D0900408

Research Techniques for High Pressure and High Temperature

Research Techniques for High Pressure and High Temperature

Edited by

Gene C. Ulmer

Springer-Verlag
New York · Heidelberg · Berlin

1971

Library of Congress Catalog Card Number 71-173285.

Printed in the United States of America.

ISBN 0-387-05594-0 Springer-Verlag New York • Heidelberg • Berlin
ISBN 3-540-05594-0 Springer-Verlag Berlin • Heidelberg • New York

PREFACE

Within the last two decades, the experimental technology for the study of high temperature solid-vapor and liquid-vapor equilibria has mushroomed so fast that both academic and industrial researchers desirous of working in this field — be they physical chemists, metallurgists, ceramists, petrologists, crystal chemists, or members of any of the several branches of materials science — find themselves in the situation that in order to learn the art of the latest techniques, a period of apprenticeship or residency needs be spent at an institution or laboratory currently engaged in this type of solid-vapor or liquid-vapor research. The techniques for control of the vapor phase at total pressures of one atmosphere or greater have not been well defined in the literature. Therefore, the purpose of this volume will be to serve as a laboratory manual for the control, calibration, and measurement of high temperature-high pressure equilibria.

The avowed aims of this treatment of experimental techniques are:

(1) to give, in terms understandable at the graduate student level, the laboratory procedures necessary to the design and utilization of good experimental technique,

(2) to list the limitations, dangers, and technical pitfalls inherent or intrinsic to the described techniques,

(3) to give theory and specific data <u>only</u> where they are essential to the experimental design,

(4) to give with each chapter references that are extensive enough to serve as a bibliography of the state-of-the-art of technique development within the last decade.

The authors have attempted to delete jargon and make the most general description that is consistent with understanding the technique. With these aims fulfilled, the experimental details within this volume should be applicable to academic, institutional, and industrial research in all branches of chemical, physical, earth, and materials sciences.

ACKNOWLEDGMENT

This volume was initiated, planned, and the table of contents finalized in the hallways at three of the annual Washington meetings of the American Geophysical Union. Special thanks are due to Drs. H.P. Eugster, G. Kullerud, and E.F. Osborn whose encouragement was appreciated. The editor would also like to mention Drs. P.M. Bell, L.S. Walter, and D.C. Presnall who served as "editorial sounding boards." Special thanks are due to Mrs. R. Krakow for copy editing, to Mr. R. Texter for drafting, and to Mrs. D. Wyszynski for typing. Only with the extreme cooperation and patience displayed by this editorial staff did the idea of this manuscript become a reality. Springer-Verlag, the authors, and editorial staff are to be recognized for their combined efforts to keep the costs of this book low enough that the average student might own a copy of this volume.

TABLE OF CONTENTS

LIST OF CONTRIBUTORS

Dr. H. L. Barnes	Department of Geochemistry and Mineralogy The Pennsylvania State University
Dr. P.M. Bell	Geophysical Laboratory Carnegie Institute Washington, D.C.
Dr. A.L. Boettcher	Department of Geochemistry and Mineralogy The Pennsylvania State University
Dr. J.R. Holloway	Department of Chemistry Arizona State University
Dr. J.S. Huebner	U.S. Geological Survey Washington, D.C.
Dr. D.M. Kerrick	Department of Geochemistry and Mineralogy The Pennsylvania State University
Dr. G. Kullerud	Geophysical Laboratory Carnegie Institute Washington, D.C.
Dr. R. Nafziger	U.S. Bureau of Mines Albany, Oregon
Dr. D.C. Presnall	Divisions of Geosciences University of Texas at Dallas
Dr. M. Sato	U.S. Geological Survey Washington, D.C.
Dr. G.C. Ulmer	Department of Geology Temple University

Dr. L.S. Walter — Planetology Branch
NASA-Goddard Space Flight Center
Greenbelt, Maryland

Dr. W.B. White — Materials Research Laboratory
Department of Geochemistry and Mineralogy
The Pennsylvania State University

Dr. D.W. Williams — Electricity Council
Research Center, Capenhurst
Chester, England

Dr. E. Woermann — Institut für Kristallographie
Aachen, Germany

CHAPTER 1

An Introduction to High Pressure-High Temperature Technology

Gene C. Ulmer

I. Historical Background

High temperature and high pressure technologies are both new and old subjects. A romanticist would claim that mankind's control of temperature began with the gift of fire from Prometheus, while a historian would point to Egyptian glass fabrication and Hittite metal smelting as examples of early high temperature technology. Even as far back as 212 B.C., Archimedes employed a very sophisticated high temperature technology by using a solar imaging lens to defend Syracuse, his home town, by setting fire to the sails of the attacking Roman ships. Pressure technology, while not dating as far back, is at least 300 years old and begins perhaps with Otto von Guericke (1602-1686) who is credited with making the first air compressor pump and devising the famous Magdeburg hemispheres which were held together so firmly by vacuum that several horses were needed to separate them. Despite this antiquity, high temperature-high pressure technology has advanced so dramatically in the last two decades that even the 350-odd pages of this manual can do little more than introduce the subject.

II. Present Extent of Technology

A. Technology

A glance at Figure 1 shows the range of modern pressure-temperature (P-T) technology with the scope of this book shown by shading. The chapters in this book do not follow any temperature or pressure sequence, but rather were arranged in terms of a progression in the complexity of the technology with an additional intent to group subject matter. Accordingly, Chapters 2 and 3

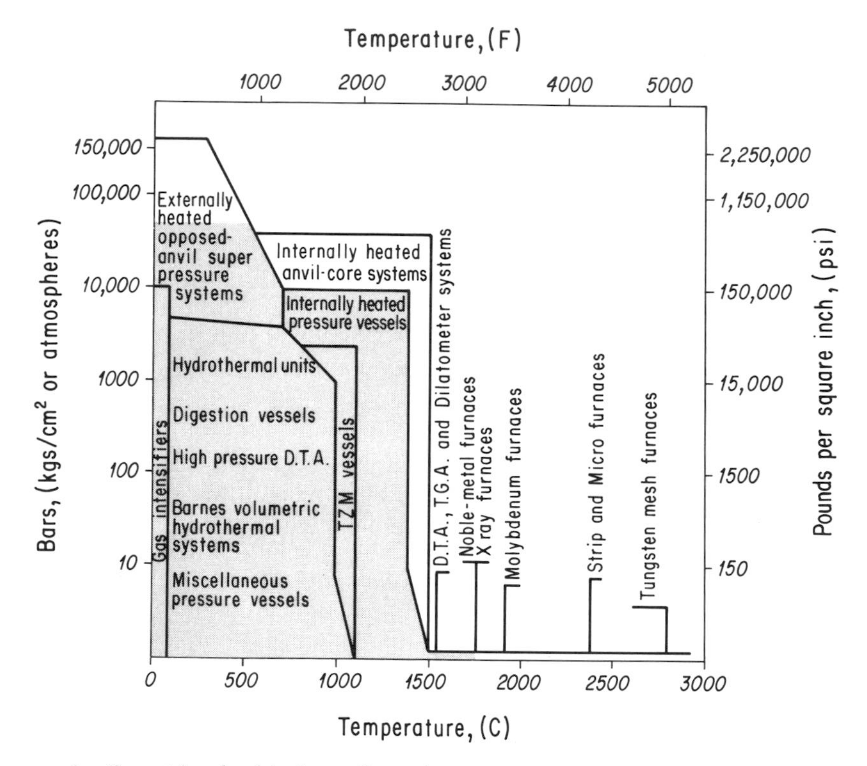

Figure 1. Practical limits for the various high temperature-high pressure techniques. Shading shows the technology discussed within this book.
(Basic diagram kindly furnished by Tem-Pres Research, The Carborundum Company, State College, Pa.)

deal with oxidation equilibria at high temperature; Chapter 4 deals with oxygen equilibria at high temperature and pressure; Chapters 4, 5, and 6 discuss P-T control and measurement in the hydrothermal technique, again primarily for redox equilibria; Chapters 7, 8, and 9 give the details for calibration (Table 7.1) and control of very high pressure equilibria techniques; Chapter 10 deals primarily with liquid-vapor equilibria; and finally Chapters 11 and 12 treat high temperature-high pressure techniques for sulfur equilibria with Chapter 11 also giving a very good general review of the "quenching-technique."

B. Active Experimentalists

Even though this book presents at least some detailed discussion on all but the highest pressure and highest temperature techniques (see Figure 1), the number of contributors to this book could have easily been doubled. Since 1967 the "growing pains" and revisions in the table of contents have been continuous. Many potential contributors were enthusiastic, but for their own valid reasons (usually "prior publication" or "heavy schedules") could not contribute: H.P. Eugster, H.R. Shaw, J.L. Munoz, W.C. Burnham, P. Orville, P.J. Wyllie, M. Carapezza, K. Schwedtfeger, and A. Piwinskii were potential contributors in this category. Furthermore, a glance through the bibliographies for each chapter points up still other names and several are noticeable by their repetition: E.F. Osborn, A. Muan, G.C. Kennedy, E.U. Franck, B.M. French, and D.R. Wones to name but a few. In addition to the bibliographies in this volume, extensive bibliographies of high pressure research are available in the literature. (See for example Bell, 1967 and ASME, 1964) A clearing house of high pressure data has in fact been established at Brigham Young University; the High Pressure Data Center (1969) established there by Dr. H.T. Hall is an excellent source of recent data. Section III details still other active workers by their literature references.

C. Fields of Research

While this book has a decidedly geochemical-geophysical persuasion, the scope of Pressure-Temperature-Composition (P-T-X) equilibria reaches into all fields of earth and materials sciences, as the bibliographies at the end of each chapter again show. Physical chemistry, metallurgy, solid state physics, ceramics, and fuel engineering are just a few of the fields of research as intimately concerned with P-T-X equilibria as is geochemistry or geophysics.

III. General Literature

The growing interest in the last decade for high pressure and high temperature technology is perhaps best proven by the dates of publication for the general literature.

Bradley (1969) and Tsiklis (1965, translated into English by Bobrowsky, 1968) have both made the job of this volume considerably easier. In _High Pressure Methods in Solid State Research_, Bradley has very adequately described the techniques at pressures higher than those covered by this volume (see Figure 1); and in _Handbook of Techniques in High Pressure Research and Engineering_, Tsiklis has given the recent Russian developments in the field.

In the area of high temperature, the U.S. National Bureau of Standards (1968) has issued Volume 2 of _Precision Measurement and Calibration-Temperature_ which again allowed the shortening of this volume. Materials behavior and measurement techniques at high temperature have been treated in several earlier works (for example Campbell, 1956 and Kingery, 1959).

Several other references in the field of high pressure-temperature are worthy of mention; although these references are aimed at theory and/or experimental data interpretation rather than at design and technology: Eyring (1967, Vol. 1; 1969, Vol. 2) is editing a serial publication entitled _Advances in High Temperature Chemistry_; Margrave (starting in 1969) is editing a monthly journal entitled, _High Temperature Science_; Toropov (1969) has edited a compilation of Russian refractory and chemical studies, _Chemistry of High Temperature Materials_; and Zharkov and Kalinin (1971) present a great deal of necessary theory in their book, _Equations of State for Solids at High Pressures and Temperatures_.

Literature on the application of high pressure-high temperature research is too voluminous to treat here, but a few general review articles are mentioned in that they also discuss experimental design. Wyllie (1963, 1966) and Simmons (1968) have both given excellent reviews of geophysical applications of high pressure research. Williams (Chapter 7, this volume) and Edgar are co-authoring a manuscript, _Introduction to Experimental Petrology_, which is now in preparation and will deal with petrologic applications of P-T research. Zeitlin and Boggio (1965) have an excellent review article entitled High Pressure Technology: Industrial Applications in which they discuss equipmental design for many industrial applications. Bibliographies of high pressure research are referenced in section II B above.

IV. Limitations to High Pressure-Temperature Technology

Perhaps mostly in subconscious recognition of the inherent danger in exceeding design limitations, liquid and gaseous pressure vessels are non-semantically referred to as bombs. Recalling the Custom's Declaration problem encountered by one researcher taking hydrothermal equipment into Europe, a facetious title for this book could have been "Blue Prints to Build a Bomb." The apparent ridiculousness of such a title can immediately be offset with grim anecdotes of explosive failures of high pressure equipment at even the most experienced of labs.

While each of the chapters below point out the specific precautions (beyond the common-sense ones) to be employed for the particular technology, this book is no substitute for experience. The virgin "do-it-yourselfer" is prewarned to use this book in the spirit in which it is presented, *i.e.*, as a guide and not *the* blueprint.

The limitations for various equipment designs are suggested in Figure 1, but the age of the equipment, the skill of the operator, the absence of embrittling gases and corrosive liquids, and the lack of exothermic effects of redox equilibria are but a few of the assumed factors in drawing up Figure 1. When in doubt, non-ricochetting, plywood-clad steel plate shielding is a very cheap insurance policy. The publishers, the editorial staff, and the authors, either singly or collectively, accept no responsibility for catastrophic occurrences.

V. Abbreviation, Symbology and Units

Unless otherwise stated, the symbols, abbreviations, and units used throughout this book were conventional, *i.e.*,

pX or fX - partial pressure or fugacity of species X, respectively

1063 C	-	temperatures given in degrees centigrade without degree symbol
percentages	-	in weight percent
H	-	Heat Function
G	-	Gibbs Free Energy
K_n	-	Equilibrium constant of process n
units	-	standard metric system with the exception listed below
Kb	-	kilobar, pressure unit with pressure conversion factors listed in Chapter 4, Table 4.1.

BIBLIOGRAPHY:

American Society of Mechanical Engineering (1964). Annotated bibliography on high-pressure technology, compiled by A. Zeitlin. Washington: Butterworth.

Bell, P.M. (1967). Geophysical research at pressures above 30 kilobars. Trans. Amer. Geophys. Union, 48:702-706.

Bradley, C.C. (1969). High pressure methods in solid state research. New York: Plenum Press.

Campbell, I.E. (1956). High temperature technology: The electrochemical society series. New York: John Wiley and Sons.

Eyring, L. (1967 and 1969). Advances in high temperature chemistry, Vol. 1 and 2. New York: Academic Press.

High Pressure Data Center (1969). Bibliography on high pressure research in chemistry and physics, with bi-monthly supplements; Director, H.T. Hall: Provo, Utah: Brigham Young University under contract with U.S. Natl. Bureau Stand., No. CST-116.

Kingery, W.D. (1959). Property measurements at high temperatures, New York: John Wiley and Sons.

Margrave, J.L. (1969-). The journal of high temperature science, monthly, New York: Academic Press.

Simmons, G. (1968). High pressure geophysics-equipment and

results. J. Geol. Education, 16:21-29.

Toropov, N.A. (1969). Chemistry of high-temperature materials, translated by C.B. Finch. New York: Plenum Press, Consultants Bureau.

Tsiklis, D.S. (1965). Handbook of techniques in high-pressure research and engineering, translated by A. Bobrowsky (1968). New York: Plenum Press.

U.S. National Bureau of Standards (1968). Precision measurement and calibration-temperature, Vol. 2. Washington, D.C.: U.S. Government Printing Office.

Wyllie, P.J. (1963). Applications of high pressure studies to the earth sciences, in High pressure physics and chemistry, Vol. 2, edited by R.S. Bradley. London: Academic Press.

Wyllie, P.J. (1966). High pressure techniques, in Methods and techniques in geophysics, Vol. 2, edited by S.K. Runcorn. London: Interscience Publishers.

Zeitlin, A. and Boggio, F.G. (1965). High pressure technology: industrial applications. Mechanical Engineering, 87: No. 11, 14-21.

Zharkov, V.N. and Kalinin, V.N. (1961). Equations of state for solids at high pressures and temperatures, translated by A. Tybulewicz. New York: Plenum Press.

CHAPTER 2

Gaseous Buffering for the Control of Oxygen Fugacity at One Atmosphere

Ralph H. Nafziger

Gene C. Ulmer

Ed Woermann

I. Introduction

The importance of controlling fugacities[1] when studying reactions in closed systems which involve elements whose oxidation states are otherwise readily changed at high temperatures is well known. Perhaps the most important of these transition elements is iron, due in large part to its widespread technological uses and natural occurrences. Osborn and Muan (1963), for example, illustrated the importance of oxygen fugacity (fO_2) on liquidus temperatures for liquids in the system $FeO\text{-}Fe_2O_3\text{-}SiO_2$. Crystallization sequences under various fO_2 conditions for liquids in this system were discussed by Muan (1955). Comprehensive discussions concerning the system Fe-O are given elsewhere (Darken and Gurry, 1945, 1946; Muan, 1958; Osborn and Muan, 1963; Muan and Osborn, 1965). Nickel, cobalt, manganese, chromium, vanadium and their oxide compounds have all become important materials in present day technology (<u>e.g.</u>, stainless steels, super-strength alloys, magnetic alloys, oxide magnets, cermets, and ferrites). Oxygen fugacity

---

[1] Throughout this chapter, the term fugacity is used in preference to pressure. Since only at high temperatures and low total pressures are these two terms essentially equivalent, the usage of the more thermodynamically rigorous term fugacity does not imply ideality of the gas phase.

control in reactions involving such variable oxidation state elements is important. This has been emphasized in the literature (see for examples, Hahn and Muan, 1961, for nickel; Massé *et al.*, 1966, for cobalt; Davies and Richardson, 1959, Morris and Muan, 1966, and Glasser, 1958, for manganese; Johnson and Muan, 1968, and Graham and Davis, 1971, for chromium; Katsura and Hasegawa, 1967, for vanadium). In addition, copper and other base metals (*e.g.*, lead, bismuth, and tin) exist in two or more oxidation states in compounds of importance to metallurgical processes. Among reactive metals, titanium has found a secure place in the aerospace industry and small amounts of oxygen are detrimental to this metal. There are a great many phases present in the system Ti-O which have different structural properties and whose stabilities often depend on the atmosphere composition (*i.e.*, oxygen fugacity) at high temperatures of interest (*e.g.*, Porter, 1965). As another example, the separation of rutile and iron from ilmenite depends largely on a low oxygen fugacity (Ramdohr, 1964). Refractory metals such as molybdenum, tungsten, and tantalum are finding increasing uses in very high temperature applications and the importance of oxygen fugacity on the stability of refractory metal oxides is beginning to be realized (*e.g.*, Chang and Phillips, 1969).

Geologically, oxygen fugacity is recognized as an important controlling parameter in determining crystallization sequences and residual liquid compositions in iron-containing silicate systems which have direct applicability to basic igneous rocks as demonstrated by Osborn (1959, 1962). Another example is the system $MnO-Mn_2O_3-SiO_2$ (Morris and Muan, 1966) which contains rhodonite and tephroite, two minerals which are not stable at liquidus temperatures under highly oxidizing conditions. Oxygen fugacity is an important intensive variable which controls the ferric-ferrous equilibrium in many magmas (Fudali, 1965; Carmichael and Nicholls, 1967). Furthermore, oxygen fugacity can determine the compositions of crystallizing phases in both magnetite-bearing assemblages and metallic iron-bearing assemblages in the system Fe-O-MgO-SiO_2 (Speidel and Nafziger, 1968). Thus, oxygen fugacity can control not only the phases present but the composition of these phases as well.

Cosmochemically, oxygen fugacity also plays an important role. The stability of observed phases in chondritic meteorites has been studied with regard to oxygen fugacity in the systems MgO-Fe-O-SiO_2 and MgO-Ni-O-SiO_2 (Nafziger and Muan, 1967; Campbell and Roeder, 1968, respectively). Natural meteorite samples and recovered lunar samples have both been investigated for their fO_2-T relationships (Walter, 1969; Tuthill and Sato, 1970). The presence of several phases hitherto unobserved terrestrially and the existence of certain phase assemblages and phase compositions are attributed to highly reducing lunar conditions at the Apollo 11 lunar landing site (Anderson et al., 1970). For example, divalent chromium in olivines recovered in both Apollo 11 and 12 landings has been interpreted as indicative of highly reducing conditions during the crystallization of lunar materials (Bell, 1971).

Inasmuch as the control and measurement of oxygen fugacity is important in a wide variety of disciplines, this chapter has as its objective to describe methods for the calculation of this parameter in a number of gaseous systems and to outline advantages of each system and under what conditions each system should be employed. This is followed by a description of the apparatus and techniques which are used in laboratories for the aforementioned control and measurement. (See also Chapters 3, 4, 5, 8, 11, and 12).

II. Calculation of Oxygen Fugacity

Two early systems investigated which involve a transition element were iron oxide-SiO_2 and MgO-iron oxide-SiO_2 (Bowen and Schairer, 1932 and 1935 respectively). Samples were heated in iron crucibles in a nitrogen atmosphere to prevent as much ferric iron from forming in the phases as possible. The phases are thus in equilibrium with metallic iron at low oxygen fugacities. However, measurement and control of the latter cannot be accomplished under these conditions. In order to measure and control gas partial pressures in a gas mixture, some means of regulating the amounts of separate gases must be achieved. This is most conveniently attained through the use of flowmeters to accurately and constantly proportion the gases. These will be described in detail later in this chapter.

Controlled oxygen fugacities at high temperatures and at one atmosphere total pressure may be attained with the use of a variety of readily available room-temperature oxygen-containing gases and gas mixtures. This method depends on the high-temperature dissociation equilibria of suitable gases to produce a desired oxygen fugacity. The gas with a desired oxygen fugacity can then be used to control oxidation-reduction reactions in the involved condensed phases. Of particular importance is the relation between achieved oxygen fugacity at the given temperature and the room-temperature gas species or gas mixing ratio used. This relation is chiefly a function of the high-temperature thermodynamic properties of the involved gas species (especially standard free energies of formation, and hence corresponding equilibrium constants). Compilations of the necessary constants are readily available in the literature (e.g., Coughlin, 1954; Sawamura and Matoba, 1960, 1961; Elliott and Gleiser, 1960; Wicks and Block, 1963; Stull et al., 1965, et seq.).

A. Oxygen Fugacities from Single Gas Species

The most simple case is that involving the high-temperature dissociation of oxygen-containing gases. An example is

$$CO_2 = CO + 1/2\ O_2 \tag{1}$$

which readily proceeds at high temperatures. The equilibrium constant $\left[K_{(1)}\right]$ is expressed by

$$K_{(1)} = \frac{fCO \cdot fO_2^{\frac{1}{2}}}{fCO_2} \tag{2}$$

where f refers to the fugacity of the respective gas species. At one atmosphere total pressure, $K_{(1)}$, and therefore fO_2, is a function of temperature. Hence, the oxygen fugacity cannot be varied independently of temperature and this represents the chief limitation of using single gas species to attain desired oxygen fugacities at high temperatures. Oxygen fugacities of high-temperature dissociation reactions may be rigorously calculated as follows: The carbon monoxide reaction

$$2\ CO = 2\ C + O_2 \qquad (3)$$

has an equilibrium constant $K = \frac{fO_2 \cdot a_C{}^2}{fCO_2}$. (4)

Letting $fO_2 = y$, then $a_C = 2y$ and $fCO = 1 - 3y$ or

$$K = \frac{y(2y)^2}{(1 - 3y)^2} . \qquad (5)$$

Rearranging terms $y^3 - 9/4\ Ky^2 + 3/2\ Ky - K/4 = 0.$ (6)

Stull, *et al.* (1965, *et seq.*) gives -log K as a function of temperature for this reaction (and also for CO_2, COS, H_2O, N_2O, N_2O_3, and SO_2 dissociations). For other dissociations (*e.g.*, NO and NO_2), the standard free energy of the reaction (ΔG^o) can be obtained (*e.g.*, Wicks and Block, 1963) and K calculated from this, using the $\Delta G^o = RT \ln K$ relation. Substituting the value of K in equation (6) permits the calculation of y; therefore, fO_2 can be obtained as a function of temperature. In order to considerably simplify the above tedious calculation of a cubic equation, it is often assumed that y is small and therefore 3y is small since $K \simeq 10^{-9}$ at high temperatures ($\simeq$1200 C). Therefore equation (5) reduces to

$$K \simeq \frac{4y^3}{1} \quad \text{or} \quad y = fO_2 \simeq \sqrt[3]{K/4} . \qquad (7)$$

The value of y can then be compared to unity to determine if the above assumption was valid.

B. <u>Oxygen Fugacity from Dilution of Oxygen by Inert Gases</u>

Another method for controlling oxygen fugacities at high temperatures is to dilute the oxygen or oxygen-containing gas species with an inert gas such as nitrogen (Muan, 1959), helium, or argon (Speidel, 1967). Dilution may also be accomplished by means of a vacuum which of course decreases the total pressure and hence the oxygen fugacity proportionally. Representative values of total pressure and corresponding log oxygen fugacities for evacuating air are given in Table 2.1. When the gas phase oxygen fugacity is used to control the oxidation-reduction reaction in the condensed phase, a transfer of oxygen between gas and condensed phases must take place, unless by some extremely fortuitous circumstance equilibrium is attained without such a transfer.

TABLE 2.1

Oxygen Fugacity as a Function of Total Pressure for Air Evacuation

Total Pressure		
Atm	Torr (mm Hg)*	Log fO_2 (atm)
1	760	-0.68
0.5	385	-0.98
0.2	152	-1.38
0.1	76	-1.68
10^{-2}	7.6	-2.68
10^{-3}	0.76	-3.68
1.3×10^{-5}	10^{-2}	-5.56
1.3×10^{-6}	10^{-3}	-6.56

*To convert to microns (μ), multiply these values by 10^3.

Whereas temperature can be varied independently of oxygen fugacity by these oxygen dilution methods, Muan (1958) has pointed out that this equilibration requires a certain volume of gas which is proportional to $1/fO_2$. Accordingly, a practical lower limit of approximately 10^{-3} atm fO_2 for this method was mentioned by this author. Lower oxygen fugacities would require prohibitively large volumes of gas. Impurities in the gases would therefore become more significant. In addition, oxygen is present only as a diatomic molecule in these dilution methods which limits the rate of oxygen transfer between gas and condensed phases (Muan, 1958). Dilution of oxygen with oxygen-containing gases such as CO_2 (Hahn and Muan, 1960) presents essentially similar limitations because the dissociation of CO_2 is negligible compared with the oxygen concentration of the gas phase. Johnston and Walker (1925) used CO_2-air mixtures and obtained satisfactory constant mixing ratios of these gases. However, these authors were not interested in high temperatures and hence gas dissociation was of little concern. At high temperatures, the dissociation of CO_2 in such an atmosphere may provide significant oxygen in addition to that from the air so that the resulting oxygen fugacity is influenced by the two contributions. Providing the natural CO_2 content of the air remains relatively stable, an air-CO_2 synthetic mixture could conveniently provide oxygen fugacities between the limits set by air and pure CO_2 (for example, $10^{-0.68}$ to 10^{-3} atm at 1400 C).

C. Oxygen Fugacity from Oxygen-Containing Gas Mixture

1. CO_2-CO Mixtures

Perhaps the most versatile method for attaining desired oxygen fugacities at high temperatures is the use of a mixture of two or more gases which react with one another at high temperature to release or consume oxygen. This permits independent variation of oxygen fugacity with temperature and, since at least one of the gas species must contain oxygen, this serves as a buffering reservoir to continuously replenish gas-phase oxygen lost to the condensed phases assuming that gas phase reactions are sufficiently rapid. In this way, equilibration times are reduced even though the oxygen fugacity may be low (Muan, 1958). (See also Chapters 3, 5, and 12).

Consider first the mixing of two gases such as CO_2 and CO. The reaction of interest is equation (1) with its corresponding equilibrium constant given by equation (2). The dissociation of CO_2 at one atmosphere total pressure can be expressed with the assumption of gas ideality (fugacity equals pressure at one atmosphere) by the equations which follow:

$$fCO_{2(i)} = fCO_2 + fCO + fO_2 = 1, \tag{8}$$

where (i) refers to the initial room-temperature gas. Upon rearranging terms and multiplying through by $fCO_{2(i)}$:

$$fCO_{2(i)} \left[fCO_2 + fCO\right] = fCO_{2(i)} \left[1 - fO_2\right] \tag{9}$$

But since $fCO_{2(i)} = 1$, equation (9) can also be rearranged:

$$fCO_2 = fCO_{2(i)} (1 - fO_2) - fCO \tag{10}$$

and since from equation (1), $N_{CO} = 1/2\, N_{O_2}$, where N is the mole fraction of the gas species, then

$$fCO_2 = fCO_{2(i)} (1 - fO_2) - 2\, fO_2 \tag{11}^*$$

Similarly for CO:

$$fCO = fCO_{(i)} \quad (1 - fO_2) + 2\, fO_2. \tag{12}$$

Substituting (11) and (12) into (2):

---

*By Henry's Law, $p_{gas} = N_{gas} \cdot P_{total}$ where p_{gas} is the partial pressure of a gas, N_{gas} is the mole fraction of the gas, and P_{total} is the total pressure of the system. Again invoking the ideality assumption of p = f and one atmosphere total pressure $f_{gas} = N_{gas} \cdot f_{total}$ and $f_{gas} = N_{gas} = p_{gas}$.

$$K_{(1)} = fO_2^{\frac{1}{2}} \left[\frac{fCO_{(i)}\ (1 - fO_2) + 2\ fO_2}{fCO_{2(i)} (1 - fO_2) - 2\ fO_2} \right]. \tag{13}$$

At room temperature with no dissociated gases, for CO_2-CO mixtures:

$$fCO_{2(i)} + fCO_{(i)} = 1 \text{ (total pressure)}. \tag{14}$$

Let $R = \frac{fCO_{2(i)}}{fCO_{(i)}}$ (initial gas mixing ratio).

Then $fCO_{2(i)} = \frac{R}{1 + R}$ and $fCO_{(i)} = \frac{1}{1 + R}$.

$$\text{Thus } \frac{K_{(1)}}{fO_2^{\frac{1}{2}}} = \frac{\frac{1 - fO_2}{1 + R} + 2\ fO_2}{\frac{R(1 - fO_2)}{1 + R} - 2\ fO_2} = \frac{2\ R \cdot fO_2 + fO_2 + 1}{R(1 - 3\ fO_2) - 2\ fO_2}. \tag{15}$$

Let $A = \frac{K_{(1)}}{fO_2^{\frac{1}{2}}}$; $B = 2\ fO_2$; $C = 1 + fO_2$; and $D = 1 - 3\ fO_2$.

Then $A = \frac{BR + C}{DR - B}$ or $ADR - AB = BR + C$ or $ADR - BR = C + AB$.

Hence $R(AD - B) = C + AB$ or $R = \frac{AB + C}{AD - B}$

or

$$R = \frac{1 + fO_2 + \frac{K_{(1)} \cdot 2\ fO_2}{fO_2^{\frac{1}{2}}}}{\frac{K_{(1)}}{fO_2^{\frac{1}{2}}} (1 - 3\ fO_2) - 2\ fO_2} \tag{16a}$$

and volume percent $CO_{2(i)} = V = \frac{100\ R}{R + 1}$, (16b) since $R = \frac{-V}{V - 100} = \frac{AB + C}{AD - B}$.

Then $V = \frac{-100\ C - 100\ AB}{B - AD - C - AB} =$

$$\frac{-100\,(1 + fO_2) - 100\,\frac{K_{(1)} \cdot 2\,fO_2}{fO_2^{\frac{1}{2}}}}{2\,fO_2 - \frac{K_{(1)}}{fO_2^{\frac{1}{2}}}\,(1 - 3\,fO_2) - (1 + fO_2) - \frac{K_{(1)} \cdot 2\,fO_2}{fO_2^{\frac{1}{2}}}} \qquad (17)$$

$$\text{or volume percent } CO_{2(i)} = \frac{100\,(1 + fO_2 + 2\,fO_2 \cdot \frac{K_{(1)}}{fO_2^{\frac{1}{2}}}}{(1 - fO_2)\,(1 + \frac{K_{(1)}}{fO_2^{\frac{1}{2}}})}. \qquad (18)$$

Equation (18) therefore permits the calculation of the necessary volume percent CO_2 in a CO_2-CO gas mixture for any selected oxygen fugacity at a given $K_{(1)}$. The above calculation is somewhat tedious especially for a variety of conditions which are often desired in experimental work. Therefore, computer tables have been prepared (Deines et al., 1971) whereby the user may select an oxygen fugacity and temperature and simply read off the required room-temperature volume percent CO_2 in a CO_2-CO mixture. These tables give values in intervals of 5 C and 0.05 atm log fO_2.

The above calculations assume that: (1) the separate gas species are devoid of impurities; (2) no precipitation or condensation takes place for these species under the pertinent experimental conditions; (3) the gases behave ideally at high temperatures; (4) equilibrium is achieved among the phases; and (5) extraneous reactions which are not listed, such as formation of organic polymer species do not occur, due chiefly to the high temperatures of interest (e.g., Cairns and Tevebaugh, 1964).

It has been common practice to perform the above calculations with several simplifying assumptions (Muan and Osborn, 1965; Porter, 1965). If it is assumed that $fO_2 << fCO_{2(i)}$ and $fO_2 << fCO_{(i)}$, then equation (13) becomes

$$K_{(1)} \simeq fO_2^{\frac{1}{2}} \left[\frac{fCO_{(i)}}{fCO_{2(i)}}\right] \quad \text{or} \quad fO_2 \simeq \left[K_{(1)} \frac{fCO_{2(i)}}{fCO_{(i)}}\right]. \qquad (19)$$

Such assumptions are valid at temperatures and mixing ratios commonly encountered in phase equilibrium determinations where oxygen fugacity control is desired. However, with the advent of applicable computer techniques, and vastly reduced calculation time, these assumptions become unnecessary and a rigorous treatment can be obtained [Speidel and Heald, 1967; Deines et al., 1971; and also equation (7) of Chapter 3].

The use of gas mixtures to attain desired high-temperature oxygen fugacity buffers does not preclude all limitations. As noted in Table 2.2, mixtures of various gases at room temperature limit oxygen fugacities which can be obtained conveniently. Use of specific gas mixtures outside of the imposed limits requires inordinately large relative gas volumes which adds to uncertainties in obtaining the final high temperature equilibrium oxygen fugacity values. This limitation, however, can in some circumstances be circumvented as described later in this chapter.

Another inherent disadvantage in the use of gases which contain carbon atoms is the precipitation of that element under certain experimental conditions. This results, for example, from decreasing the CO_2 content in a CO_2/CO mixing ratio at a given temperature to a point at which graphite precipitates. French and Eugster (1965) show that according to the reaction

$$C + O_2 = CO_2 \qquad (20a)$$

the oxygen fugacity at which carbon precipitation just begins

TABLE 2.2

Recommended Single-Species Gases and Gas Mixtures for Achieving Desired Oxygen Fugacities at High Temperatures at One Atmosphere Total Pressure*

Approximate fO_2 Range (log atm)	Approximate Temperature Range C	Gas Species
0	Any	O_2
- 0.68	Any	Air
-0.68 to -3	Any	O_2-inert gas**
-4.3 to -3	1100-1400	CO_2
-4.2 to -2.6	500-2000	CO
-4.6 to -1.1	500-2000	H_2O
-3.3 to -2.1	500-2000	N_2O
-3.4 to -2.6	500-2000	N_2O_3
-6.8 to -1.7	500-2000	SO_2
-20.0	900-1000	CO_2-H_2***
-10.0	1000-2000	
- 8.0	1100-2000	
- 5.0	1350-2000	
- 3.0	1500-2000	
-20.0	900	CO_2-CO***
-10.0	1000-1800	
- 8.0	1150-2000	
- 5.0	1350-2000	
- 3.0	1550-2000	
-20.0	575-1000	H_2O-H_2***
-10.0	1000-2000	
- 8.0	1150-2000	
- 5.0	1500-2000	
- 3.0	1800-2000	

*Although theoretically, any fO_2 may be obtained with any two mixed gases provided enough stages are used in the gas mixer and/or capillary flowmeters are small enough to allow reasonable manometer settings, the above values are those which can be attained conveniently in a single-stage gas mixer.

**For example, Ar, He, etc., or other O_2-containing gases, for example, CO_2-air.

***Practical log mixing ratio limits range from +2.2 to -2.5.

at one atmosphere is given by:

$$\log fO_2 = \frac{\Delta G^o_{(20)}}{2.303\ RT} + \log fCO_2 \qquad (20b)$$

where ΔG^o is the standard Gibbs free energy of formation, R is the gas law constant, T is the temperature in K and fCO_2 is the equilibrium CO_2 fugacity as given by equation (11) earlier. Muan and Osborn (1965) graphically show the "forbidden" region of CO_2/CO mixtures in a T-fO_2-mixing ratio plot for one atmosphere (their Figure 11). The exact mixing ratios of CO_2-CO at which carbon precipitates are given in the computer tables of Deines et al. (1971). Typical approximate temperature, volume percent CO_2 (in CO_2-CO), and corresponding log fO_2 atmosphere values at which graphite precipitates are: 1300 C, 0.32, -16.7; 1100 C, 4.76, -17.8; and 900 C, 26.14, -19.3. Precipitation of sulfur in S-containing gases could similarly be envisaged.

2. H_2O-H_2 Mixtures

When H_2O-H_2 gas mixtures are used, similar calculations to those described previously for CO_2-CO will give gas compositions necessary to obtain desired oxygen fugacities. In this case, the reaction of interest is

$$H_2O = H_2 + 1/2\ O_2, \qquad (21)$$

$$\text{for which } K_{(21)} = \frac{fO_2^{\frac{1}{2}} \cdot fH_2}{f\ H_2O} \qquad (22)$$

The composition of the initial H_2O-H_2 mixture is given by

$$\text{volume percent } H_2O_{(i)} = \frac{100\ (1 + fO_2 + 2\ fO_2 \cdot \frac{K_{(21)}}{fO_2^{\frac{1}{2}}})}{(1 - fO_2)\ (1 + \frac{K_{(21)}}{fO_2^{\frac{1}{2}}})} \cdot \qquad (23)$$

Temperature-oxygen fugacity-volume percent $H_2O_{(i)}$ values are also tabulated by Deines et al. (1971). Larimer (1968) relates temperature-oxygen fugacity-H_2-H_2O mixing ratio by the equation:

$$\log H_2/H_2O = -1/2 \log fO_2 - 12{,}850/T\ (K) + 2.86 \qquad (24)$$

Similar assumptions as described previously for CO_2-CO mixtures must also be applied in this case. [See also Chapter 3, in particular equation (9).]

The H_2O-H_2 gas mixture was a favorite of earlier workers (see e.g., Emmett and Shultz, 1930; Chipman and Marshall, 1940; Darken and Gurry, 1946). However, certain precautions which involve added complexities to the gas mixing apparatus must be taken to insure accuracy and precision of the resulting high-temperature oxygen fugacities and this has caused large-scale abandonment of this gas mixture in favor of others which do not require such precautions. With H_2O-H_2 mixtures, hydrogen is either mixed with steam or passed through "saturator" towers filled with glass beads and distilled water. Since the vapor pressure of steam or water vapor over liquid water is pressure dependent, atmospheric pressure fluctuations must be considered in very accurate work. Furthermore, the necessary constant temperature of the mixing apparatus must be controlled to ± 0.01 C by a well-stirred surrounding oil bath. The steam-hydrogen mixture is then led into the furnace; the tubes carrying the gas mixture must be heated to prevent condensation and alteration of the mixing ratio. A check on the exit gas composition may be accomplished by a dew-point determination, mass spectrometry, or condensing and weighing the water and measuring the volume of hydrogen per unit time.

3. CO_2-H_2 Mixtures

The derivation of equilibrium oxygen fugacity for a mixture of CO_2-H_2 is more complicated than that discussed previously for CO_2-CO and H_2O-H_2 mixtures. At high temperature, the reaction of interest for CO_2-H_2 mixtures is:

$$CO_2 + nH_2 = CO + nH_2 + \left(\frac{1 - n}{2} \right) O_2 \tag{25}$$

$$\text{for which } K_{(25)} = \frac{fCO \cdot fH_2O^n \cdot fO_2^{(\frac{1-n}{2})}}{fCO_2 \cdot fH_2O^n} . \tag{26}$$

Where rigorously permissable values are: n is equal to or less than one and n is greater than zero. For each permissable value of n, a unique set of solutions are obtained for fO_2 as a function of temperature and initial CO_2/H_2 ratios. This is not to be confused with multiples of the same equation which do not have unique values; for examples, $CO_2 + 1/2\ H_2 = CO + 1/2\ H_2O + 1/4\ O_2$ and its multiple, $2\ CO_2 + H_2 = 2\ CO + H_2O + 1/2\ O_2$, are the same equation and would produce the same calculated fO_2 value for each temperature and initial CO_2/H_2 ratio. On the other hand, consider the equations, $CO_2 + 1/2\ H_2 = CO + 1/2\ H_2O + 1/4\ O_2$ and $CO_2 + 1/3\ H_2 = CO + 1/3\ H_2O + 1/3\ O_2$, each of which is correctly balanced, and each of which would give a different calculated fO_2 at a given temperature and a given initial CO_2/H_2 ratio.

With this dilemma of being able to write non-multiple varieties of legitimately balanced equations for reactions between CO_2 and H_2, it would seem that the utility of CO_2-H_2 mixtures for unique and accurate fO_2 control would be limited. This, however, is fortunately not the case. Consider the total pressure equation for reaction (25) at one atmosphere constant pressure:

$$P_{Total} = pCO_2 + pH_2 + pCO + pH_2O + pO_2 = 1 . \tag{27}$$

The dilemma from balancing equation (25) is that the exact partial pressure of the gases will depend on the choice of n (Henry's Law). However, if the pO_2 desired is already very small compared to the total pressure, then the choice of n is "already made" as being very close to one, since there is very little free oxygen present at equilibrium. Pragmatically, the pO_2 (or fO_2) control with CO_2-H_2 mixtures ranges from 10^{-20} to 10^{-3} atmosphere (see Table 2.2) and even the pO_2 of 10^{-3} atmosphere is only $10^{-3}/1$ or 0.1% of the total pressure. At worst then, the utilized range of choices of n influence the partial pressures of CO_2, CO, H_2, and H_2O by one part in a thousand and experimentally, gas impurity itself is typically one or more parts in a hundred (see Table 2.3).

The conventional solution (Muan, 1958; Muan and Osborn, 1965) to the dilemma of equation (25) is to choose the value of n which simplifies the algebra, namely, n = 1:

$$CO_2 + H_2 = CO + H_2O \tag{28}$$

$$K_{(28)} = \frac{pCO \cdot pH_2O}{pCO_2 \cdot pH_2} \tag{29}$$

and

$$P_{Total} = pCO + pH_2O + pCO_2 + pH_2 = 1. \tag{30}$$

The use of this equation at first thought is philosophically perplexing in that with no oxygen expressed, the equation is being used to calculate fO_2 values for the given temperatures and the given initial CO_2/H_2 ratios. However, as pointed out above, while the choice of n determines a rigorous and exact calculation of the fO_2 values, the differences in fO_2 at a given temperature and a given initial CO_2/H_2 ratio are experimentally indistinguishable for the different choices of the value of n. Hence, the convention of n = 1 has persisted, even for the latest computer tabulations of fO_2 obtainable from CO_2-H_2 mixtures (Deines _et al._, 1971).

Muan and Osborn (1965) have shown that for equation (28), the initial mixing ratio can be expressed in terms of equilibrium partial pressures as (their equation 5-10):

$$\frac{pCO_{2(i)}}{pH_{2(i)}} = \frac{1 + \dfrac{pCO_2}{pCO}}{1 + \dfrac{pH_2}{pH_2O}} \tag{31}$$

which can be recast with the equations of this chapter as:

$$\frac{fCO_{2(i)}}{fH_{2(i)}} = \frac{1 + \dfrac{fO_2^{\frac{1}{2}}}{K_{(1)}}}{1 + \dfrac{K_{(21)}}{fO_2^{\frac{1}{2}}}} \tag{32}$$

and since volume percent $CO_2 = \dfrac{100\ fCO_{2(i)}/fH_{2(i)}}{\left[fCO_{2(i)}/fH_{2(i)}\right] + 1}$ (33)

the volume percent $CO_2 = \dfrac{100\left[1 + fO_2^{\frac{1}{2}}/K_{(1)}\right]}{2 + \dfrac{fO_2^{\frac{1}{2}}}{K_{(1)}} + \dfrac{K_{(21)}}{fO_2^{\frac{1}{2}}}}$. (34)

Since the evaluation of $K_{(1)}$ and $K_{(21)}$ has been discussed earlier in this chapter, equation (34) can be used to calculate the volume percent CO_2 in a CO_2-H_2 mixture for obtaining any desired fO_2 at any desired temperature. (See tables of Deines et al., 1971.)

As with the CO_2-CO mixtures, the limitation of carbon precipitation from CO_2-H_2 mixtures must be considered. Calculations for carbon precipitation limits with equation (20a) and (20b) give typical temperatures, volume percentages of CO_2 (in CO_2-H_2 mixtures) and corresponding log fO_2 atmosphere values of: 1300 C, 7.36, -16.7; 1100 C, 4.76, -17.8; and 900 C, 26.14, -19.3. These and other carbon precipitation values are shown graphically by Muan and Osborn (1965) and are tabulated by Deines et al. (1971).

French (1966) has shown that once graphite has precipitated

(or equilibrated) with a C-H-O gas, CH_4 is an important species even at one atmosphere. Accordingly, an additional limitation to the use of CO_2-H_2 mixtures may be the appearances of CH_4 or its polymers at temperatures of interest. Even without graphite present:

$$CO + 3H_2 = CH_4 + H_2O \quad (35)$$

$$2\ CO + 2\ H_2 = CH_4 + CO. \quad (36)$$

Both reaction (35) and (36) have negative valued ΔG^o below 900 K. As this manuscript goes to press, Deines et al. (1971) are finishing a computer calculation to determine if there is a significant pCH_4 before carbon precipitates from CO_2-H_2 mixtures.

III. Control and Measurement of fO_2 at High Temperature

After appropriate gases and/or gas mixtures have been chosen to measure and control the desired oxygen fugacity, and the latter has been calculated, construction of the necessary equipment and application of careful experimental technique is required.

A. Single Gases

Much simplification of design can be achieved if single gases or commercially premixed gases are used. All that is necessary in this case is some type of reducing valve and a means of transferring the gas from the cylinder regulator to the reaction chamber at an accurate and reproducible rate. Materials such as copper, stainless steel, rubber, and various types of plastic are satisfactory for most gases. Premixed gases preclude independent variation of the mixing ratio but are useful in equilibrium studies of condensed phases with gases at constant mixing ratios where it is desired to vary oxygen fugacity as a function of temperature (e.g., Muan and Osborn, 1956). Some premixed gas mixtures available commercially include: CO-CO_2, N_2-CO_2, H_2-CO_2, air-CO_2, O_2-CO_2, CO-air, CO-N_2, CO-H_2, N_2-O_2, noble gases (He, Ne, Ar, Kr) with CO, CO_2, H_2, N_2, and O_2.

The gas critical pressure however precludes the delivery of gas tanks with very high pressure. For example, any mixture containing CO_2 cannot be pressurized beyond about 650 psi in order to avoid CO_2 liquifying and concomitant unmixing of the premixed gases. Some evidence of slight gas unmixing, even short of critical pressures, has been reported (R. Roy, personal communication).

B. Gas Mixers

1. Design

Johnston and Walker (1925) were among the first to devise a relatively simple and satisfactory apparatus to mix two gases (air and CO_2) in constant proportions. Each gas was passed through manometer-flowmeters equipped with overflow tubes immersed in water to insure a uniform line pressure for each gas. Flow rates for the gases through the flowmeters therefore depended on the relative depths of the overflow tubes in the water reservoir. These tubes were adjustable. After passing through their respective flowmeters, the air and CO_2 were directed toward opposite sides of a thin rubber diaphragm in a mixing chamber. Carbon dioxide diffuses through the rubber at a greater rate than air and this affords a means of mixing the gases in constant proportions. The gas mixture composition is dependent on temperature and on the thickness and cross-sectional area of the rubber diaphragm. An overflow tube on the CO_2 side of the diaphragm serves to eliminate excess amounts of this gas. The gas mixture is passed through a chamber filled with glass beads to promote thorough mixing and thence through a final flowmeter which insures a constant rate of flow into the reaction chamber. The rubber diaphragm is only necessary for very low concentrations of CO_2 ($\leq$0.1%). For higher concentrations of any gas and for simplicity, a T-tube suffices. Darken and Gurry (1945) used this basic design and employed capillaries for manometer-flowmeters to provide a constant pressure drop for each gas. The capillaries were calibrated by the displacement of gas-saturated oil. A subsequent modification (see, for example, Muan, 1955) includes the use of separate water reservoirs for each gas and the final gas mixture, thereby permitting independent adjustments of the pressure drop for each gas by regulating the water level in each reservoir. This type of gas mixer is illustrated schematically in Figure 1 for CO_2-H_2 mixtures. For each gas, including the final mixture, there is a U-tube manometer, a calibrated capillary, and a "line pressure regulator" which consists of a manostatic bubbler (or overflow valve) with adjustable reservoir. In operation, for each desired setting of the manometer, the reservoir is adjusted in such a way that bubbles are streaming continuously, but slowly, up through the liquid of the manostatic chamber. This slow bubble overflow must be maintained, of course, throughout the experiment to insure the desired constant line pressure of gas flow to the manometer, the capillary, and hence the mixing chamber. The mixing chamber should contain either glass beads, short pieces of glass tubing, or glass wool to insure gas mixing. For any reservoir setting, the flow rate for each gas is dependent on manometer fluid and on the capillary size. Dibutyl-phthalate, even though it will react with Tygon and most rubber, makes an excellent low vapor pressure manometer-fluid, which when coupled with calibrated capillaries (2 mm i.d. and less) permits the selection of flow rates up to 20 cc/sec. For very long experiments at a constant mixing ratio, the evaporation surface of the "line pressure regulator" of Figure 1 should be enclosed to minimize reservoir fluid losses which could cause manometer level changes. If water is used in this line pressure regulator, a hundred milligrams of $CuSO_4 \cdot 5H_2O$ per liter is recommended as an algal poisoning agent.

In another design, the line pressure regulator is permanently closed at the top, fitted with a rubber tube, and fitted with a stopcock located between the reservoir and manostatic bubbler. In order to adjust reservoir levels, this stopcock is opened and air pressure (provided, for example, by blowing) is

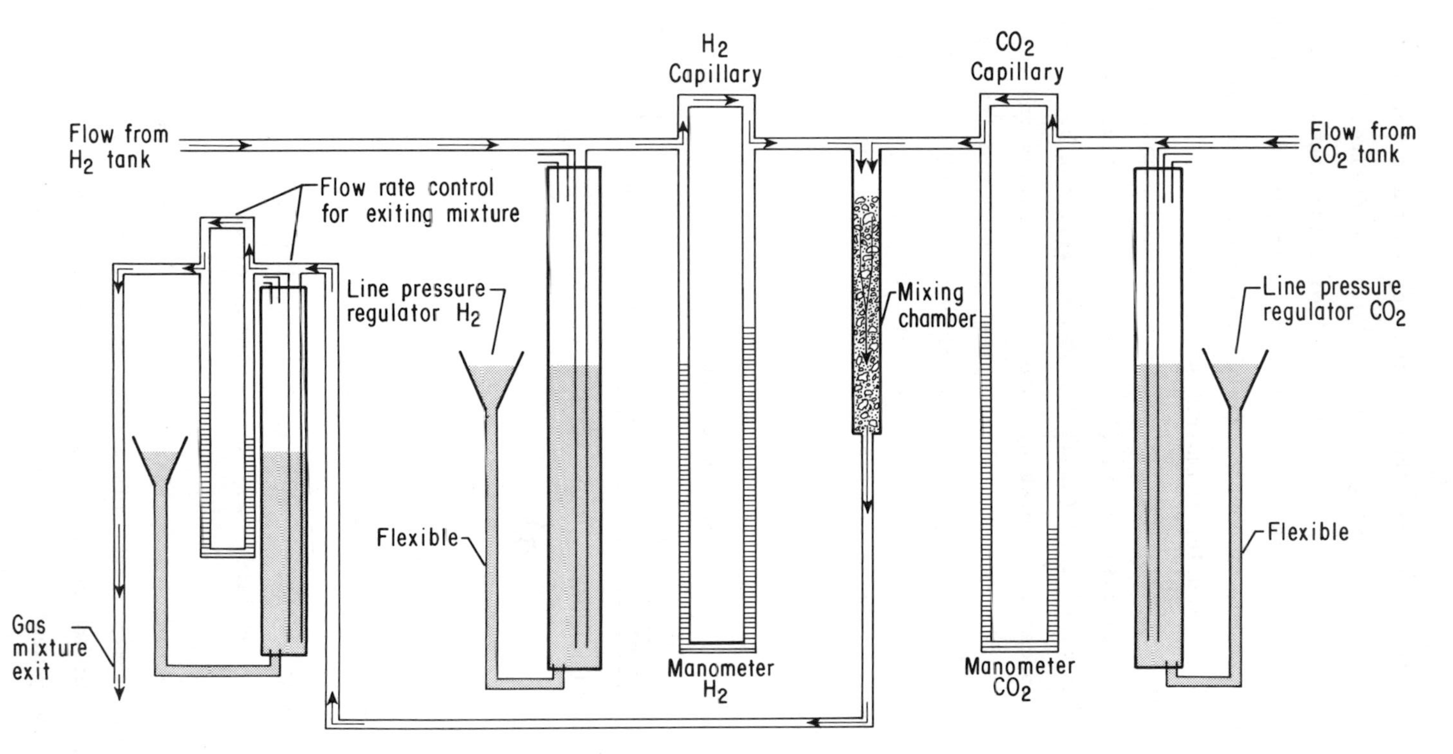

Figure 1. Schematic of a binary gas mixer.

applied to the rubber tube at the top to force the water into the manostatic bubbler slightly above the desired level. The precise level is then adjusted by allowing the excess water to drain back. The stopcock is then closed. For this to be accomplished there must be, of course, a higher level of water in the reservoir than in the bubbler chamber. Advantages of this type of mixer include: (1) it is relatively compact and eliminates the need for separate reservoirs; (2) it is less expensive to construct; (3) manometer fluid (usually organic) will not be forced by excess pressure to back up into the capillaries and contaminate them; and (4) it is more convenient to make precise reservoir adjustments. Stopcocks can be placed at strategic points (for example, between each capillary and the mixing chamber) to shut off one gas and meter the other alone without the use of clamps.

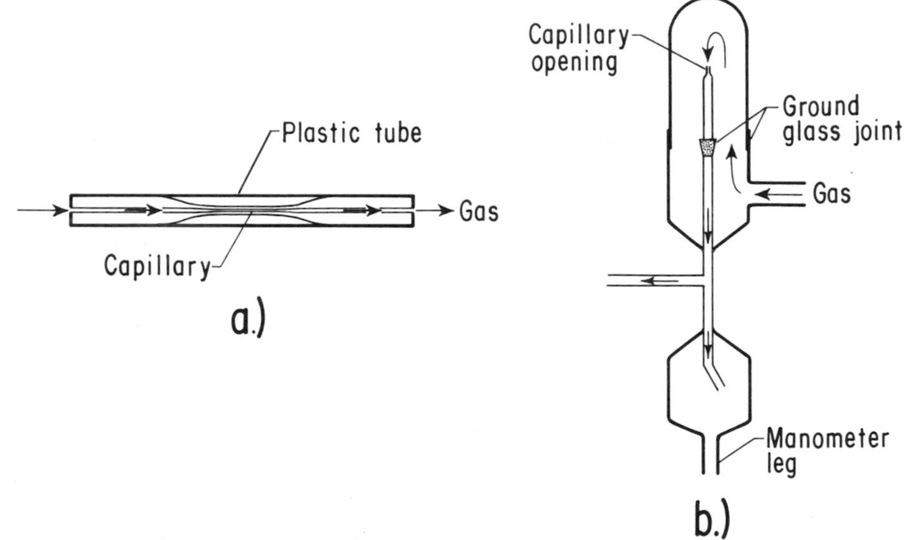

Figure 2. Capillary designs for gas mixing. a.) Drawn-out variety. b.) Encapsulated variety.

Variations in capillary design are shown in Figure 2. In Figure 2 a.) commercial glass capillary tubing is heated and drawn out to the approximate desired diameter. A plastic tube may be placed over the glass to protect the fragile capillary from breaking. The capillary opening in Figure 2 b.) is usually easier to construct, although the construction of the associated permanent glass parts requires considerable skill.

It was mentioned previously that problems are often encountered when large volumes of gas are required to achieve the desired mixing ratios. Limits of accurate mixing ratios can be extended with the use of multistage mixers whereby a mixer such as is shown in Figure 1 is used to mix two gases and the resultant gas mixture is fed into one side of a second gas mixer wherein it is mixed with a gas whose desired volume percent is large with respect to the other gas (for example, 100:1). Several stages can thus be employed.

Numerous types of gas flowmeters are now commercially available, chiefly from gas-supply and chemical-supply houses. One micro variety employs needle valves to control flow rates, although precautions must be taken to minimize relatively large back pressures. Complete portable gas mixers are also available on the market with and without total flow control.

Gas mixer tubing is usually constructed of glass which is satisfactory for a wide variety of gases. As indicated in Figure 1, the tubes connecting the water reservoirs with the regulators should be of flexible rubber or plastic, as should the capillary connections to permit rapid replacement. With proper caution concerning leaks, the capillary connections can also be constructed of ground glass. It should be noted that Tygon-type plastic tubing [modified plasticized polyvinyl chloride $(CH_2:CHCl)_2$] is impervious to hydrogen, but CO_2 diffuses through such tubing at the approximate rate of 0.001 cm^3/sec/linear meter for a wall thickness of 1.5 mm (Ulmer, 1964).

The gas mixing apparatus can be shock mounted on a board or frame. A suitable arbitrary scale should be placed adjacent to the manometer to aid in achieving desired settings. If the gas mixture exit in Figure 1 is to be attached to a furnace wound on a ceramic tube, a ground glass connection can be arranged by sealing Pyrex glass to a good grade of mullite ceramic tube. Alternatively, a gas-tight rubber sleeve (gooch tubing) can be constructed to join the ceramic and the glass, if the ceramic is not too hot.

2. Calibration

Prior to use, it is necessary to calibrate a set of capillaries with the gas to which each will be subjected. The procedure is to install the capillary to be calibrated in the gas mixer where it will eventually be used. The other side of the mixer is blocked off at its capillary location (or this side of the mixer may be isolated by closing a stopcock, if the gas mixer is so equipped). The gas mixer exit line is detached from the reaction chamber and connected to the calibration instrument. For flow rates of approximately ≥ 3 cm^3 sec^{-1} or more, a commercial precision wet test meter is satisfactory. However, for lower flow rates a Bunsen tower has been used with good results. This may simply be a glass analytical burette divided into units of 0.01 cm^3 to which suitable glass tubing has been connected as shown in Figure 3. Various capacity towers may be constructed depending on desired flow rates. A satisfactory fluid may be prepared by dissolving two tablespoons of a powdered soap (e.g., "Alconox") into 2/3 gallon (2.5 liter) of distilled water to which approximately 1 1/4 pt. (0.59 liter) of glycerin has been added. It is important to saturate the fluid for 1-2 hours with the gas which will be used to calibrate the capillary. In addition, a thermometer should be placed near the tower and volume corrections (glass expansion) made for temperatures above (or below) the burette calibration [e.g., for a 250 ml glass burette, a correction of 0.006 cm^3 must be made per degree centigrade difference from 20 C (subtract for temperatures higher than this and add for lower temperatures)]. The gas is introduced into the apparatus and the fluid reservoir raised to allow a few bubbles to form and rise up the column (the back pressure of up to six bubbles does not affect the flow rate). The differential

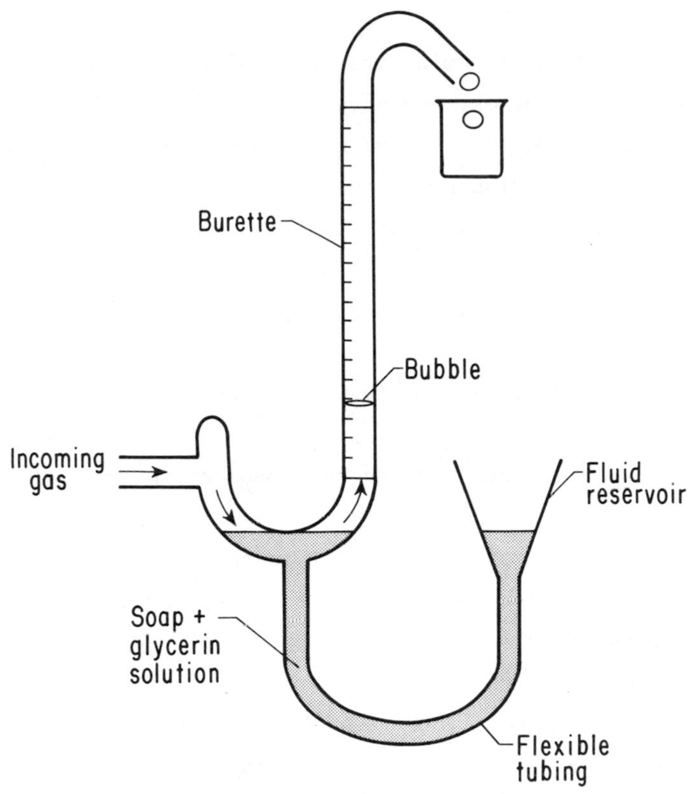

Figure 3. Bunsen bubble tower for low gas flow measurements.

manometer level on the gas mixer is noted and a stopwatch establishes the time required for one bubble to traverse the indicated volume. This measurement should be repeated several times and an average value established for each manometer setting.

Several different pressure settings on the gas mixer will result in various flow rates for this gas, and a calibration curve [$\Delta P_{manometer}$ vs. flow (cm^3 sec^{-1})] plotted for this capillary. Calibration curves for several capillaries and for each gas are thus obtained. Usually these curves are nearly linear. When these calibration curves show distinct regions of different slope, i.e., different flow regimes, turbulences are indicated and the capillary should be reconstructed. Darken and Gurry (1945) have established that approximately 0.9 cm sec^{-1} is an optimum linear flow rate of a gas mixture through an 18 inch long, 12 inch heated zone, cylindrical furnace. This rate minimizes thermal diffusion and temperature uncertainties, and provides for a slightly positive furnace exit pressure (2-3 inches water pressure). These authors provide an extensive discussion on the derivation of this rate. In order, therefore, to calculate the minimum flow rate for a gas mixture through a cylindrical furnace, the inside diameter must be known and the flow rate calculated from:

$$\text{Flow rate} = \pi r^2 (0.9)\ \text{cm}^3\ \text{sec}^{-1} \tag{37}$$

or

$$\text{Flow rate} = \text{Cross-sectional area}\ (\text{cm}^2)\ (0.9),$$

where r is the radius of the cylinder in cm.

Of course, the essential components of any gas mixer must be gas-tight and leak checking may be accomplished by closing off one side of the mixer. A steady rise in the manometer pressure should be noted if no leaks are present. It is advisable to check the entire gas mixer-furnace system for leaks prior to experimental work. This can be most conveniently accomplished by checking one or more equilibrium reactions which have been extensively reported in the literature. One reaction is $FeO = Fe + 1/2\ O_2$ which for example, is in equilibrium at 1303 C and $\log fO_2 = -10.72$ atm. Other values for this reaction have been compiled and are shown in Figure 4 along with other oxygen fugacity data for the Fe-O systems (Darken and Gurry, 1946; Muan, 1958; Osborn and Muan, 1965; Vallet and Raccah, 1965). If leaks are present, an electrolytic iron pellet or iron wire equilibrated in the furnace with a known oxygen fugacity derived from the gas mixture will show wüstite (either as "pimples" on the surface in reflected light microscopy or by a weight gain at an oxygen fugacity lower than that given for the Fe-FeO boundary at the temperature of interest. Similarly, a wüstite pellet will not be entirely reduced to metallic iron at the proper fO_2 if there is an air leak. Final gas mixture compositions

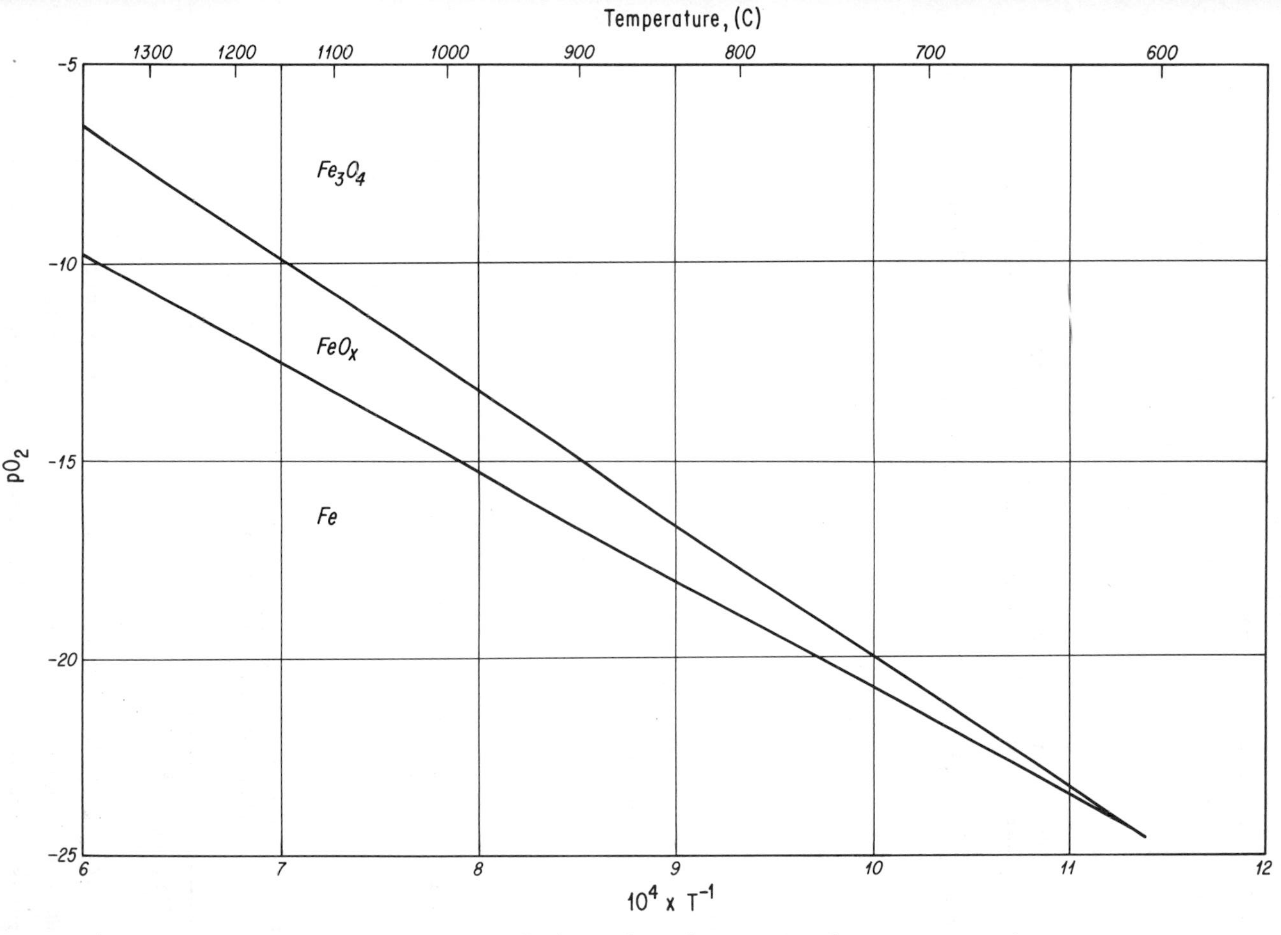

Figure 4. Data for oxygen fugacities of phase boundaries in the system Fe-O.

may also be checked by gas chromatography or by more conventional means (see, for example, Muan, 1955).

The importance of gas purity should also be stressed. Table 2.3 lists several common gases, the means for purifying these before entrance into the gas mixer, and typical purity limits.

TABLE 2.3

Means of Purification and Typical Impurity Limits for Common Buffering Gases Used to Control Oxygen Fugacities

Gas	Purifying Agent	Typical Impurity Limits	Remarks
H_2	Platinized asbestos or drierite ($CaSO_4$)	0.2%	Pure H_2 alone, may reduce ceramic furnace walls. Extra dry: electrolytic grade.
CO	Activated Al_2O_3	0.5%	C.P. grade
CO_2	Drierite, heated Cu gauze, activated Al_2O_3	0.2-0.4%	Bone dry grade
O_2	Ascarite + activated Al_2O_3	0.1%	Extra dry
Air	Same		
Inert Gas*	Anhydrous Mg perchlorate, heated Cu; or Ti-Zr turnings + ascarite	0.1%	Extra "oxygen getters" commonly used in the furnace itself
N_2O		2%	
N_2	Mg perchlorate; Cu gauze (heated 600 C + drierite)	0.3%	Extra dry

*(Ar, He, etc.)

Errors in the final equilibrium oxygen fugacity are derived from sources such as capillary calibration, flowmeter reading, impure gases, slow reaction rates in the furnace, and furnace porosity.

IV. Summary

As a final summary for this chapter, consider the following example: Calculate the room-temperature CO_2 and CO flow rates required to obtain an equilibrium oxygen fugacity of $10^{-8.00}$ atm at 1300 C in a furnace with a reaction zone 2.5 cm in diameter. The volume percent of CO_2 in a CO_2-CO mixture required to attain the above oxygen fugacity at the specified temperature is 87.92 (Deines _et al._, 1971). Therefore the room-temperature CO_2-CO mixing ratio is 87.92/12.08 or 7.28. The cross-sectional area of the furnace is $(3.14)\ (1.25\ cm)^2 = 4.91\ cm^2$. Therefore the minimum gas mixture flow rate is $4.91\ cm^2 \times 0.9\ cm\ sec^{-1}$ or 4.42 $cm^3\ sec^{-1}$. In order to attain this desired total flow rate at the appropriate mixing ratio, the CO_2 flow rate (x) must be 3.88 $cm^3\ sec^{-1}$, and that for CO (y) must be 0.54 $cm^3\ sec^{-1}$, _i.e._, $x + y = 4.42$; $x/y = 7.28$, or $x = 7.28\ y$; $y = 4.42/8.28 = 0.54\ cm^3\ sec^{-1}$; and $y = x/7.28$ or $x(1 + \frac{1}{7.28}) = 4.42$; $x = 3.88\ cm^3\ sec^{-1}$. The corresponding number readings for each gas in the mixture may be obtained from the appropriate capillary calibration curve and the gas mixer set as described previously.

BIBLIOGRAPHY:

Anderson, A.T., Jr., Crewe, A.V., Goldsmith, J.R., Moore, P.B. Newton, J.C., Olsen, E.J., Smith, J.V., and Wyllie, P.J. (1970). Petrologic history of the moon suggested by petrography, mineralogy, and crystallography. Science _167_:587-590.

Bell, P.M. (1971). Analysis of olivine crystals in Apollo 12 rocks. Annual Report of the Director, Geophysical Laboratory Yearbook, _69_:228-229.

Bowen, N.L. and Schairer, J.F. (1932). The system $FeO-SiO_2$. Amer. J. Sci., _224_:177-213.

Bowen, N.L. and Schairer, J.F. (1935). The system $MgO-FeO-SiO_2$. Amer. J. Sci., _229_:151-217.

Cairns, E.J. and Tevebaugh, A.D. (1964). CHO gas phase compositions in equilibrium with carbon, and carbon deposition boundaries at one atmosphere. J. Chem. and Eng. Data, 9:453-462.

Campbell, F.E. and Roeder, P. (1968). The stability of olivine and pyroxene in the Ni-Mg-Si-O system. Amer. Mineral., 53:257-268.

Carmichael, I.S.E. and Nicholls, J. (1967). Iron and titanium oxides and oxygen fugacities in volcanic rocks. J. Geophys. Res., 72:4665-4687.

Chang, L.L.Y. and Phillips, B. (1969). Phase relations in refractory metal-oxygen systems. J. Amer. Ceram. Soc., 52:527-533.

Chipman, J. and Marshall, S. (1940). The equilibrium $FeO + H_2 = Fe + H_2O$ at temperatures up to the melting point of iron. J. Amer. Chem. Soc., 62:299-305.

Coughlin, J.P. (1954). Contributions to the data on theoretical metallurgy XII. Heats and free energies of formation of inorganic oxides. U.S. Bureau of Mines Bull., 542.

Darken, L.S. and Gurry R.W. (1945). The system iron-oxygen. I. The wüstite field and related equilibria. J. Amer. Chem. Soc., 67:1398-1412.

Darken, L.S. and Gurry R.W. (1946). The system iron-oxygen. II. Equilibrium and the thermodynamics of liquid oxide and other phases. J. Amer. Chem. Soc., 68:798-816.

Davies, M.W. and Richardson F.D. (1959). The nonstoichiometry of manganous oxide. Trans. Faraday Soc., 55:604-610.

Deines, P., Nafziger, R.H., Ulmer, G.C., and Woermann, E. (1971). T-fO_2 tables for selected gas mixtures in the C-H-O system. The Pennsylvania State Univ., Coll. of Earth and Mineral Sci. Bull., in press.

Elliott, J.F. and Gleiser, M. (1960). Thermochemistry for steelmaking, Vol. I. Reading, Mass.: Addison-Wesley.

Emmett, P.H. and Shultz, J.F. (1930). Equilibria in the Fe-H-O system. Indirect calculation of the water gas equilibrium constant. J. Amer. Chem. Soc., 52:4268-4285.

French, B.M. (1966). Some geological implications of equilibrium between graphite and a C-H-O gas phase at high temperatures and pressures. Rev. Geophys., 4:223-253

French, B.M. and Eugster, H.P. (1965). Experimental control of oxygen fugacities by graphite-gas equilibrium. J. Geophys. Res., 70:1529-1539.

Fudali, R.F. (1965). Oxygen fugacities of basaltic and andesitic magmas. Geochim. Cosmochim. Acta, 29:1063-1975.

Glasser, F.P. (1958). The system $MnO-SiO_2$. Amer. J. Sci., 256: 398-412.

Graham, H.C. and Davis, H.H. (1971). Oxidation/vaporization kinetics of Cr_2O_3. J. Amer. Ceram. Soc., 54:89-93.

Hahn, W.C., Jr. and Muan, A. (1960). Studies in the system Mn-O: The Mn_2O_3-Mn_3O_4 and Mn_3O-MnO equilibria. Amer. J. Sci., 258:66-78.

Hahn, W.C., Jr. and Muan, A. (1961). Activity measurements in oxide solid solutions: the systems NiO-MgO and NiO-MnO in the temperature interval 1100-1300°C. Internat. J. Phys. Chem. Sol., 19:338-348.

Johnson, R.E. and Muan, A. (1968). Phase diagrams for the systems Si-O and Cr-O. J. Amer. Ceram. Soc., 51:430-433.

Johnston, J. and Walker, A.C. (1925). Preparation and analysis of constant mixtures of air and carbon dioxide. J. Amer. Chem. Soc., 47:1807-1817.

Katsura, T. and Hasegawa, M. (1967). Equilibria in the V_2O_3-VO_2 system at 1600°K. Bull. Chem. Soc. Japan, 40 #3:561-569.

Larimer, J.W. (1968). Experimental studies on the system Fe-MgO-SiO_2-O_2 and their bearing on the petrology of chondritic meteorites. Geochim. Cosmochim. Acta, 32:1187-1207.

Massé, D.P., Rosén, E., and Muan A. (1966). Activity-composition relations in CO_2SiO_4-Fe_2SiO_4 solid solutions at 1180°C. J. Amer. Ceram. Soc., 49:328-329.

Morris, A.E. and Muan, A. (1966). Phase equilibria in the system MnO-Mn_2O_3-SiO_2. J. Metals, 18:957-960.

Muan, A. (1955). Phase equilibria in the system FeO-Fe_2O_3-SiO_2.

J. Metals, 7, Trans. Metal. Soc., AIME, 203:965-976.

Muan, A. (1958). Phase equilibria at high temperatures in oxide systems involving changes in oxidation states. Amer. J. Sci., 256:171-207.

Muan, A. (1959). Stability relations among some manganese minerals. Amer. Mineral., 44:946-960.

Muan, A. and Osborn, E.F. (1956). Phase equilibria at liquidus temperatures in the system $MgO-FeO-Fe_2O_3-SiO_2$. J. Amer. Ceram. Soc., 39:121-140.

Muan, A. and Osborn, E.F. (1965). Phase equilibria among oxides in steelmaking. Reading, Mass: Addison-Wesley.

Nafziger, R.H. and Muan, A. (1967). Equilibrium phase compositions and thermodynamic properties of olivines and pyroxenes in the system MgO-"FeO"-SiO_2. Amer. Mineral., 52:1364-1385.

Osborn, E.F. (1959). Role of oxygen pressure in the crystallization and differentation of basaltic magmas. Amer. J. Sci., 257:609-647.

Osborn, E.F. (1962). Reaction series for subalkaline igneous rocks based on different oxygen pressure conditions. Amer. Mineral., 47:211-226.

Osborn, E.F. and Muan, A. (1963). Physical chemistry of steel-plant refractories. Electric furnace steelmaking, Vol. II: Theory and fundamentals, E.C. Sims, editor. New York: Interscience.

Porter, V.R. (1965). Studies in the titanium oxygen system and the defect nature of rutile. Ph.D. thesis, The Pennsylvania State Univ., University Park, Pa.

Ramdohr, P. (1964). Opaque minerals in stony meteorites. Carnegie Inst. Washington Year Book, 63:217-218.

Sawamura, H. and Matoba, S. (1960, 1961). Recommended equilibrium values for the reactions $O + H_2 = H_2O$, $O + CO = CO_2$, $C + CO_2 = 2\ CO$, and $C + O = CO$. Subcommittee for Physical Chemistry of Steelmaking, 19th Committee, 3rd division, Japan Soc. for Promotion of Sci.

Speidel, D.H. (1967). Phase equilibria in the system MgO-FeO-Fe_2O_3: the 1300°C isothermal section and extrapolations to other temperatures. J. Amer. Ceram. Soc., 50:243-248.

Speidel, D.H. and Heald, E.F. (1967). Gaseous equilibria in portions of the system C-H-O-S. The Pennsylvania State Univ., Coll. of Earth and Mineral. Sci. Bull., 83.

Speidel, D.H. and Nafziger, R.H. (1968). P-T-fO_2 relations in the system Fe-O-MgO-SiO_2. Amer. J. Sci., 266:361-379.

Stull, D.R. et al. (1965 and yearly supplements). JANAF (Joint Army, Navy, Air Force) thermochemical tables. Dow Chem. Co., Midland, Mich.

Tuthill, R.L. and Sato, M. (1970). Phase relations of a simulated lunar basalt as a function of oxygen fugacity, and their bearing on the petrogenesis of the Apollo 11 basalts. Geochim. Cosmochim. Acta, 34:1293-1302.

Ulmer, G.C. (1964). Oxidation-reduction reactions and equilibrium phase relations at 1300°C at oxygen pressures from 0.21 to 10^{-14} atmospheres for the spinel solid solution series $FeCr_2O_4$-$MgCr_2O_4$ and $FeAl_2O_4$-$MgAl_2O_4$. Ph.D. thesis, The Pennsylvania State Univ., University Park, Pa.

Vallet, P. and Raccah, P. (1965). Contribution to the study of the thermodynamic properties of solid ferrous oxide. Mem. Sci. Rev. Met., 62:1-29.

Walter, L.S. (1969). Determination of oxygen activities of chondritic meteorites. Geol. Soc. Amer. Abstracts, Part 7: Atlantic City, Meeting, 232-233.

Wicks, C.E. and Block, F.E. (1963). Thermodynamic properties of 65 elements, their oxides, halides, carbides, and nitrides. U.S. Bureau of Mines Bull., 605.

CHAPTER 3

Electrochemical Measurements and Control of Oxygen Fugacity and Other Gaseous Fugacities with Solid Electrolyte Sensors[1]

Motoaki Sato

I. Introduction

Recent advances in the theory and technology of solid electrolyte sensors have opened up a very promising field of direct monitoring of thermodynamic behavior of a chemical system which involves a non-condensed phase directly or indirectly. At present only oxygen and sulfur sensors are practical to use, but it is very likely that sensors for other gases will soon be developed. A fluorine sensor, for example, has already reached experimental stage. By combining appropriate solid electrolyte, electrode material, and design, sensors for hydrogen, chlorine, and even carbon dioxide may soon come into existence.

The advantages of the solid electrolyte sensors are many; to name a few: (1) the response is usually very specific, rapid and continuous, (2) the measurement can be made directly in or with the system to be investigated without disturbing the system, (3) a single sensor can often cover exceedingly wide ranges of fugacity and temperature, (4) the nature of the sensor output, *i.e.*, voltage, is such that high precision is easily obtainable, (5) not only equilibrium values but also the amount of departure from equilibrium or the rate of approach to equilibrium can be observed, and (6) fugacity or activity rather than partial pressure or concentration is measured.

The above advantages arise mainly from the facts that the solid electrolyte sensors are reversible electrochemical cells and that solid electrolytes are normally very selective ionic

---

[1] Publication authorized by the Director, U.S. Geological Survey.

conductors. An electrochemical cell is a transducer of available chemical energy to electrical energy; it translates chemical potential (i.e., fugacity or activity; not partial pressure or concentration) to electrical potential. A reversible cell can achieve the translation without net chemical changes because the electrode reactions responsible for the electrical potential proceed in both directions at the same rate (if no current is allowed to flow); hence, there is virtually no disturbance to the system under investigation. A solid electrolyte differs from aqueous and molten salt electrolytes in that it normally allows the passage of only anion or cation, and is frequently specific to only one kind of anion or cation. This specificity restricts the number of possible electrode reactions and hence the chemical potentials to be translated. It becomes possible in many cases to deal with the fugacity of only one component in the presence of others which might otherwise interfere with the measurement.

The solid electrolyte sensors have many proven and feasible areas of application in addition to the primary fugacity or activity determinations. Their ability to monitor the fugacity of a specific component directly and continuously makes an automated control of gas mixing possible. It is also possible to use the sensor itself as a quantitative pump for a specific component by deliberately allowing an electric current through the sensor. The direction of pumping is reversible. Such pumping is useful for titration, solubility studies, and investigation of non-stoichiometric properties of a solid compound. Reaction rate study is another area of application. These features will be discussed in detail in later sections.

II. Oxygen Fugacity Sensor

A. General Background

The oxygen fugacity sensor is the most developed of all solid electrolyte sensors. The modern foundation for this sensor was laid down by the classic work of Kiukkola and Wagner (1957). Numerous papers were published on this subject in the following years, but only occasional reference will be made to selected papers for brevity. The sensor is essentially an oxygen concentration cell which is composed of a solid electrolyte for O^{2-} ion, an oxygen reference electrode, and an electrode of unknown oxygen fugacity (sample). By measuring the e.m.f. and the temperature of the cell and knowing the fO_2 value of the reference at this temperature, the unknown fO_2 value can be determined.

To illustrate, suppose we have two compartments separated with an impermeable wall and one compartment is filled with air and the other with a fuel gas, say a mixture of carbon monoxide and a little carbon dioxide, both at 1 atm total pressure and at $T^{o}K$. If a hole is punched in the wall, the combustion reaction:

$$2\ CO + O_2 = 2\ CO_2 \tag{1}$$

will take place provided the temperature is sufficiently high. The molar free energy of reaction at the incipient stage of the reaction is:

$$\Delta G_r = \Delta G^o_r + RT\left[2\ \ln\ (fCO_2/fCO) - \ln\ fO_2(air)\right]. \tag{2}$$

At equilibrium $\Delta G_r = 0$, and the oxygen fugacity of the CO_2-CO gas mixture is defined by reaction (1) as:

$$\ln\ fO_2\ (CO_2\text{-}CO) = (\Delta G^o_r/RT) + 2\ \ln\ (fCO_2/fCO). \tag{3}$$

From (2) and (3) we obtain the relationship:

$$\Delta G_r = RT\ \ln\left[fO_2\ (CO_2\text{-}CO)/fO_2\ (air)\right]. \tag{4}$$

Reaction (1) can also proceed without having a hole punched in the wall, if the wall allows migration of negatively charged oxygen ions and suitable metallic conductors are in contact with the opposite surfaces of the wall as shown in Figure 1. As soon as the conductors are connected to form a closed circuit, oxygen molecules colliding with the wall in the first compartment can be ionized to O^{2-} ions by receiving electrons from the conductor and migrate into the wall. Simultaneously, the O^{2-} ions in the opposite side of the wall can give up their electrons to the conductor, convert to neutral oxygen, and immediately react with carbon monoxide. As long as the chemical potential or fugacity

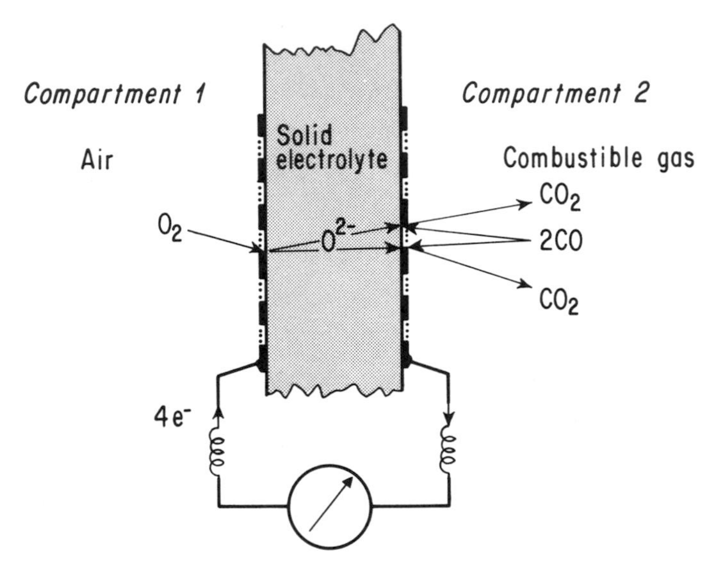

Figure 1. Diagram illustrating the principle of a solid electrolyte oxygen sensor. If the external circuit is closed, oxygen will move from left to right through the solid electrolyte in the ionized form, while electrons move in the reverse direction through the external conductor. The open-circuit voltage is proportional to the logarithm of the ratio of oxygen fugacities.

of oxygen in compartment 1 remains higher than that in compartment 2 the electrons continue to flow from right to left (and an electric current in the opposite direction) through the conductor and the combustion reaction proceeds. This is the basic principle of the solid electrolyte fuel cell, which provides electric power by quiet combustion of a fuel gas. The implication here is that, in order to have an electric current through a conductor of a finite resistance, there must be a finite potential difference between the two conductor-electrolyte contacts to begin with. Such an electrical potential difference indeed exists and can be measured with a potentiometer. This open-circuit potential, or e.m.f. value is related to the molar free energy of reaction ΔG_r in (4), simply as:

$$E = - \Delta G_r/nF$$
$$= (RT/4F) \ln [fO_2 (CO_2\text{-}CO)/fO_2 (air)] \quad (5a)$$
$$= 0.04960 T \log [fO_2(CO_2\text{-}CO/fO_2 (air)] \text{ mV (millivolts)} \quad (5b)$$

where n is the number of electrons involved in the cell transfer, $O_2 + 4e^- = 2 O^{2-}$, R gas constant (8.31439 abs. joules/degree-mole, and F the faraday constant (96,484 abs. coulombs) (Latimer, 1952). The polarity indicated by the above equations is such that the electrode (gas + conductor in this case) with a higher oxygen fugacity corresponds to the positive electrode. If the air electrode is taken as the reference electrode, the CO_2-CO electrode then has a negative potential. The general form of the above equations is often referred to as the Nernst relationship. This relationship applies only if there is no appreciable current drain during the voltage measurement and if the electrolyte has only negligible electronic conduction. The presence of electronic conduction is a characteristic problem with solid electrolytes. These two conditions are very important in the practical application of the solid electrolyte sensors and will be discussed in more detail later.

From the above discussion, it is evident that the fugacity in a gas mixture can be determined with such a cell by measuring the e.m.f. and the temperature, if the reference fugacity is known, and if a suitable solid electrolyte is found. It is also evident that only thermodynamic quantities are involved in this method, and that partial pressure or concentration must be approximated or inferred from other sources of information such as the degree of deviation of the gas from ideality at the temperature and pressure of the measurement. The determination of the oxygen fugacity of a condensed phase is similarly possible. Because the second law of thermodynamics dictates that the chemical potential of an element be identical in all phases at equilibrium, the chemical potential or fugacity of oxygen in a gas phase in equilibrium with a condensed phase, no matter how small the volume of the gas, must be identical with that of the condensed phase. If the gas is in direct contact with the sensor, it is apparent that the same principle applies as in the case of a gas mixture. Even when a condensed phase alone is in direct contact with the sensor, the same principle should still apply. Such an arrangement may be regarded as a special case where the volume of the gas phase is infinitesimally small. The determination of oxygen fugacity in a molten metal or slag, into which a solid electrolyte sensor is immersed, corresponds to this case. As will be discussed later, however, it is preferable to avoid such a direct contact for a number of technical reasons.

The sensor design varies depending on the shape of the electrolyte, configuration of the cell, and type of oxygen fugacity reference. Variations also exist with the choice of the solid electrolyte, metallic conductors, and oxygen fugacity reference material. The application of the sensor for direct measurements of oxygen fugacity has been tried in diverse fields such as chemical thermodynamics, metallurgy, and geology to obtain data on thermodynamic parameters, reaction rates, and change in oxidation states, mostly at high temperatures. Other areas of application include the control of the oxygen fugacity of a furnace atmosphere, quantitative addition or subtraction of oxygen to study the range of homogeneity of an oxide, and determination of the optimum mixing ratio of a fuel gas-air mixture.

B. Solid Electrolyte

1. General Characteristics

The most critical part of the oxygen fugacity sensor is the solid electrolyte which is specific to O^{2-} ion. The first step in constructing a good workable sensor is to obtain a solid electrolyte of a desired shape and of quality which meets the following requirements: (1) oxygen ion conductivity predominates over other types of conductivity throughout a wide range of oxygen fugacity, (2) the ionic conductivity is large enough to permit accurate e.m.f. measurements with existing potentiometric equipment over a wide temperature range, (3) the reversible electrode reaction for the ionization of molecular oxygen can take place easily at the electrolyte-conductor interface, (4) the material is thermally stable (i.e., no phase change) and chemically inert up to high temperatures, and (5) it has very low permeability for neutral gases including molecular oxygen. The reason behind the first requirement may need a little elaboration here. The presence of cationic and other anionic conductivity may result in interference by other elements. The presence of electronic conduction not only lowers the accuracy of the measurements, but also results in the leakage of oxygen across the electrolyte. The electronic conduction generally becomes appreciable when oxygen fugacity is very low or very high. When this happens, the e.m.f. of the cell no longer follows the Nernst relationship and becomes smaller than expected from equation (5). The voltage measured is no longer the open-circuit voltage but becomes a voltage with a resistive load of a finite value. It is equivalent to having the opposite poles of the cell short-circuited with a resistor, allowing a flow of current and hence a leakage of oxygen across the electrolyte. This oxygen leakage may seriously contaminate a system under investigation.

Few substances can satisfactorily fulfill all these requirements. In order to have appreciable oxygen ion conductivity, oxygen ion must be abundant and easily mobile within the structure of the material. This requirement limits the selection to oxides and oxide solid solutions which have appreciable defects in the crystal structure. The requirement that the electronic conduction be negligible (less than one percent) eliminates oxides of transition elements which can assume more than one oxidation state and thus allow hopping of electrons among the cationic sites, contributing to electronic conduction. Scores of oxides and oxide solid solutions of non-transition elements have been examined for this purpose, but so far only a few have been found to be satisfactory for practical applications.

2. Zirconia Electrolytes

The solid electrolytes most commonly used today for oxygen fugacity sensors are the solid solutions of ZrO_2-CaO containing 15 mole % (7.5%) of CaO (lime, calcia) and of ZrO_2-Y_2O_3 containing 10 mole % (15%) of Y_2O_3 (yttria). Pure ZrO_2 (zirconia), is monoclinic below about 1000 C and tetragonal at higher temperatures up to its melting point of approximately 2700 C (Garvie, 1970). The monoclinic-tetragonal transition involves a large, disruptive volume change and hence thermal cycling across the transition temperature renders the oxide mechanically unstable. Pure zirconia is also an electric insulator with a conductivity of about 10^{-8} ohm^{-1} cm^{-1} at 1000 C. When a suitable amount of one or more oxides of metals having valence states of +2 or +3 is added, the resulting solid solution assumes a fluorite-type cubic structure which is stable from low temperatures up to the melting point, which is in the range of 2200 to 2600 C. This cubic form is often referred to as "stabilized" zirconia and has been utilized extensively as a high-temperature refractory material. In addition to the increase in structural stability, the electrical conductivity is also increased drastically to as high as 10^{-1} ohm^{-1} cm^{-1} (at 1000 C). The conductivity is almost entirely anionic and explained as due to the presence of a significant population of vacant anion sites allowing easy migration of O^{2-} ions in the structure. Under an electrical field the O^{2-} ions exchange positions with the vacant anion sites and move toward the positive pole, thus sustaining a flow of negative charge. The vacant anion sites are due to the occupation of some of the cation sites by cations of +2 or +3 valence state instead of Zr^{4+} ions, requiring fewer oxygens to maintain charge balance. Apparently it is easier to have anion vacancies than to have excess cations to maintain the electrical charge balance in this particular structure.

The conductivity varies with temperature, composition, and oxygen fugacity. The temperature dependence is the most straightforward of these variables. The logarithm of the conductivity increases linearly with the decrease in the reciprocal of the absolute temperature for a given composition, similar to normal diffusion rates. Figure 2 shows examples of this behavior. This

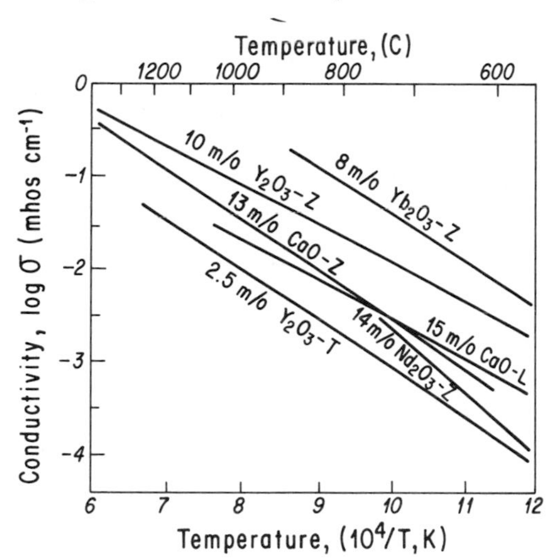

Figure 2. The variation of conductivity (σ ohm^{-1} cm^{-1}) of various solid electrolytes for O^{2-} ion as a function of temperature. The compositions chosen are for maximum or near maximum ionic conductivity. The symbol m/o denotes mole percent of the oxide as written. The data are from Strickler and Carlson (1964); Subbarao et al. (1965); Etsell and Flengas (1969); and Spacil and Tedmon (1969). L = La_2O_3, T = ThO_2, Z = ZrO_2.

behavior is, as a matter of fact, regarded as supporting evidence for the diffusion-controlled conduction mechanism discussed above. The composition dependence is not so straightforward. Although it has been clearly demonstrated that the conduction is by O^{2-} ions, conductivity does not necessarily increase with increase in anion vacancy sites (larger concentration of the stabilizing oxide) which presumably favors higher O^{2-} ion mobility, but it shows a maximum at a certain intermediate composition which has not been predicted theoretically. Examples of this behavior are shown in Figure 3. Nor is it possible to explain why a certain oxide

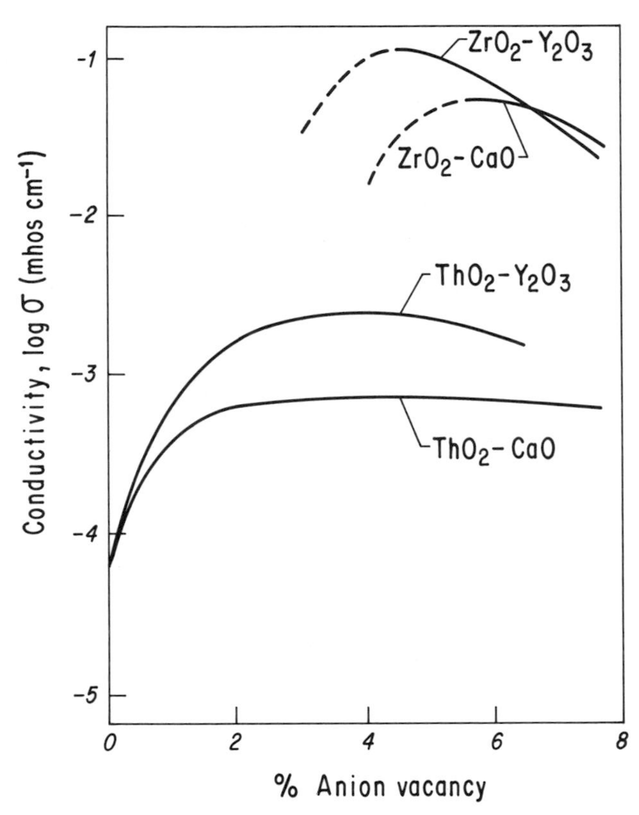

Figure 3. Variation of conductivity (σ) as a function of anion vacancy concentration at 1000 C for several electrolytes (after Jakeš, 1969).

produces a higher conductivity than another. As shown in Figure 2, considerable differences exist in the maximum conductivity (at a given temperature) among various stabilizing oxides. The stabilizing oxides (sometimes called "doping" oxides in analogy to semiconductor doping for creation of defects) are most commonly CaO and Y_2O_3, but other alkaline earth oxides (MgO and SrO) and rare earth oxides have also been tested. Oxides of transition elements are to be avoided because these elements enhance electronic conduction by assuming more than one valence state and thus allow hopping of electrons (mixed valence type conduction). At present there is no theoretical basis to predicting which element will give the highest conductivity, and the search for a better electrolyte is largely empirical.

The change in conductivity as a function of oxygen fugacity is a problem of a different nature. When oxygen fugacity is high, an oxide ionic conductor normally exhibits an exponentially increasing p-type electronic conductivity (conduction by positive holes; see, for example, Weise, 1964) with increasing log fO_2, as schematically illustrated in Figure 4. Schmalzried (1962) gives an equation

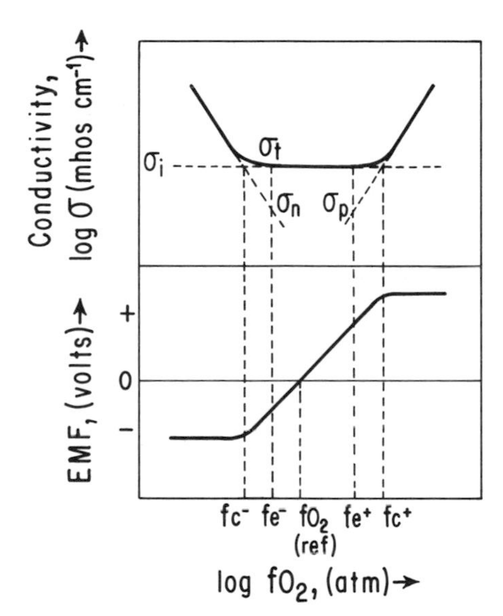

Figure 4. Schematic diagrams showing the variation of conductivity (upper half) and cell e.m.f. (lower half) as a function of oxygen fugacity for a typical oxide ion solid electrolyte. Total conductivity (σ_t) increases at low and high oxygen fugacities because of increase in n-type and p-type electronic conduction (σ_n and σ_p), respectively. The limiting fO_2 values (f_e+ and f_e-), where $\sigma_i = 99\% \sigma_t$, are about 8 log units inside of the critical fO_2 values (f_c+ and f_c-) where $\sigma_i = 50\% \sigma_t$. Accurate response of the cell e.m.f. to a change in oxygen fugacity is obtained between f_e+ and f_e-.

which describes the e.m.f. of a cell with a solid electrolyte that shows this kind of mixed conductivity. The variation of the e.m.f. of such a cell as a function of log fO_2 is also schematically shown in Figure 4, where it is assumed that the reference fO_2 is within the range of predominantly anionic conduction. For the purpose of oxygen fugacity measurements, a solid electrolyte becomes unsuitable to use when the electronic conductivity exceeds 1% or so of the total conductivity, because the error produced by this factor becomes comparable to the limits of error tolerated in routine potential and temperature determinations. It is important to know, therefore, the oxygen fugacity range of each electrolyte in which its electronic conductivity remains below this level. The approximate limits of fO_2 range for a few commonly used electrolytes are shown as a function of temperature in Figure 5. Certain important oxygen buffer curves are also plotted in the figure for reference. It turns out that both lime- (calcia) and yttria-stabilized zirconia remain >99% ionic conductors even at 1 atm of oxygen, which is a rather unique characteristic among solid electrolytes. They show n-type conductivity (conduction by negative electrons) however, at low oxygen fugacities. The "limiting" fO_2 value where electronic conductivity exceeds the level of about one percent depends on temperature as is evident in the figure, and the slope may not be close to those of solid oxygen-buffers (see Chapter 5 for definition). Thus, lime-stabilized zirconia is safe to use even at a few log fO_2 units below that of the wüstite-iron (WI) buffer at low temperatures, but as the temperature is raised its limiting fO_2 approaches the fO_2 value of the WI buffer. Above about 1400 C, this electrolyte is unsuitable to use with this buffer or any system with a lower oxygen fugacity. The electronic conductivities of zirconia electrolytes at low oxygen fugacities are thought to be caused by the partial reduction of ZrO_2 with a loss of some oxygen leading to the mixed valence mode of conduction by zirconium ions. When this happens, the ivory white ceramic becomes grayish in color. As shown in Figure 5, yttria-stabilized zirconia

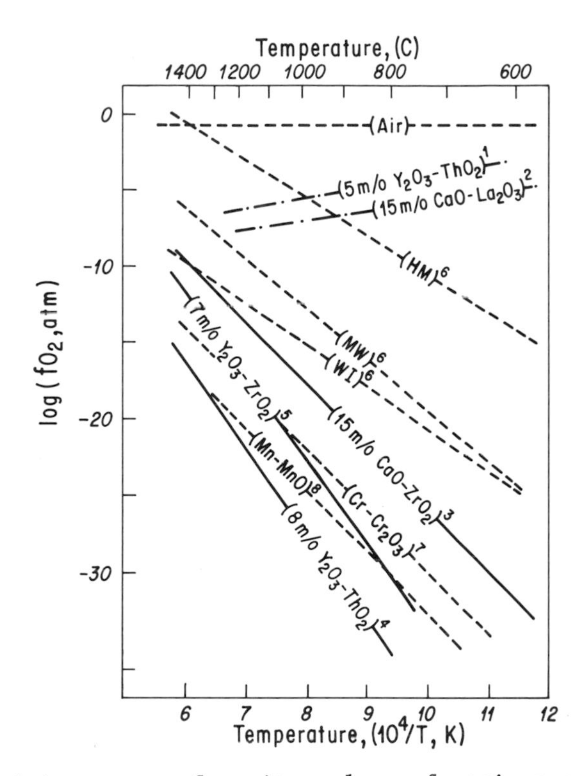

Figure 5. Limiting oxygen fugacity values of various solid electrolytes as a function of temperature. Electronic conductivity becomes appreciable above the broken line and below the solid line of the specified electrolytes. The values shown are approximate ($\pm$1 log atm unit). References: 1. Lidster and Bell (1969); 2. Etsell and Flengas (1969); 3. Schmalzried (1962); Baker and West (1966); 4. Tretyakov and Muan (1969); and 5. Swinkels (1970). The dashed lines indicate buffer fugacities with M for magnetite, W for wustite, H for hematite, and I for iron. References: 6. Eugster and Wones (1962); 7. Jeannin *et al.* (1963); and 8. Alcock and Zador, (1967).

has a limiting fO_2 value several log units lower than that of the lime-stabilized zirconia. In addition, 10 mole % yttria-stabilized zirconia has a conductivity almost one order of magnitude higher than that of 15 mole % lime-stabilized zirconia at relatively low temperatures. As shown in Figure 2, 8 mole % Yb_2O_3 (ytterbia)-stabilized zirconia has the highest conductivity, particularly at high temperatures, but the lower cost and buffer availability make the yttria-stabilized zirconia more attractive. Besides, at high temperatures the conductivity becomes less important than other factors (such as limiting fO_2 value, cost, mechanical and thermal properties) in the selection. Most zirconia electrolytes have large enough conductivity at these temperatures for presently available potential readout systems. It is at low temperatures (<600 C) that the conductivity assumes a great importance, because the internal resistance of the sensor cell becomes large compared to the input resistance of many potential measuring devices. The lesser temperature-dependence of the conductivity of yttria-stabilized zirconia makes this material much preferable to lime-stabilized zirconia for low-temperature measurements. Zirconia electrolytes stabilized with MgO (magnesia), SrO (strontia), La_2O_3 (lanthania) and other rare earth oxides, either singly or mixed, have been explored for one reason or another. Some of them are said to have shown superior thermal shock resistance, and some superior conductivity. These electrolytes are still in the testing stage of development. For more information, reference may be made to the papers by Baker and West (1966), Jakeš (1969), and Spacil and Tedmon (1969).

3. Thoria and Other Electrolytes

For very low oxygen fugacities, ThO_2 (thoria) doped with Y_2O_3 (yttria) is frequently used as the solid electrolyte because of its extremely low limiting fO_2 value at the reduction side, although its ionic conductivity is a good deal lower than those of zirconia electrolytes. It is not simple to use thoria electrolytes for reasons that will be mentioned later. In contrast to zirconia, pure ThO_2 has intrinsically a cubic fluorite-type structure up to its melting point (~3300 C). It can accomodate a fair amount of Y_2O_3 in its structure at high temperatures (10~12 mole % Y_2O_3 at 1400 C), and the conductivity increases in proportion to the amount of Y_2O_3 added in solid solution. Normally, ThO_2 with 5-8 mole % Y_2O_3 is used for the electrolyte. The function of yttria in this case is not to stabilize the cubic structure but simply to create oxygen vacancies. Thoria electrolytes also differ from zirconia electrolytes in that p-type electronic conduction becomes significant at moderate oxygen fugacity (in the neighborhood of 10^{-6} atm) and sharply increases with increase in oxygen fugacity. This behavior is normal for an ionic conductor as schematically illustrated in Figure 4, but makes the use of the thoria electrolyte more difficult than the zirconia electrolyte, because direct exposure to oxygen, air, or any other system with a relatively high oxygen fugacity has to be avoided during the measurements. The p-type electronic conduction is believed to be caused by excess oxygen entering the vacancies and creating electron holes. A thoria electrolyte sintered in air is pale brown. The color disappears if the electrolyte is heated in a reducing atmosphere for a period of time, as the excess oxygen is then driven off. The fractional amount of this excess oxygen is reportedly less than 4×10^{-5} (Bauerle, 1966). This is small enough to be ignored in many cases, but if the system to be investigated is unbuffered and relatively small, deoxygenation of the electrolyte by such a treatment prior to loading a sample is recommended in order to avoid possible oxidation of the sample by the evolving oxygen. A thoria electrolyte tube weighing 30 g could produce nearly 1 cm^3 of free oxygen at STP. Various techniques for using the thoria electrolyte are discussed in a later section. The upper and the lower limiting oxygen fugacities of yttria-doped thoria, which are temperature-dependent, are graphically shown in Figure 5.

Solid solutions of thoria with CaO and La_2O_3 have also been tried experimentally as the solid electrolytes. The ThO_2-CaO (1.4%) was reported (Worrell, 1966) to be superior to the ThO_2-Y_2O_3

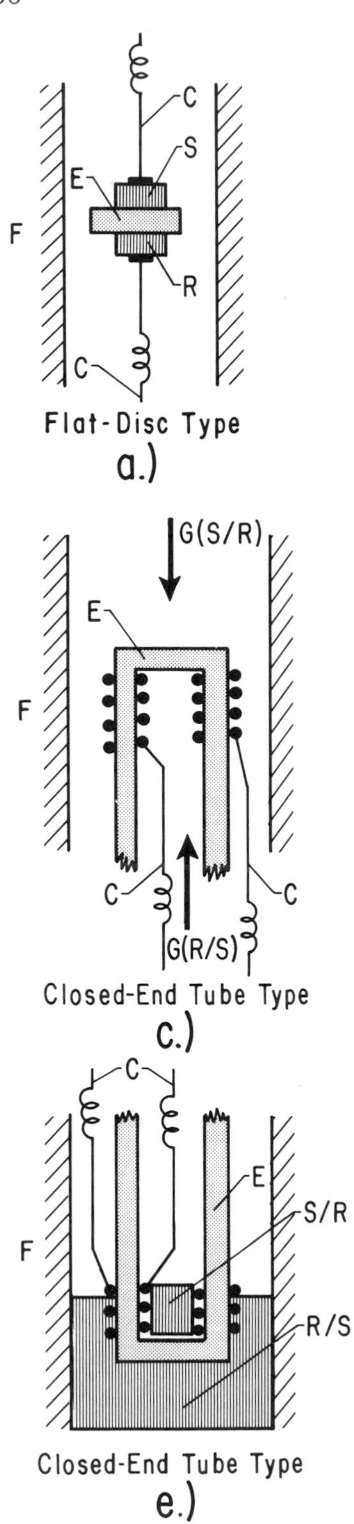

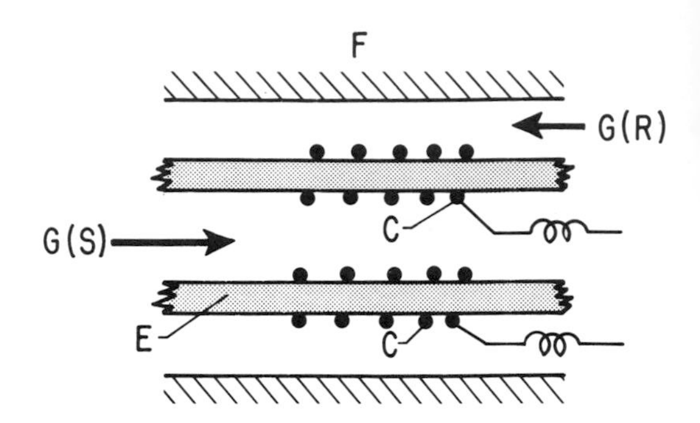

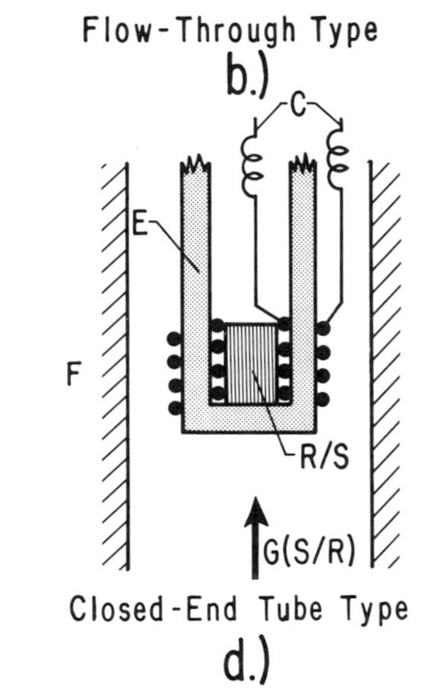

EXPLANATION

C = Conductor
E = Electrolyte
F = Furnace
G = Gas
R = Reference
S = Sample

electrolyte when iron was in direct contact with the electrolytes. The ThO_2-La_2O_3 was reported (Kiukkola and Wagner, 1957) to have shown significant electronic conduction.

Other oxide solid solutions investigated with respect to electrical conductivity include those of CeO_2 (ceria) with a trivalent oxide such as La_2O_3, Y_2O_3, and Nd_2O_3 (Neuimin and Pal'guev, 1962), and those of La_2O_3 (lanthania) with CaO (Etsell and Flengas, 1969). In these solid solutions, electronic conduction is always appreciable and their possible usefulness appears to be limited to fuel cell application.

C. Sensor Design

There are three basic varieties of design for the solid electrolyte oxygen sensors, classified on the basis of the shape of the electrolyte. These are flat-disc type, flow-through tube type, and closed-end tube type, as shown in Figure 6.

1. Flat-Disc Type

The flat-disc type was used by early investigators who had to prepare the electrolyte in their own laboratory, and its applicability is limited to experimental work involving only metallic or semiconductive oxygen buffers. A typical arrangement is to have two such buffers pressed to flat-disc shapes and tightly sandwich the electrolyte. Inert metallic conductors such as platinum are attached to the exposed sides of the conductive

---

Figure 6. Various types of the solid electrolyte oxygen sensors classified in terms of the shape of the electrolyte and the aggregate state of the reference and the sample. a.) Flat-disc type. b.) Flow-through type. c.) Closed-end tube type with two gas phases. d.) Closed-end tube type with a condensed phase and a gas phase. e.) Closed-end tube type with two condensed phases. Condensed phases that react with the electrolyte must be contained in a suitable crucible. When a metal crucible is used, it can function also as a part of the conductor.

buffer discs as shown in Figure 6 a.). The cell transfer is made, presumably, directly from one buffer to the other *via* the electrolyte. If the buffers are not electronically conductive, this arrangement becomes awkward because the metallic conductors must be placed between the buffer and the electrolyte and becomes vulnerable to interference by the external atmosphere. Even using electronically conductive buffers in a highly purified inert gas atmosphere or a reasonably high vacuum, reliable results seemed to have been obtained only with buffers which attain equilibrium rapidly and reversibly (such as Ni-NiO, Fe-"FeO", and $Cu\text{-}Cu_2O$). Purification of an inert gas is an extremely tricky operation which requires complete removal of water vapor, oxygen, carbon gases, and hydrogen (often produced by the reaction of trace water vapor with the oxygen getter). In order to prevent diffusion of oxygen from air, ultra-high vacuum connections must be used throughout the system. The difficulty in obtaining leak-tight seals is a familiar problem to a mass spectroscopist. A small internal positive pressure alone is not a satisfactory means of preventing the back-diffusion of air. Oxygen moves along steep chemical potential gradients that exist at the seals. Some investigators have stressed the importance of avoiding the use of glass or ceramic parts to prevent moisture contamination. In addition to employing an all metal (such as copper) system, the use of high grade diffusion pump oil in the outlet gas trap is probably necessary. In summary, the flat-disc type is not recommended for general use, except perhaps for high pressure work in a tight-fitting cylinder where gaskets or other seals can isolate the buffer discs from the external atmosphere and from each other.

2. Flow-Through-Tube Type

The flow-through-tube type is suitable for monitoring a continuous gas flow. Cylindrical platinum (sometimes silver) electrodes are placed near the center of the tube, one in contact with the interior wall and the other with the exterior wall, and this part of the tube is kept at a fixed temperature (normally about 750 C) in a small regulated tubular furnace as shown in Figure 6 b.). The gas to be monitored is passed through the interior of the tube and the exterior is exposed to air which functions as the fO_2 reference (0.21 atm). Oxygen fugacity monitoring instruments of this design are commercially available. Normally a small suction pump is provided to facilitate the gas flow. The sensor of this type can advantageously be used where the temperature of the gas to be monitored is too low to allow proper functioning of a direct insertion type sensor, or where the fluctuation of the temperature is too severe for the electrolytes. Examples of the applications of the flow-through type sensor are monitoring of the oxygen consumption in human exhalation, measuring optimum combustible gas mixing ratio, determining fumarolic gases, and monitoring metallurgical furnace atmospheres. This design is generally not suitable for the study of condensed phases.

3. Closed-End-Tube Type

The closed-end-tube type is the most versatile and reliable of the three and is the most suitable for investigating condensed systems which are slow to equilibrate. Tubes with either flat or hemispherical bottoms made of yttria-stabilized zirconia, lime-(calcia) stabilized zirconia, and yttria-doped thoria are commercially available in the United States. These tubes come in sizes as small as 1/4 inch in outside diameter and as long as a few feet. They are slip-cast and sintered to obtain a very low permeability (helium leakage rate less than 10^{-7} cc/sec STP). A number of variations in design are possible and most have been tested. The oxygen reference can be either inside or outside the tube if both the reference and the unknown are gas phases (Figure 6 c.)). If the reference is a gas phase and the unknown a condensed phase or the reverse, configuration d.) is suitable. If both the reference and the unknown are condensed phases, obviously there are two alternatives in the configuration of e.). As the unknown often represents samples of less availability, small buffering capacity or slower equilibration rate, it is often necessary that the unknown be placed inside the tube. The internal volume of the electrolyte tube can be reduced by inserting a Vycor glass or silica glass rod which fits snugly to the interior of the tube and has a small bore for the metal conducting wire. The top of the electrolyte tube can be sealed with a combination of a fluorocarbon high-vacuum connector and a silicone rubber stopper to isolate the sample from the exterior atmosphere.

4. Double-Tube Type

Special varieties of the closed-end-tube type are the double-tube arrangements shown in Figure 7. The first design (Figure 7 a.)) may be called "opposed double-tube" design. It is most suitable for use with a sample which is slow to equilibrate or has a small buffering capacity, i.e., its fO_2 changes appreciably with a minute addition or subtraction of oxygen. This design has been used in the author's laboratory at the U.S. Geological Survey for several years for the determination of oxygen fugacities of basalts, mafic mineral separates and oxide mixtures which are slow to react. A basic problem in the use of solid electrolytes at high temperatures is that their permeability to molecular gases increases sharply with temperature, a typical behavior for ceramic material. Even the best commercially available electrolyte tubes show evidence of leakage of reactive gases such as O_2, H_2, H_2O, CO_2, and CO at high temperatures. The leakage could be tolerated so long as the sample has a large fO_2 buffering capacity and also is quick to react, or alternately, is a rapidly flowing gas. Otherwise, a substantial error could be introduced by the leakage.

In an attempt to determine fO_2 values of basaltic melts using the single tube arrangement, the author observed that the e.m.f. of a melt tended to drift slowly toward zero, and the ultimate steady fO_2 value of the melt varied significantly depending on the reference gas used; a relatively higher fO_2 value was obtained with a high-fO_2 reference gas and lower fO_2 with a low-fO_2 gas.

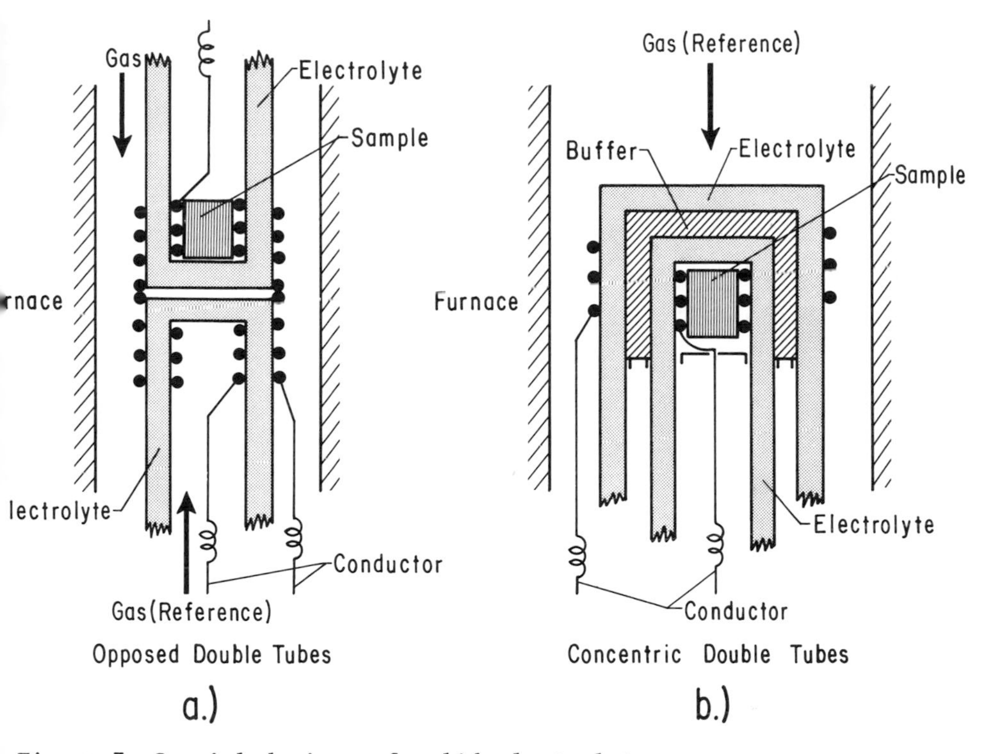

Figure 7. Special designs of solid electrolyte oxygen sensors. a.) Opposed double tube type designed to reduce contamination of the sample due to gas diffusion. b.) Concentric double tube designed to make use of thoria electrolyte for low oxygen fugacity measurements while air is used as the reference. The nomenclature is the same as in Figure 6 except for buffer mixtures which typically are Co-CoO or Ni-NiO.

A basaltic melt does not buffer the fO_2 and a minor leakage could produce significant errors. Also active diffusion of reactive gases produces so-called "mixed potential" and results in erroneous fO_2 values even with a sample that has a large buffering capacity unless it is also rapidly reactive. The magnetite-hematite buffer shows this mixed potential effect because of its slow reversibility. In addition to normal gas diffusion problems, electronic conduction within the electrolyte, though normally small, would also contribute to the migration of oxygen across the electrolyte wall when the e.m.f. difference is substantially large, as discussed earlier. Migration of oxygen could also be caused by the input leakage current of the potentiometer. The best remedy for these problems is obviously to diminish the difference in fO_2 and hence e.m.f. across the wall of the sample electrode tube. This is the basic reasoning behind the opposed double-tube design. The method utilizes two opposing flat-ended electrolyte tubes. Each tube is capped with a platinum basket. One tube is placed above the other end to end in a vertical furnace chamber. The sample is placed at the bottom of the upper tube, while an fO_2 reference gas such as air is circulated through the interior of the lower tube. The furnace atmosphere is regulated manually or, more preferably, automatically with a servomechanism in such a way that the e.m.f. between the inside and outside of the sample-containing electrolyte tube reads very close to zero. This means that the fO_2 of the furnace atmosphere closely approximates that of the sample. The fO_2 value of the sample is calculated from the difference of the e.m.f. values of the sample and the reference electrodes (the furnace atmosphere electrodes are grounded) and the temperature. It is convenient to measure the temperature at the interior top of the reference electrode tube with a built-in thermocouple, if the vertical temperature distribution is symmetrical about the contact of the two tubes. This arrangement tends to reduce the temperature error arising from the horizontal temperature gradient that often exists to a marked extent in the furnace. The opposing double-tube arrangement would also make the use of the thoria electrolyte easy, although the fO_2 matching feature cannot be incorporated. Thoria electrolyte is used as the sample electrode tube and a zirconia electrolyte as the reference electrode tube. The furnace atmosphere is regulated at an fO_2 which is a few log fO_2 units above the lower limiting fO_2 value of the zirconia electrolyte.

Another double-tube type arrangement similar to the method just described was recently devised by Tretyakov and Muan (1969) in order to measure the fO_2 values of the Mn-MnO buffer using air as the reference. The arrangement is characterized by concentric double-tube configuration as shown in Figure 7 b.). The larger tube is made of the ZrO_2-CaO electrolyte and contains the sample buffer mixture. Another buffer of an intermediate fO_2 value (Co-CoO mixture) is sealed between the two electrolyte tubes. The tubes were placed in a vertical furnace upside-down to avoid the direct contact of the sample with the thoria tube. This arrangement simplifies the use of the thoria electrolyte and may find further application for solid buffers of extremely low fO_2 values. An effort must be made, however, to reduce the temperature gradient across the two concentric electrolyte tubes.

D. Oxygen Fugacity Reference

1. Gaseous Reference

In principle, any standard oxygen buffer or gas mixture for furnace fO_2 control should be usable as the oxygen fugacity reference. There are a number of restrictions, however. Suitable references are either a flowing gas with a stable fO_2 value or a solid oxygen buffer which equilibrates rapidly and reversibly and has a high melting point. The most versatile and easily obtainable reference is air which contains 20.95% by volume of oxygen in a normal environment. Since oxygen behaves almost as an ideal gas at low pressures and high temperatures, the fO_2 value of air is close to 0.21 atm at 1 atm total pressure. Where the elevation is high, a barometric pressure correction becomes important. Another easily obtainable reference is pure oxygen itself. This also requires the pressure correction at high altitudes, although again the fugacity coefficient can be ignored. These two reference gases have the advantage that at low pressure the fO_2 is independent of temperature. The e.m.f. of air in reference to pure oxygen is, however, a function of temperature according to equation (5) and is given as:

$$E(air) = 0.0496\ T \log 0.210 = -0.0337\ T\ mV \qquad (6)$$

where T is the absolute temperature (the polarity is such that the potential of the pure oxygen electrode is taken as zero). When air is used as the reference the above E(air) value must be added to the measured e.m.f. to compute the fO_2 value of the unknown. Alternatively, -0.68 can simply be added to the log fO_2 value computed on the basis of the measured e.m.f. The results are the same. Air or oxygen are superior to reactive gas mixtures and solid buffers in terms of simplicity of use and instant e.m.f. response with change of temperature even at relatively low temperatures (300~700 C). Since oxygen diffusion coefficients of the solid electrolytes are significant (Simpson and Carter, 1966), these gases should not be used in direct contact with an electrolyte tube containing a low-fO_2 sample of limited quantity, small buffering capacity or slow equilibration rate, as discussed earlier. If continuous rise in fO_2 of the sample is detected, these gases must be replaced by more reducing reference gas mixtures, or cell design changed to the opposed double-tube type. Air and oxygen should not be used directly with the thoria electrolytes.

Gas mixtures such as CO_2-CO, CO_2-H_2, and H_2O-H_2 can also be used as reference gases. (See also Chapter 2 for more details in the use of those gases.) These gases have the advantage of reducing the oxygen contamination of a sample when the latter has a low oxygen fugacity. Also they can be used directly with thoria electrolytes. The oxygen fugacities of these gas mixtures are dependent on both temperature and composition. The fO_2 values at various temperatures must be calibrated electrochemically against air or oxygen or thermodynamically computed for each batch of the mixture in use. Equilibration of these gas mixtures is rather sluggish at temperatures below about 500 C and the adjustment of the flow rate becomes critical. Pre-heating is recommended for the CO_2-H_2 mixture. The CO_2-CO mixture is probably the best to use with an anhydrous system. The e.m.f. response of the electrolytes with this mixture is rapid for both compositional change and temperature change. Neither CO_2 nor CO appears to diffuse through the electrolytes at a significant rate, though no quantitative experimental data are available. The mixture must, of course, be carefully vented out or bubbled through a scrubbing solution (such as potassium permanganate solution) to eliminate releasing the poisonous CO into the laboratory after circulation through the furnace chamber. On the basis of the data by Zeise (1954) the fO_2 value for CO_2-CO mixtures are computed from:

$$\log fO_2\ (CO_2\text{-}CO) = 4.505 - \frac{14700}{T} + \log\left(\frac{pCO_2}{pCO}\right) \pm 0.005 \quad (7)$$

and the e.m.f. against a pure oxygen electrode is given by:

$$E\ (CO_2\text{-}CO) = -1458 + \left[0.047 + 0.0992 \log\left(\frac{pCO_2}{pCO}\right)\right] T \text{ mV.} \quad (8)$$

The range of pCO_2/pCO practical to use is from 100 to 0.01.

The mixtures H_2O-H_2 and CO_2-H_2 have a few problems. Firstly, it is not desirable to use pH_2O higher than the saturation vapor pressure at room temperature (0.031 atm), because water tends to condense in the coldest portions of the gas train and can short the potential measuring circuit. Secondly, hydrogen diffuses through practically everything at high temperatures, and if the fO_2 of the reference gas is significantly lower than that of the sample, the sample tends to be reduced by this diffusing hydrogen. The best way to use the H_2O-H_2 reference is to change the pH_2 by mixing H_2 with an inert gas such as argon or nitrogen as diluent, and bubble this mixture through pure water kept at a constant temperature. The log fO_2 value for the H_2O-H_2 mixtures between 700 and 1400 C are calculated from the data by Zeise (1954) as:

$$\log fO_2(H_2O\text{-}H_2) = -2.947 - \frac{13008}{T} + \log\left(\frac{pH_2O}{pH_2}\right) \pm 0.005 \quad (9)$$

and the e.m.f. against a pure oxygen electrode is:

$$E(H_2O\text{-}H_2) = -1290 + \left[0.292 + 0.0992 \log \left(\frac{pH_2O}{pH_2}\right)\right] T \text{ mV}. \quad (10)$$

The water-gas mixture CO_2-H_2 is not recommended because hydrogen must be less than 3% to avoid water condensation in the gas train and the mixture also tends to become non-uniform in storage due to gravity separation. The computation of the log fO_2 values is rather complex.

For graphical estimation of the fO_2 value of this mixture at a given temperature and mixing ratio, the reader may refer to the diagram by Muan and Osborn (1964, Figure 34). For the CO_2-CO and H_2O-H_2 mixtures, reference may be made to Muan and Osborn (1964), Darken and Gurry (1953, Figure 14-4), and Mackowiak (1965, Figure 9.2). For graphic solution of the e.m.f. values (reference: oxygen) for these mixtures, refer to the diagrams by Möbius (1965). Computer calculated tables for all these gas mixtures are given by Deines *et al.* (1971) and the calculations are derived in Chapter 2 of this volume.

2. Solid Reference

Solid oxygen buffers have advantages where simplicity of the sensor design is a prime consideration. This is particularly true with insertion-type sensors. The buffer can be sealed in the bottom of an electrolyte tube; only a conducting wire is required to come out of the interior of the tube — no gas feeder nor outlet tube is required. The slowness of the equilibration of most buffers at relatively low temperatures, however, makes them of secondary importance at these temperatures. The most commonly used and probably most reliable buffer is the Ni-NiO mixture. At temperatures above about 700 C, this buffer obtains equilibrium within a matter of minutes and gives reproducible results through repeated temperature cycling. No significant reaction with the electrolytes has been observed even when this buffer is in direct contact with the electrolyte wall.

Its log fO_2 value adopted by Huebner and Sato (1970) is:

$$\log fO_2 \text{ (Ni-NiO)} = 9.31 - \frac{24810}{T} \pm 0.02 \quad (11)$$

and from this the e.m.f. against pure oxygen is given as:

$$E \text{ (Ni-NiO)} = -1231 + 0.462\,T \text{ mV.} \quad (12)$$

This buffer can be used with the thoria electrolytes.

Other buffers frequently used are the Fe-"FeO", Co-CoO, and Cu-Cu_2O. These buffers equilibrate fairly rapidly. In choosing solid buffer references, attention has to be paid to the reactivity of the buffers with the electrolyte. Iron oxides, for example, tend to react with yttria to form a compound $YFeO_2$ (Worrell, 1966). Placing solid buffers directly in contact with the electrolyte is still a controversial issue and it appears wise to avoid a direct contact of buffer and electrolyte to prevent possible contamination of the electrolyte by transition elements. Such contamination enhances electronic conductivity within the electrolyte.

E. Electrode Conductors

The selection of the metallic conductor for the solid electrolyte cell is more involved than it might at first appear. Required properties for suitable conductors are: that (1) they be chemically inert and not appreciably miscible with the electrode material, (2) they have a high melting point and low vapor pressure, and (3) they not contaminate the electrolyte. Platinum is most commonly used and is quite satisfactory for use with a gas phase at temperatures up to 1769 C. It can be used even with sulfur-rich gases except at low temperatures and high fS_2, in which case platinum sulfide forms. Platinum also becomes brittle in gases rich in hydrogen. Gold may be used advantageously for such gases if the temperature is below 1060 C. Platinum is not really suitable for use with condensed phases rich in iron. As is well known, an appreciable amount of iron dissolves in platinum at high temperatures, and unless excess metallic iron exists in the sample, the activity of iron in the latter is altered and an erroneous fO_2 measurement results. The error would be normally toward a higher fO_2 value because of the disproportionation reaction:

$$3Fe^{2+} \text{ (in sample)} \rightarrow Fe \text{ (in Pt)} + 2Fe^{3+} \text{ (in sample)}. \quad (13)$$

This effect could be minimized by increasing the amount of the sample relative to platinum, or pre-saturating the platinum at the temperature of interest with the extra sample material. If the sample contains basic silicates and melts at the temperature of measurement, platinum is unsuitable to use, because a basic silicate melt creeps over the surface of pure platinum extensively at high temperatures. A basic silicate melt cannot be allowed to be in direct contact with the electrolyte tube, because it adheres to the electrolyte and cracks the tube upon cooling. A crucible is necessary to contain the melt inside the electrolyte tube. It is most convenient if the crucible is made of an inert metal and functions simultaneously as the electrode conductor. A platinum crucible is not suitable for this purpose for the reason discussed above. A suitable material for use below about 1175 C is the alloy of silver-palladium studied by Muan (1963). The alloy containing 40% of Pd was found to be satisfactory for use as a crucible-conductor for most terrestrial basalts up to about 1175 C if not kept at this temperature over an hour. If the recovery of the sample or the electrolyte tube is important, heating above 1160 C is not recommended. This material is not suitable for use with an air reference electrode because this author found it deteriorated in air at 400~600 C. The cause has not been investigated in detail, but it is probably due to oxide formation and incongruent dissolution of the alloy into the oxide. It is necessary to use platinum with the air reference electrode and the Ag-Pd alloy with the silicate melt. This arrangement gives rise to a thermoelectric power (= thermocouple potential) between the two dissimilar metals which is superposed on the isothermal cell e.m.f. In order to obtain the isothermal cell e.m.f., this thermocouple potential has to be subtracted from the potential measured between the two cold ends of the conducting wires. This procedure may appear puzzling because in the sensor arrangement the two conductors are not directly connected. Calibration work with reference buffers shows, however, that the correction is necessary. This phenomenon can be understood if the energetic state of electrons in the conductors at different temperatures is considered in the over-all cell transfer mechanism starting from the copper leads of the potentiometer. The thermocouple-potenial correction factor can be obtained by actually making a thermocouple with the two dissimilar conducting wires and measuring its thermal e.m.f. values at a number of temperatures. For example, on the basis of 53 measurements, the thermal e.m.f. of the pair $Ag_{60}Pd_{40}$ (negative) to Pt (positive) between 500 C and 1200 C, when the cold end is at 25 C, is found to be given approximately ($\pm$0.1 mV) by:

$$E(Ag_{60}Pd_{40}\text{-}Pt) = -0.125 + 0.0108\ (t-25) + 1.33 \times 10^{-5}\ (t-25)^2\ \text{mV} \quad (14)$$

where t is in degrees Celsius (centigrade). This correction amounts to about 30.9 mV at 1200 C. Since $Ag_{60}Pd_{40}$ is normally used with a sample of low oxygen fugacity (negative potential) and its thermal e.m.f. against Pt is negative, the potential value indicated by the potentiometer is more negative than the true isothermal cell e.m.f. Hence the above correction value (given as an absolute value) has to be added to the potentiometer reading of the $Ag_{60}Pd_{40}$ terminal.

A promising alloy for use with silicate melts at temperatures up to 1600 C and over a wide range of oxygen fugacity is an alloy of platinum and gold. Crucibles of the alloy containing 5% Au were tested and found satisfactory for melting simulated lunar basalt high in titanium and iron (Tuthill and Sato, 1970). Although some iron was lost to the crucibles, the melt remained at the bottom without showing a tendency to creep. Though more tests are needed to ascertain the suitability of this alloy for use with oxygen sensors, its application in this field appears to have a bright future. For applications below 800 C and for systems poor in sulfur, silver is used as the electrode conductor in some commercial sensors chiefly because of economy. Its value for laboratory use is not certain. Rhodium-platinum alloys do not seem to have particular advantages except for their high melting points. They are not suitable for use in gases containing carbon monoxide, because rhodium forms volatile compounds (carbonyls and carbonyl hydrides — c.f., Sidgwick, 1950, pages 547 and 1528) with this gas.

F. E.m.f. Readout and Recording Devices

An important qualification for the e.m.f. readout device for the solid electrolyte oxygen sensor is to have a very high input resistance. This is necessary because it is desirable to have no appreciable current through the cell at any time. Except at very high temperatures the resistance of the solid electrolyte is so high that a small current drain results in a large potential drop across the electrolyte. Also allowing electric current through the cell means permitting migration of oxygen into or out of (depending on the direction of the current) the sample and this disturbs the system to be investigated. For these reasons the conventional potentiometer which uses a Kelvin-Varley bridge and a mechanical galvanometer (relatively low input resistance) is unsuitable for use in measurements which involve relatively low temperatures and/or samples with slow reversibility or small buffering capacity. It may be argued that the input resistance of a null type potentiometer is infinite at balance so it should be usable, but in reality, an exact balance does not happen and in the process of balancing appreciable current is allowed to flow. It should be noted here also that high resolution of potential is not of prime importance for the solid electrolyte sensor. Resolution of 0.1% or ± 1 mV, whichever is greater, is normally sufficient. Other factors of experimental error such as uncertainties in the temperature determination and barometric pressure fluctuations make a higher precision in potential measurement less important than a higher input resistance.

The best e.m.f. readout device available today is the vibrating reed (or vibrating capacitor) electrometer, which has an input resistance of 10^{14} to 10^{16} ohms and very low zero drift (as low as 20 microvolts per day). This unit is rather expensive, however. There are less expensive electrometers with comparably high input resistance available (see, for example, Archambault, 1971). These include special high-Z tube electrometers, FET (field effect transistor)-input electrometers, and Varactor bridge electrometers. The input resistances of these electrometers range between 10^{14} and 10^{12} ohms and the zero drift is about a few millivolts per day. They are available both as analog and digital display models equipped with analog output for recorder operation. In using these high input resistance electrometers, it is important to give much attention to the shielding problem. The use of a shielded cable is essential to avoid picking up stray potentials. Also a grounded shield plate is necessary inside the furnace to avoid induction potential from the furnace current, particularly at relatively low temperatures where the cell resistance is high. A platinum plate shielding the electrolyte tubes for the entire length of the furnace is a desirable feature. Also non-inductive winding of the furnace element is an effective method to reduce the induction potential. Turning off the furnace current during the measurement is not desirable, because the temperature of the cell would drop rapidly, particularly at high temperatures. When the temperature change is fast, a significant temperature gradient is developed across the zirconia electrolyte tube because zirconia has a very low thermal conductivity. Use of a smooth direct current supply for the furnace would provide an alternate solution. Such a direct current supply can easily be obtained by inserting large-capacity silicon rectifiers and ripple-filtering condensors between the furnace current regulator and the furnace winding.

Recording can be either analog or digital. Simultaneous recording of the e.m.f. and cell temperature is desirable. If the recorder is a potentiometric multichannel recorder, or a digital recorder with input resistance greater than 2×10^4 ohms, the thermocouple output (cold junction) can be directly fed to the recorder.

G. Applications

1. General

The applications of the solid electrolyte oxygen sensors are numerous. A wide range of problems in thermodynamics, metallurgy, geology, chemical engineering, and combustion control can be investigated. Detailed discussions of these applications alone would fill an entire book. Only brief descriptions of principles of major applications are given below.

2. Gas Monitoring

The monitoring of fO_2 values of furnace atmospheres, volcanic gases, combustion systems, and metallurgical furnace gases are the most direct and practical applications of the solid electrolyte oxygen sensors. For preparation of minerals and semiconductor materials sensitive to oxygen fugacity, direct monitoring of fO_2 with the sensors provides easier and more accurate control of the furnace atmosphere. A mixture of more than two gases such as N_2-H_2O-H_2 can be used without tedious calibrations. The production of H_2 by the electrolysis of water in such cases provides a simpler but finer control of oxygen fugacity (Sato, 1970).

Dynamic change in oxidation state of basaltic magma in lava lakes and active vents can be studied with solid electrolyte sensors even at magmatic temperatures (Sato and Wright, 1966). Sensors with the closed-end type design are suitable for this purpose. At temperatures above 600 C, the Ni-NiO reference works well if tightly sealed from the exterior. For lower temperatures (down to about 300 C), the use of a yttria-stabilized zirconia tube and internal air flow is much more suitable. For fumarolic gases, a flow-through type sensor with an auxiliary heater is effective.

Monitoring of oxygen fugacity is an efficient way of regulating combustible mixtures at the optimum mixing ratios (Möbius, 1965;

Lawrence et al., 1969). When neither oxygen nor the fuel is in excess, the e.m.f. of the sensor monitoring the combustion products (exhaust gas) changes very rapidly and oscillates with slightest variations in the mixing ratio (Figure 8). This effect is very similar to pH titrations. The optimum mixing ratio is found where the slope is steepest.

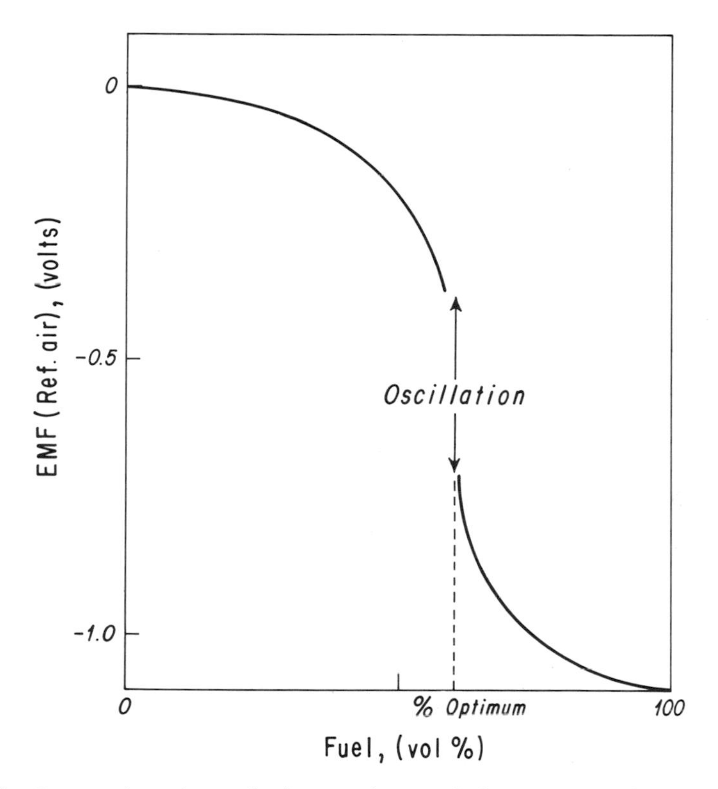

Figure 8. Determination of the optimum mixing ratio of a fuel gas and air by monitoring the fO_2 of the combustion products. At the optimum mixing ratio, the slope is steepest and the e.m.f. becomes extremely sensitive to the slightest irregularities in mixing ratio of the fuel/air supply.

3. Oxygen Content in Molten Metals

The determination of the oxygen content in steel and other metals is a very important procedure in process metallurgy. The quality of steel, for example, is closely related to its oxygen content. Equilibrium oxygen fugacities of molten metals have been investigated by the solid electrolyte sensor technique to correlate their oxygen content with the nature of the atmosphere of the melting furnace. The most common electrode arrangement adopted in the past is to use the molten metal as the electrode conductor by immersing a zirconia electrolyte tube directly into the molten metal (Figure 9) (Belford and Alcock, 1965; Fischer and Ackermann, 1965, 1966; Wilder, 1966; El-Naggar et al., 1967). The oxygen reference, either the Ni-NiO mixture or air flow, constitutes the internal electrode with a platinum conductor. The electrical connection to the molten metal is made by dipping a conducting wire

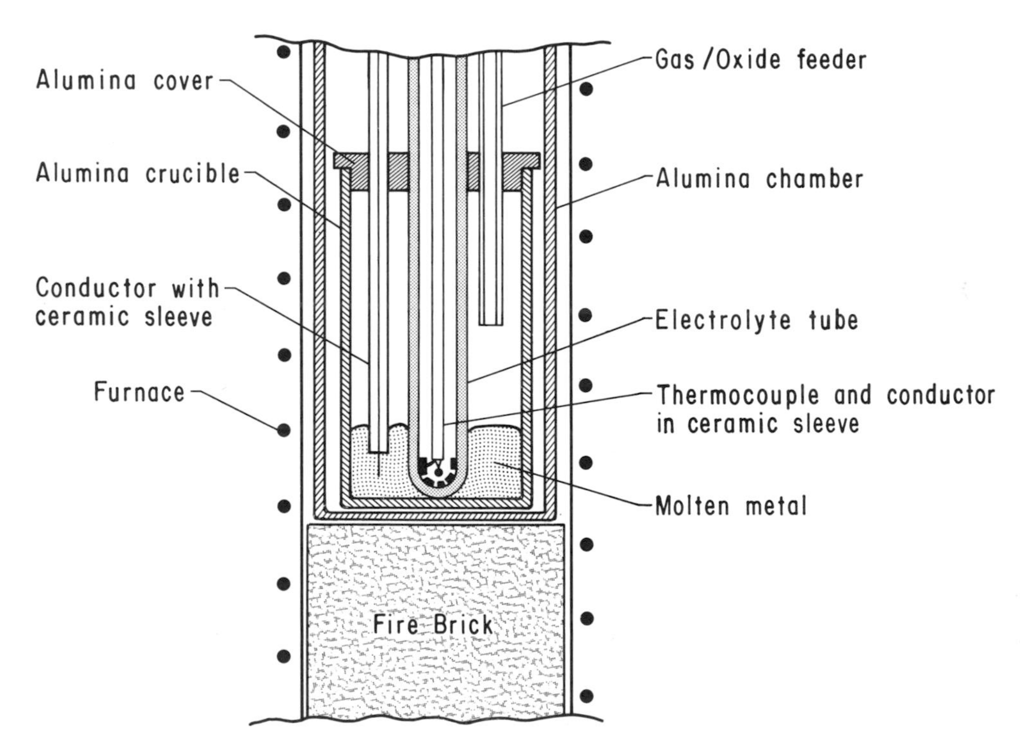

Figure 9. A sensor arrangement for the determination of the fugacity coefficient of dissolved oxygen in molten metals. A simplified version of the design used by Wilder (1966). (Courtesy of Trans. Metal. Soc., AIME)

such as Pt, Ta, or a cermet (refractory metal-ceramic mixture) into the molten metal. The conducting wire must be non-reactive and immiscible with the molten metal. If the external conductor is different from the conductor used for the reference electrode, the thermocouple effect of the pair must be calibrated and subtracted from the e.m.f. reading as discussed before. The oxygen content of the molten metal is either separately analyzed by the vacuum fusion method, or controlled by the addition of an oxide of the metal or by coulometric titration of oxygen through the electrolyte, a method which will be described later. Examples of the e.m.f. of the solid electrolyte cell vs. oxygen concentrations in molten metals are shown in Figure 10. With iron or steel

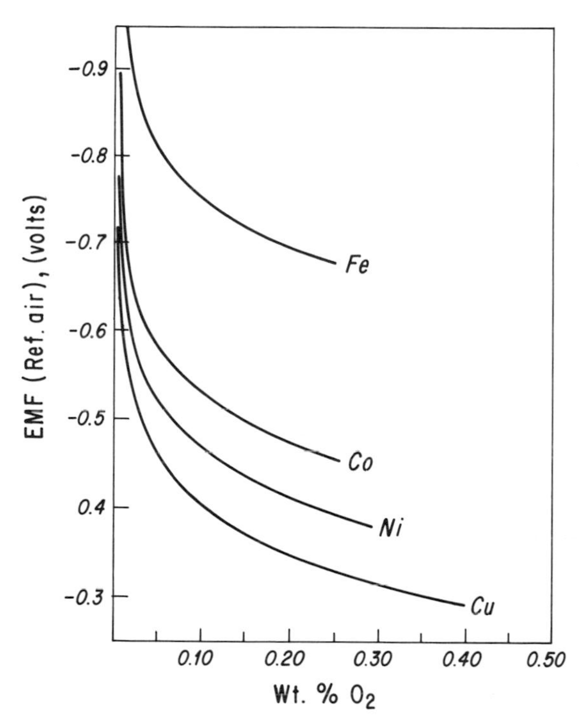

Figure 10. The dependence of the e.m.f. value of the solid electrolyte cell (in reference to air) on the oxygen content in molten iron, cobalt, nickel, and copper at 1600 C (after Fischer and Ackermann, 1966). (Courtesy of Arch. Eisenhüttenwesen)

the above arrangement requires extreme caution at high oxygen concentrations because molten wüstite, if formed, would attack the electrolyte tube quickly.

The solid electrolyte sensor method is a very attractive one for use in acid open hearth operation. The conventional sampling-and-analysis method for the oxygen content determination is not satisfactory for the control of current heats (heated batches), as the analytical procedure is often too slow. Direct insertion of the sensor into the molten steel is necessary in this case because a layer of slag covers the molten metal and the hearth atmosphere is not well equilibrated with the metal. The thermal shock encountered in the process of plunging the sensor into the molten steel through the slag is far more than an ordinary zirconia electrolyte tube can resist. An inexpensive, shock-resistant disposable sensor tip described by Fitterer (1967) may provide an answer for this problem. The disposable tip is made up of an electrolyte disc and a quartz tube which holds the disc tightly at the bottom.

4. Thermodynamic Data

Thermodynamic applications are based on the determination of equilibrium oxygen fugacity of a system under study. These include the determinations of: (1) standard free energies of formation of metal oxides; (2) activities of components of binary metal solutions; (3) activities of components of binary oxide solutions; (4) standard enthalpies and entropies of formation of metal oxides; and (5) thermodynamic properties of compounds and solutions other than oxides. The determination of standard free energy of formation of a metal oxide is possible if the oxide is virtually immiscible with the metal, has a reversible reaction relationship, and the equilibrium fO_2 is above the lower limiting fO_2 value of the electrolyte. For a metal that forms an oxide M_iO_j reversibly and directly by oxidation:

$$iM + \frac{j}{2} O_2 = M_iO_j . \qquad (15)$$

If the oxide and the metal are mutually immiscible, their activities are at unity. Hence the equilibrium constant at absolute temperature T and the standard free energy of formation of the oxide are given simply as:

$$\log K_T = -\frac{j}{2} \log fO_2 \tag{16}$$

and

$$\Delta G_f^o (M_iO_j) = -RT \ln K_T = \frac{2.303j}{2} \cdot RT \log fO_2$$

$$= 2.288 \text{ jT} \log fO_2 \text{ cal/mole.} \tag{17}$$

For a higher oxide, the equilibrium fO_2 of the oxide with an adjacent oxide (whose ΔG_f^o value is known) must be determined and similar calculations followed.

Papers which deal with this application include Kiukkola and Wagner (1957; CoO, NiO and Cu_2O), Blumenthal and Whitmore (1961; FeO, MnO and Mn_3O_4), Matsushita and Goto (1966; FeO, Fe_3O_4, SnO, PbO, Cu_2O, CuO and Ta_2O_5), Worrell (1966; NbO, NbO_2 and Ta_2O_5), Alcock and Zador (1967; MnO), Tretyakov and Muan (1969; FeO and MnO), Pugliese and Fitterer (1970; Cr_2O_3), and Huebner and Sato (1970; NiO, Mn_3O_4 and Mn_2O_3).

The determination of the activities of the components in a binary metal solution can be accomplished if metal A and its oxide A_iO_j has an equilibrium fO_2 at least several log units below that of the other pair and if the oxide of A is mutually immiscible (less than a few percent) with the metal solution. The equilibrium constant is given by:

$$\log K_T = -\frac{j}{2} \log fO_2^* \tag{18}$$

also

$$\log K_T = i \log a_A - \frac{j}{2} \log fO_2 \tag{19}$$

where fO_2^* is the oxygen fugacity of the pair, pure A - pure A_iO_j. By equating the two equations we obtain

$$\log a_A = \frac{j}{2i} (\log fO_2^* - \log fO_2). \tag{20}$$

The activity of metal B in the solution can be obtained by using the Gibbs-Duhem equation. Examples of this method are found in Matsushita and Goto (1966; Pb-Sn), Lidster and Bell (1969; Fe-Cr, Ni-Cr), and Pugliese and Fitterer (1970; Cr-Ni).

To be able to determine the activity of a metal oxide A_iO_j in solution with another metal oxide B_xO_y, it is necessary that the pair A-A_iO_j be reversible, have an equilibrium fO_2 which is at least several log units higher than that of the B-B_xO_y pair, and metal A be mutually immiscible with the oxide solution. If these requirements are all met, then the measurement of the equilibrium fO_2 of the mixture of metal A and the oxide solution (A_iO_j, B_xO_y of a particular composition gives the activity of A_iO_j for that composition as:

$$\log a_{A_iO_j} = \frac{j}{2} \log fO_2 - \log fO_2^* \tag{21}$$

where fO_2^* is again the equilibrium oxygen fugacity of the pair, pure A - pure A_iO_j. The activity of B_xO_y is obtained by using the data on the activity of A_iO_j and the Gibbs-Duhem equation. Other thermodynamic functions such as activity coefficients, free energy of mixing, enthalpy and entropy of mixing, and excess free energy of mixing can be derived from the activity data. An example of this type of investigation is found in Matsushita and Goto (1966).

Standard enthalpies and entropies of formation of metal oxides can be determined if the determinations of the equilibrium fO_2 values of the metal-metal oxide pairs are carried out over an appropriate range of temperature and the ΔG_f^o (M_iO_j) values are obtained as a function of temperature in the first order form:

$$\Delta G_f^o = a + bT. \tag{22}$$

Because of the relationship:

$$\Delta G_f^o = \Delta H_f^o + T \Delta S_f^o \tag{23}$$

then it is apparent that $\Delta H_f^o = a$, and $\Delta S_f^o = b$.

An interesting application of the solid electrolyte oxygen sensor, which was pioneered by Larson and Elliott (1967), is the determination of the thermodynamic functions of metal sulfides and related compounds. This method makes use of the fact that reactions of the types listed below can be used to determine the standard free energies of reaction (ΔG_r^o), if the variables T, fO_2 and fSO_2 or fS_2 are known or measured:

$$MO + SO_2 = MS + 3/2\ O_2;$$

$$\Delta G_r^o = -2.303\ RT\ (3/2 \log fO_2 - \log fSO_2) \qquad (24a)$$

$$MO + 1/2\ S_2 = MS + 1/2\ O_2;$$

$$\Delta G_r^o = -2.303\ RT\ (1/2 \log fO_2 - 1/2 \log fS_2) \qquad (24b)$$

$$M + SO_2 = MS + O_2;$$

$$\Delta G_r^o = -2.303\ RT\ (\log fO_2 - \log fSO_2). \qquad (24c)$$

The fO_2 values can be measured with the solid electrolyte sensor. A flow of SO_2 of a known partial pressure (<u>e.g.</u>, 1 atm) can be considered to have a fixed fSO_2. The presence of condensed sulfur with the sulfide or in a gas train at a controlled temperature fixes fS_2. The choice of reaction depends on the relative stabilities of the metal oxides and the sulfide of interest. If the oxides of a metal are very stable relative to the sulfide, (24b) is preferred, whereas for a reverse case (24c) is a better choice. From the ΔG_r^o value of a chosen reaction and the known ΔG_f^o value of the oxide, if involved, the standard free energy of formation of the sulfide at a given temperature may be obtained. The enthalpy and entropy of formation can be obtained by determining the ΔG_f^o values at various temperatures as discussed earlier for oxides. This method could be applied to similar compounds.

5. Phase Equilibrium Studies

In metallurgy and geology, phase equilibrium studies are of prime importance in obtaining basic information of the behavior of condensed phases. The determinations of equilibrium fO_2 values of condensed phase assemblages have varied applications in this field. For example, the temperature of first-order phase transition of a metal or metal oxide can be investigated if the entropy change of the transition is appreciable. The transition appears as a distinct break in the slope of (1) ΔG_j^o vs. T curve, (2) equilibrium log fO_2 vs. 1/T curve, or (3) e.m.f. (of the solid electrolyte sensor) vs. T curve for the metal-metal oxide assemblage. Because many reactions have nearly constant enthalpies of reaction, these curves are

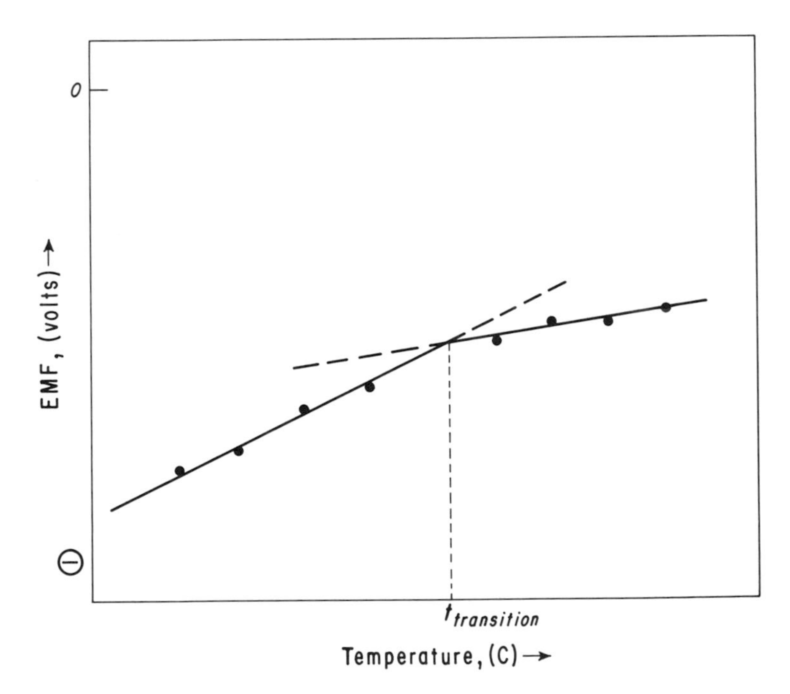

Figure 11. A schematic diagram showing that an e.m.f. versus temperature plot may be used for locating a first-order transition temperature of a metal or oxide if a proper buffer mixture ($M-M_iO_j$) is prepared. Such a transition temperature shows up as an inflection point in the plot.

almost linear and the transition temperature can be obtained as the intersection of two linear extrapolations as shown in Figure 11. A similar change in slope of these curves also takes place at a eutectoid temperature where a shift in fO_2-controlling reactions takes place. For example, the reaction $3Fe + 2O_2 = Fe_3O_4$ is stable only below about 560 C. Above this temperature, wüstite becomes a stable phase and the fO_2 controlling reaction becomes either $Fe + 1/2\ O_2 = FeO$ or $3FeO + 1/2\ O_2 = Fe_3O_4$ depending on the initial mixing ratio of iron and magnetite. The eutectoid temperature can be obtained as the intersection of three e.m.f. vs. T curves as

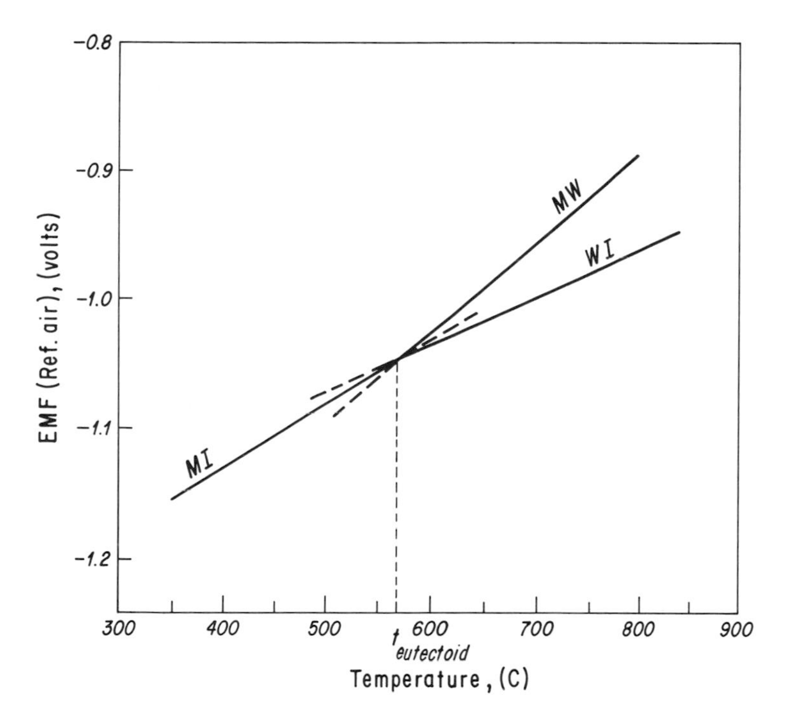

Figure 12. E.m.f. vs. temperature plots may be used for locating an eutectoid temperature. The above diagram schematically shows the relationships for the magnetite-wüstite (MW), wüstite-iron (WI), and magnetite-iron (MI) buffers. The eutectoid temperature is obtained at the intersection of the three buffer curves.

schematically shown in Figure 12. A variation of this technique, which is claimed to give a higher accuracy (Birks, 1966), is to set up a cell:

$$Fe + FeO \,|\, \text{solid electrolyte} \,|\, FeO + Fe_3O_4$$

and extrapolate the e.m.f. value down to zero from the higher temperature side. This method could be very useful for similar studies under high pressures if the data on absolute fO_2 values are not required.

6. Petrological Investigations

In petrology, oxygen fugacity is an important indicator of the environment of formation of igneous, metamorphic, and extra-terrestrial rocks, the course of differentiation of basaltic magmas, and the extent of attainment of internal equilibrium of minerals in rocks. It is highly desirable to be able to determine the equilibrium fO_2 values of rock melts or separate minerals up to about 1300 C without contaminating them. When the Fe_2O_3/FeO ratio is less than about 0.1, the fO_2 value of a basaltic melt is very susceptible to change due to contamination. A large change in fO_2 is produced by very minor leakage of reactive gases such as O_2 and H_2. The determination of intrinsic oxygen fugacity values of a mafic mineral also calls for a leakage-free measuring system. As discussed earlier, diffusion of gases through the electrolyte tube becomes appreciable at high temperatures. Also the leakage due to residual electronic conduction of the electrolyte and that due to the current drain by the voltage measuring device could become substantial sources of oxygen leakage. These difficulties can largely be alleviated by incorporating the negative oxygen fugacity feedback principle into the design of the sensor system. Such a design developed by this author is schematically illustrated in Figure 13. The main feature of this system is that the furnace atmosphere is regulated automatically in such a way that its fO_2 value is always maintained at a value very close to the fO_2 value of the sample. The addition of the reference electrode tube makes it possible to determine the fO_2 value of the

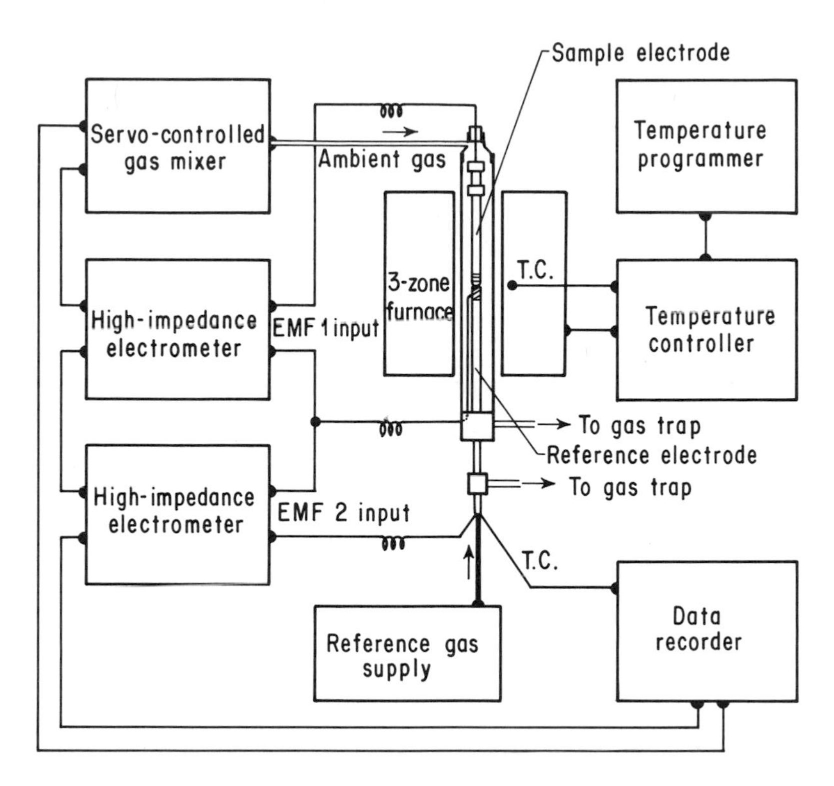

Figure 13. Schematic diagram showing the arrangement of the oxygen fugacity sensor system which incorporates the oxygen fugacity negative feedback principle to minimize contamination of the sample by diffusion of gas through the electrolyte tube. For relaible determinations of the oxygen fugacity of a poorly buffered sample at high temperatures, a complex design of this sort becomes necessary.

sample. The effect of the leakage of reactive gases across the reference electrode tube could be neglected if the furnace atmosphere flows through the furnace chamber from the top (sample tube side) to the bottom (reference tube side) and if the flow rates of both the furnace atmosphere and the reference gas (e.g., air) are reasonably rapid. The improvement in reproducibility that this design has produced is remarkable. As discussed earlier, a single-tube design often resulted in a drift of the measured fO_2 toward that of the furnace atmosphere (reference gas) even with dry basaltic melts having ferric to ferrous ratios between 1 and 0.2. The fO_2 negative feedback design has produced stable readings with these basalts (Sato, unpublished data). It should be mentioned, however, that this design, though extremely helpful, is not a universal solution. If the ferric to ferrous ratio is extremely small as in lunar basalts, for example, loss of iron to the metal conductor appears to produce significant errors (higher fO_2) at high temperatures. High concentrations of volatiles such as water and sulfur also appear to produce erratic results. When fH_2O is much higher in the sample than in the furnace atmosphere hydrogen could escape from the sample tube resulting in the oxidation of the sample. Escape of molecular sulfur from a melt and condensation in a cooler region of the sample tube could result in lowering of fO_2 of the melt due to the reaction: FeS (in melt) + $2Fe^{3+}$ + $2Fe^{3+} = 3Fe^{2+}$ + S (vapor). Individual remedies must be found for these special cases.

The use of intrinsic oxygen fugacities of mafic minerals for geologic thermometry was pioneered by Buddington and Lindsley (1964). The intrinsic oxygen fugacity may be defined as "the oxygen fugacity of a single phase of a fixed composition under given temperature and total pressure." The usefulness of the solid electrolyte method for the determination of the intrinsic fO_2 of a solid phase was demonstrated by Markin and Rand (1966) for PuO_{2-x}. Oxide and silicate minerals containing appreciable amounts of transition elements in more than one oxidation state and equilibrated at a certain temperature should show identical intrinsic fO_2 values, and hence e.m.f. values, at the temperature of equilibrium as schematically illustrated in Figure 14. If a pair of such minerals had crystallized together from a melt or equilibrated at a subsolidus temperature (as in a metamorphic

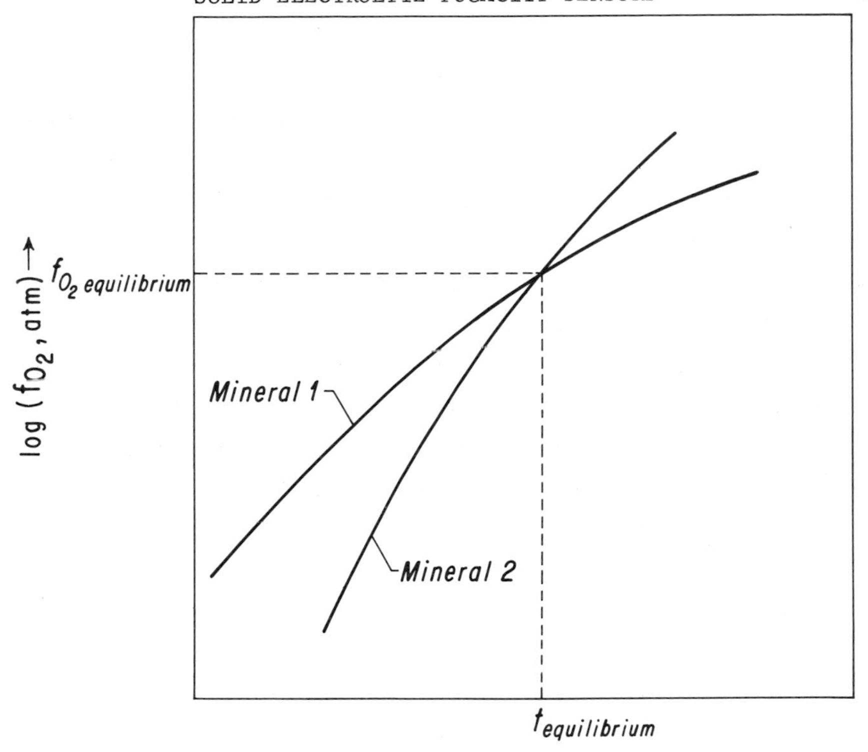

Figure 14. A log fO_2-temperature diagram to illustrate how the electrochemical determination of the intrinsic oxygen fugacities of mafic mineral separates might be used as a geothermometer. If a pair of minerals were once equilibrated at $t_{equilibrium}$ and $fO_{2\ equilibrium}$ and then quenched, a plot similar to the one shown above should be obtainable. The e.m.f. vs. temperature plot would also show a similar pattern.

suite) and subsequently quenched, the temperature of crystallization or equilibration may be determined by finding the temperature at which the e.m.f. values are identical, provided that the total pressure correction is negligible. In principle, this procedure is identical to the e.m.f. method of geothermometry suggested by Sato (1965) for sulfide minerals.

The limits of applicability of this geothermometry method are set by at least three factors. The first factor is geologic. It is a matter of finding a pair of homogenous mafic minerals which were once in turn equilibrium with each other. The second factor is the purity of the mineral separates. Inclusions of various sorts are common in natural minerals. If the minerals are fine grained, complete separation may not be technically feasible. A serious consideration must be given as to the influence of the impurity on the measured fO_2 values. The third factor is the contamination of the sample during fO_2 measurement. If a slight addition or subtraction of oxygen from a mineral by contamination during measurement causes a large change in the intrinsic fO_2 value of the mineral, the method would be unreliable with this mineral. If the measurement is done very carefully, however, reproducible results can be obtained even for a mineral such as olivine (Sato, unpublished data).

7. Quantitative Pumping of Oxygen

A somewhat distinct area of application of the solid electrolyte sensor is quantitative pumping of oxygen into or out of a given system. This can be done by deliberately passing on electric current through the electrolyte. Because of the reaction, $O_2 + 4e^- = 2O^{2-}$, one mole of oxygen passes through the electrolyte for each 4 faradays (4 x 96,529 coulombs or ampere-seconds) of electricity. This pumping application is particularly useful when a very minute quantity of oxygen is to be transferred. For example, if 10 mA of current is allowed to flow for one minute through an electrolyte tube from the interior to exterior (O^{2-} ion in the reverse direction), about 1.55×10^{-6} moles (or 4.97×10^{-5} grams) of oxygen enters a system enclosed in the tube from exterior. If the direction of the current is reversed, oxygen is removed from the system. The adjustments of the direction and magnitude of the current are achieved by applying an external voltage of appropriate polarity and value. In case the current changes with time, the integration of the recorded trace of the current on a recorder strip-chart gives the number of coulombs. The equilibrium fO_2 value of the system after pumping can be determined by measuring the open-circuit voltage of the same cell after a period of equilibration. This method has been applied to the study of fO_2 of nonstoichiometric oxides as a function of the metal-to-oxygen ratios as well as the range of homogeneity (Rizzo and Smith, 1968).

8. Reaction Rate Studies

The ability of the solid electrolyte sensor to respond instantaneously and continuously to a change in oxygen fugacity makes it a powerful tool in reaction rate studies. Several sensors placed in a stream of reactive gas mixtures, for example, could yield basic data for homogenous gas reaction rates. For example, the reaction rate constant k_1 of the phase-boundary reaction during oxidation of iron foil to wüstite by CO_2-CO gas mixtures was determined by Schmalzried (1968) using a ZrO_2-CaO cell. The study of kinetics of redox reactions with electrolyte sensors is a very rapidly expanding field of research.

III. Other Gas Fugacity Sensors

A. Sulfur Fugacity Sensor

The solid electrolyte sensor for sulfur gas is still in the developmental stage. One difficulty involved in the construction of this sensor is that there is no sulfide counterpart of zirconia or thoria electrolyte. At high temperatures most sulfides are unstable in the presence of even trace amounts of oxygen. Some of them melt or decompose at relatively low temperatures, while some are predominantly electronic conductors. In the absence of a suitable sulfide ion conductor, it is necessary to resort to an indirect approach. An arrangement similar to the electrode of the <u>second</u> kind (such as the Ag/AgCl electrode for the determination of the Cl^- ion activity; for more details see, <u>e.g.</u>, Janz, 1961) has been found workable. The first such solid electrolyte cell, which responds to varying sulfur fugacity, was constructed by Rickert and Wagner (1963). This cell uses AgI as the solid electrolyte for Ag^+ ion in the arrangement:

$$Ag|AgI|Ag_2S,\ S\ (vap) \tag{25}$$

Sulfur-containing gases can be used instead of condensed sulfur in such a cell. Basically, the above cell is silver concentration cell, in which the chemical potential of silver in silver sulfide is controlled by the chemical potential of sulfur. This is apparent from the relationship:

$$\mu_{Ag_2S} = 2\mu_{Ag} + \mu_S \tag{26}$$

Since Ag_2S has a narrow homogeneity range, μ_{Ag_2S} remains practically constant. Although sulfur exists in several molecular forms in the vapor phase, it is customary to define the sulfur fugacity in terms of the diatomic gas, S_2. Then, the overall cell reaction for the above cell becomes:

$$4\ Ag + S_2 = 2\ Ag_2S,$$

$$\Delta G_r = 2\Delta G_f^o\ (Ag_2S) - \Delta G_f^o\ (S_2) - RT\ \ln fS_2 \tag{27}$$

and by virtue of the relationship $-nFE = \Delta G_r$, and $n = 4$ in this reaction, the e.m.f. of the cell is given as:

$$E = -\frac{1}{4F}\left[2\Delta G_f^o\ (Ag_2S) - \Delta G_f^o\ (S_2) - RT\ \ln fS_2\right] \tag{28}$$

By rearranging we obtain:

$$\ln fS_2 = \frac{4FE}{RT} + \frac{1}{RT}\left[2\Delta G_f^o\ (Ag_2S) - \Delta G_f^o\ (S_2)\right] \tag{29}$$

It is apparent that the value of the sulfur fugacity for a given e.m.f. and temperature depends on the choice of the value of $\left[2\Delta G_f^o\ (Ag_2S) - \Delta G_f^o\ (S_2)\right]$. On the basis of the data given in Richardson and Jeffes (1952), the numerical value of sulfur fugacity (in log atm units) is calculated as:

$$\log fS_2 = \frac{1}{T}\left[20.155\ E\ (mV) - 9789\right] + 4.829 \tag{30}$$

below the transition temperature of Ag_2S (178 C), and:

$$\log fS_2 = \frac{1}{T}\left[20.155\ E\ (mV) - 9173\right] + 3.610 \qquad (31)$$

between 178 C and the melting point of Ag_2S (842 C). There is a small discrepancy between the two equations at the transition temperature, which results from the uncertainties in the basic thermodynamic data. The accuracy in the determined log fS_2, therefore, should be considered somewhere around ± 0.2 log units. Other solid electrolytes of Ag^+ can be used for the construction of similar sulfur sensors. For example, AgCl works just as well as AgI. These silver halides are too fragile to fabricate directly into a tubular, probe-shaped sensor. One way to overcome this difficulty is to use a porous inert material as a mechanical support. An example of such a sensor is shown in Figure 15. It consists of a flat-ended tube of porous corundum dipped in molten AgCl, silver metal powder placed inside the tube as the silver activity reference, a coating of Ag_2S film on the exterior wall of the corundum tube, and platinum conducting wires. This sensor has been used to determine the sulfur fugacity of fumarolic gases in Hawaiian volcanoes by the members of the U.S. Geological Survey.

The response of this sensor was found to be of the order of seconds even at temperatures as low as 90 C. Above the transition temperature, the response time becomes less than a second. A drawback of the sensors using silver halide as the solid electrolyte is the formation of a solid solution between Ag_2S and silver halide at temperatures above about 400 C. This results in an inaccurate determination and a slower response, because the activity of Ag_2S deviates from unity and the Ag_2S no longer exists in a well defined thin layer but diffuses inward through the halide layer.

A new electrolyte for Ag^+ ion was recently tested in our laboratory for use in the sulfur sensor and gave very satisfactory results. This electrolyte is silver beta-alumina prepared from sodium beta-alumina by the method of Yao and Kummer (1967).

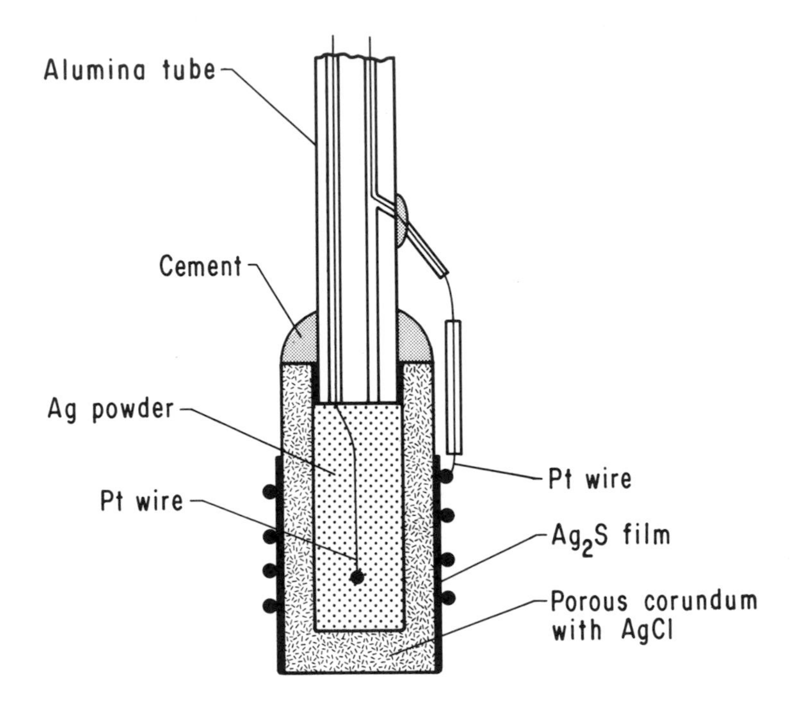

Figure 15. An example of a solid electrolyte sulfur sensor. The sensor is basically a silver concentration cell. The activity of Ag in the Ag_2S film is controlled, within a range, by the fugacity of sulfur in the gas phase.

Sodium beta-alumina is an excellent cationic conductor and maintains its high ionic conductivity even after all of its sodium is replaced by silver. The replacement is made easily by soaking sodium beta-alumina in molten silver nitrate for several minutes. Silver beta-alumina is stable in air even at 1000 C. Sulfur sensors with silver beta-alumina as the electrolyte have already been used satisfactorily from 90-800 C in drill holes in a lava lake, in glowing vents and in fumaroles of Kilauea, and also in active boccas (small glowing vents) of Etna for the determination of fS_2 in magmatic gases.

IV. Other Solid Electrolyte Sensors

A. Fluorine Sensor

Fluorides of calcium, yttrium and lanthanides show appreciable anionic conductivity and could be useful as the solid electrolyte for fluorine gas. The conductivity is generally enhanced by doping one fluoride with another fluoride whose cation has a different valence state. For example, adding N_2F or YF_3 to CaF_2 increases fluoride ion conductivity of the latter markedly by creating F^- vacancies or interstitials (Ure, 1957). The application of the fluorine sensor has been scanty. Aronson and Auskern (1966) determined the standard molar free energies of thorium borides by using CaF_2 as the solid electrolyte at 800~950 C.

B. Hydrogen Sensor

The determination of hydrogen gas fugacity could be made by constructing an oxygen sensor equipped with a H_2O-H_2 reaction chamber which contains a mixture of an oxide and a hydroxide of a metal and has a window of palladium-alloy membrane. For example,

$$\text{(Pt) Ni, NiO}\,|\,\text{CaO-ZrO}_2\,|\,\text{(Pt) CaO, Ca(OH)}_2\text{, H}_2\,\vdots\,\text{Pd}\,\vdots\,\text{H}_2 \qquad (32)$$

Because palladium allows free passage of H_2 but not H_2O from the exterior and the mixture of CaO and $Ca(OH)_2$ buffers the fugacity of H_2O inside the chamber, it should be possible to determine the fH_2 value of a gas phase by measuring the fO_2 value within the chamber and the temperature. A hydrogen sensor, however, still remains undeveloped except in theory.

BIBLIOGRAPHY:

Alcock, C.B. and Zador, S. (1967). Thermodynamic study of the manganese/manganous-oxide system by the use of solid oxide electrolytes. Electrochimica Acta, 12:673-677.

Archambault, J. (1971). The ubiquitous electrometer. Ind. Res., March:36-39.

Aronson, S. and Auskern, A. (1966). The free energies of formation of thorium borides from measurements on solid EMF cells, in Thermodynamics, Vol. 1 Proceedings in the Symposium on

Thermodynamics (1965). Vienna: International Atomic Energy Agency.

Baker R. and West, J.M. (1966). Solid electrolytes for use at steelmaking temperatures. J. Iron Steel Ind., 204:212-216.

Bauerle, J.E. (1966). Electrical conduction in thoria and thoria-yttria as a function of oxygen pressure. J. Chem. Phys., 45:4162-4166.

Belford, R.N. and Alcock, C.B. (1965). Thermodynamics and solubility of oxygen in liquid metals from E.M.F. measurements involving solid electrolytes. Part 2 - Tin. Trans. Faraday Soc., 61:No. 507.

Birks, N. (1966). Some problems in the use of solid-state galvanic cells at low temperatures: the determination of the eutectoid temperature of the iron-oxygen system. Nature, 210:407-408.

Blumenthal, R.N. and Whitmore, D.H. (1961). Electrochemical measurements of elevated-temperature thermodynamic properties of certain iron and manganese oxide mixtures. J. Amer. Ceram. Soc., 44:508-512.

Buddington, A.F. and Lindsley, D.H. (1964). Iron-titanium oxide minerals and synthetic equivalents. J. Petrol, 5:300-357.

Darken, L.S. and Gurry, R.W. (1953). Physical chemistry of metals. New York: McGraw-Hill Book Co.

Deines, P., Nafziger, R.H., Ulmer, G.C., and Woermann, E. (1971). T-fO_2 tables for selected gas mixtures in the C-H-O system. The Pennsylvania State Univ., College of Earth and Mineral Sci. Bull., in press.

El-Naggar, M.M.A., Horsley, G.B., and Parlee, N.A.D. (1967). Application of a solid electrolytic cell for measuring equilibrium pO_2 over liquid metal-oxygen solutions. Trans. Metal. Soc., AIME, 239:1994-1996.

Etsell, T.H. and Flengas, S.N. (1969). The electrical properties of lanthanum oxide-calcium oxide solid electrolytes. J. Electrochem. Soc., 116:771-778.

Eugster, H.P. and Wones, D.R. (1962). Stability relations of the ferruginous biotite, annite. J. Petrol., 3:82-125.

Fischer, W.A. and Ackermann, W. (1965). Unmittelbare elektrochemische ermittlung des sauerstoffgehaltes von eisenschmelzen. I. Untersuchungen an reineisenschmelzen. Arch. Eisenhüttenwesen, 36:643-648.

Fischer, W.A. and Ackermann, W. (1966). Unmittelbare elektrochemische ermittlung des sauerstoffgehaltes von metallschmelzen 1. Untersuchunger an eisen-, kobalt-, und kupferschmelzen. Arch. Eisenhüttenwesen, 7:43-47.

Fitterer, G.R. (1967). Further development of the electrolytic method for the rapid determination of oxygen in liquid steels. J. Metals, September:92-96.

Garvie, R.C. (1970). Zirconium dioxide and some of its binary systems, in High temperature oxides, Part II., A.M. Alper, editor. New York: Academic Press.

Huebner, J.S. and Sato, M. (1970). The oxygen fugacity-temperature relationships of manganese oxide and nickel oxide buffers. Amer. Mineral., 55:934-952.

Jakeš, D. (1969). Galvanické články s pevnými elektrolyty. Chem. listy, 63:1073-1091.

Janz, G.J. (1961). Silver-silver halide electrodes, in Reference electrodes, D.J.G. Ives and G.J. Janz, editors. New York: Academic Press.

Jeannin, Y., Mannerskants, C., and Richardson, F.D. (1963). Activities in iron-chromium alloys. Trans. Metal. Soc. AIME, 227:300-305.

Kiukkola, K. and Wagner, C. (1957). Measurements of galvanic cells involving solid electrolytes. J. Electrochem. Soc., 104:379-387.

Larson, H.R. and Elliott, J.F. (1967). The standard free energy of formation of certain sulfides of some transition elements and zinc. Trans. Metal. Soc. AIME, 239:1713-1720.

Latimer, W.M. (1952). The oxidation states of the elements and their potentials in aqueous solutions, 2nd edition. New York: Prentice-Hall.

Lawrence, S.J., Spacil, H.S. and Schroeder, D.L. (1969). The solid electrolyte oxygen sensor. Theory and applications. Automatica, 5:633-643.

Lidster, P.C. and Bell, H.B. (1969). The application of thoria-yttria electrolytes in measuring the thermodynamic properties of chromium in alloys. Trans. Metal. Soc. AIME, 245:2273-2277.

Mackowiak, J. (1965). Physical chemistry for metallurgists. New York: Elsevier Publ. Co.

Markin, T.L. and Rand, M.H. (1966). Thermodynamic data for plutonium oxides, in Thermodynamics, Vol. 1. Proceedings of the Symposium on Thermodynamics (1965). Vienna: International Atomic Energy Agency.

Matsushita, Y. and Goto, K. (1966). The application of oxygen concentration cells with the solid electrolyte $ZrO_2 \cdot CaO$ to thermodynamic research, in Thermodynamics, Vol. 1. Proceedings of the Symposium on Thermodynamics (1965). Vienna: International Atomic Energy Agency.

Möbius, H.H. (1965). Sauerstoffionenleitende festelektrolyte und ihre anwendungsmöglichkeiten - Grundlagen der gas-potentiometrischen sauerstoffbestimmung. Z. physik. Chemie, 230:396-412.

Muan, A. (1963). Silver-palladium alloys as crucible material in studies of low-melting iron silicates. Amer. Ceram. Soc. Bull. 42:344-347.

Muan, A. and Osborn, E.F. (1964). Phase equilibria among oxides in steelmaking. Reading, Mass.: Addison-Wesley.

Neuimin, A.D. and Pal'guev, S.F. (1962). Electrical conductivity of solid oxides, II. The systems CeO_2-La_2O_3, CeO_2-Nd_2O_3, CeO_2-Y_2O_3. Trudy. Inst. Elektrokhim., Akad. Nauk S.S.S.R., Ural'sk Filial. No. 3:133.

Pugliese, L.A. and Fitterer, G.R. (1970). Activities and phase boundaries in the Cr-Ni system using a solid electrolyte technique. Metallurgical Trans., 1:1997-2002.

Richardson, F.D. and Jeffes, J.H.E. (1952). The thermodynamics of substances of interest in iron and steel making. III. Sulphides. J. Iron and Steel Ind., June:165-175.

Rickert, H. and Wagner, C. (1963). Stationäre zustände und stationäre transportvorgänge in silbersulfid in einem temperaturgefälle. Ber. Bunsengesellschaft Chemie, 67:621-629.

Rizzo, F.E. and Smith, J.V. (1968). Coulometric titration of wüstite. J. Phys. Chem., 72:485-488.

Sato, M. (1965). Electrochemical geothermometer: A possible new method of geothermometry with electroconductive minerals. Econ. Geology, 60:812-818.

Sato, M. (1970). An electrical method of oxygen fugacity control of furnace atmosphere for mineral synthesis. Amer. Mineral., 55:1424-1431.

Sato, M. and Wright, T.L. (1966). Oxygen fugacities directly measured in magmatic gases. Science, No. 3740, 153:1103-1105.

Schmalzried, H. (1962). Über zirkondioxyd als elektrolyt für elektrochemische untersuchungen bei höheren temperaturen. Z. Elektrochemie, 66:572-576.

Schmalzried, H. (1966). The EMF method in studying thermodynamic and kinetic properties of compounds at elevated temperatures, in Thermodynamics, Vol. 1. Proceedings of the Symposium on Thermodynamics (1965). Vienna: International Atomic Energy Agency.

Schmalzried, H. (1968). Elektrochemische verfahren zur bestimmung kinetischer grössen, besonders von diffusionskoeiffizienten und phasengrenzreaktionskonstanten in und and festen stoffen. Arch. Eisenhüttenwesen, 39:531-533.

Sidgwick, N.V. (1950). The chemical elements and their compounds, Vols. 1 and 2. London: Oxford University Press.

Simpson, L.A. and Carter, R.E. (1966). Oxygen exchange and

diffusion in calcia-stabilized zirconia. J. Amer. Ceram. Soc., 49:139-144.

Spacil, H.S. and Tedmon, C.S., Jr. (1969). Electrochemical dissociation of water vapor in solid oxide electrolyte cells. II. Materials, fabrication, and properties. J. Electrochem. Soc., 116:1627-1633.

Strickler, D.W. and Carlson, W.G. (1964). Ionic conductivity of cubic solid solutions in the system $CaO-Y_2O_3-ZrO_2$. J. Amer. Ceram. Soc., 47:122-127.

Subbarao, E.C., Sutter, P.H., and Hrizo, J. (1965). Defect structure and electrical conductivity of $ThO_2-Y_2O_3$ solid solutions. J. Amer. Ceram. Soc., 48:443-446.

Swinkels, D.A.J. (1970). Rapid determination of electronic conductivity limits of solid electrolytes. J. Electrochem. Soc., 117:1267-1268.

Tretyakov, J.D. and Muan, A. (1969). A new cell for electrochemical studies at elevated temperatures: Design and properties of a cell involving a combination of thorium oxide-yttrium oxide and zirconium oxide-calcium oxide electrolytes. J. Electrochem. Soc., 116:331-334.

Tuthill, R.L. and Sato, M. (1970). Phase relations of a simulated lunar basalt as a function of oxygen fugacity, and their bearing on the petrogenesis of the Apollo-11 basalts. Geochim. Cosmochim. Acta, 34:1293-1302.

Ure, R.W. (1957). Ionic conductivity of calcium fluoride crystals. J. Chem. Phys., 26:1363-1373.

Weise, E.K. (1964). Conductivity (electrical) in solids, in The encyclopedia of electrochemistry, C.A. Hampel, editor. New York: Reinhold Publ. Corp.

Wilder, T.D. (1966). Direct measurement of the oxygen content in liquid copper; the activity of oxygen in dilute liquid Cu-O alloys. Trans. Metal. Soc., AIME, 236:1035-1040.

Worrell, W.L. (1966). Measurements of the thermodynamic stabilities of the niobium and tantalum oxides using a high-temperature galvanic cell, in Thermodynamics, Vol. 1. Proceedings of the Symposium on Thermodynamics (1965). Vienna: International Atomic Energy Agency.

Yao, Y.F.Y. and Kummer, J.T. (1967). Ion exchange properties of and rates of ionic diffusion in beta-alumina. J. Inorg. Nucl. Chem., 29:2453-2475.

Zeise, H. (1954). Thermodynamik auf den grundlagen der quantentheorie, quantenstatistik und spektroskopie. 3., Vol. 1. Leipzig, Germany: Hirzel-Verlag Halfte.

ACKNOWLEDGMENT:

The author is grateful to J. Stephen Huebner and Rosalind T. Helz, both of U.S. Geological Survey, for their careful review and helpful comments.

CHAPTER 4

Direct Control of the Oxygen Vapor Phase Mainly at Pressures Greater Than One Atmosphere

William B. White

I. Introduction

When the oxygen fugacity above a reacting mixture is more than a few torr, (mm of mercury) the pressure of the vapor phase can best be managed by using oxygen itself. In the range from a few torr to a few atmospheres nominal laboratory techniques of glass manifold systems, manometers to measure pressure, and routine set-ups can be used. Pressure much in excess of a few atmospheres requires special high-pressure techniques and the use of gaseous oxygen with these techniques places even more severe restrictions on their application.

Oxygen pressures have been utilized directly to pressures of many kilobars and indirectly in other high-pressure systems to pressures in excess of 40 Kb. Uses for such atmospheres include stabilization of a higher valence state such as Ni^{3+}, Cr^{4+}, or Pt^{4+} in the preparation of new compounds; the determination of phase relations in systems with unstable higher valence oxides; and growth of crystals of unstable high oxidation-state materials. In general the complexity of the apparatus increases in this same order. Figure 1 indicates by shading the elements for which these techniques would be useful.

High oxygen pressure experiments are not new. There are many examples in the chemical literature of some sort of autoclaves being used at pressures to a few tens of bars. In most cases the only objective was to prepare a material not stable at one bar pressure and few attempts were made to control the pressure or the vapor composition. For most purposes such techniques can be considered as superseded by the techniques described in the following

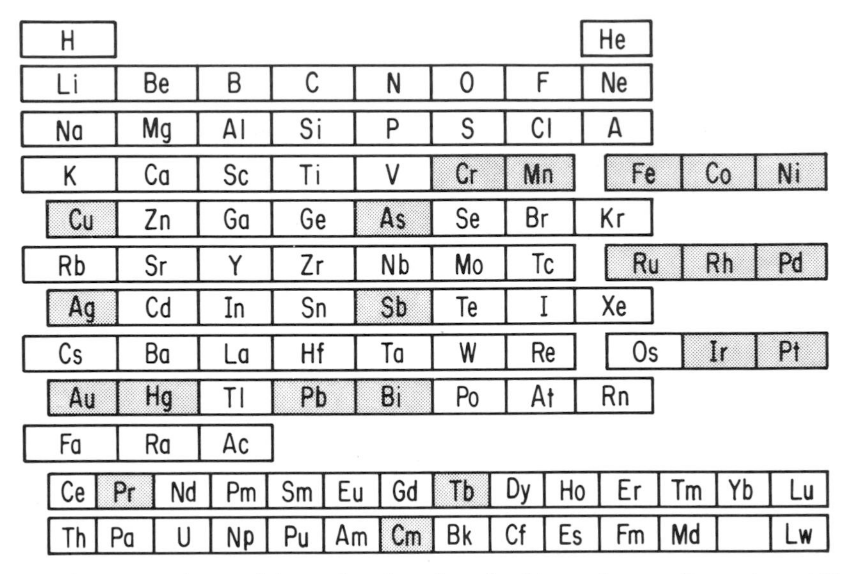

Figure 1. Periodic table showing by shading those elements with oxidation states that require high oxygen pressure for synthesis or phase equilibria.

sections.

The pressure units used in the various scientific disciplines that have had some occasion to use high-pressure techniques are many and varied. The chemical literature often expresses pressure in atmospheres or in "mm Hg," the latter now redefined as torr. Earth and materials scientists have been in the habit of using bars or kilobars as their unit, or occasionally Kg/cm^2. More recently the proper MKS unit, $Newtons/meter^2$ has come into usage. Since all units are still in common usage, a conversion table is provided in Table 1.

II. Tensiometric Measurements

At low pressures the oxygen fugacity of a redox reaction can be determined by direct vapor pressure measurements. The apparatus for doing this takes many forms since for the most part special high-pressure techniques are not required. The most recent developments of this technique have been by E.M. Otto of the National Bureau of Standards and his procedures are discussed

TABLE 4.1

Conversion Factors for Pressure Units

	Atm	Bar	Kg/cm^2	MN/m^2*	Psi	Torr
Atm	1	1.01325	1.03323	0.101325	14.6960	760
Bar	0.986923	1	1.019716	0.1000	14.5038	750.061
Kg/cm^2	0.967841	0.980665	1	0.0980665	14.223343	735.559
MN/m^2	9.86923	10.0000	10.19716	1	145.038	7500.61
Psi	0.0680460	0.0689476	0.070306958	0.0068947	1	51.7149
Torr	0.0013158	0.0013332	0.0013595	0.00013332	0.019337	1

*MN/m^2 = mega Newtons per square meter

below.

The basic apparatus is shown in Figure 2. The sample is contained in a non-reactive container (Pyrex, Vycor, and nickel have been used) which is connected by Pyrex or heavy wall tubing to a manometer. The reaction vessel is heated with a small furnace and the entire assembly is enclosed in a constant temperature chamber. Outside the constant temperature chamber, a second manometer and external valve allow pressure to be adjusted. For use at pressures

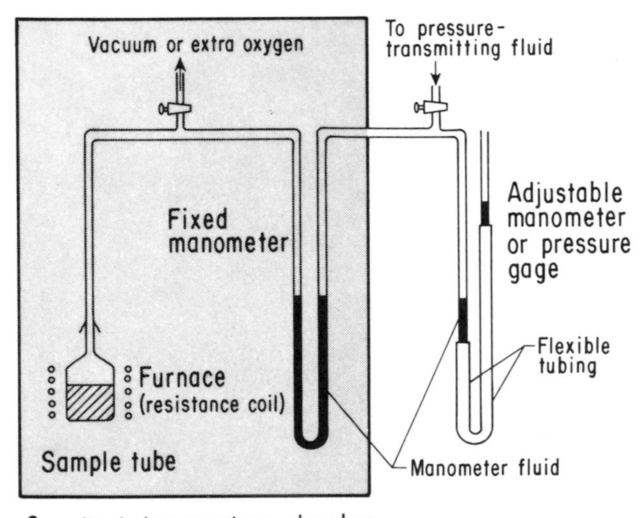

Figure 2. Basic apparatus for direct tensiometric measurement of oxygen vapor measure. (Modified from Otto, 1964, 1965)

to a few tens of bars, the internal manometers must be replaced with narrow-bore heavy-wall materials, the connecting tubing replaced with stainless steel and the external manometer replaced with a pressure gage.

This apparatus is best suited for determining univariant phase relations for reactions in which a higher oxidation state oxide decomposes into a lower oxidation state oxide plus oxygen. A quantity, typically several grams of the higher oxidation state oxide is loaded into the chamber and the system evacuated. When the sample is heated above the kinetic threshold, it begins to decompose, forming the lower oxidation state oxide (or metal) and releases oxygen into the closed system. When the oxygen pressure reaches the equilibrium value for the coexistence of the two phases at the temperature of the experiment the reaction stops and the oxygen pressure is fixed. A requirement is that some of each solid phase must remain at the equilibrium state (otherwise the system would not be univariant). The vapor pressure can thus be measured directly by the manometer and the temperature adjusted to a new and higher value and the experiment run again. The external manometer can be used to counterbalance the pressures so that the internal system is operated at constant volume.

The advantages of tensiometric measurements are that points on the univariant curve are obtained directly and further that a number of points can be obtained in a single experiment. This has clear advantages over any sort of run-and-quench technique simply in saving time. On the other hand it is necessary to analyze the final product of any experiment to insure that there were indeed two phases in coexistence throughout the run. A disadvantage is the relative lack of control over the vapor phase. Enough sample must be used to fill the entire system with oxygen to the pressure at equilibrium. Any adsorbed water or stray gaseous impurities in the starting materials contribute to the total pressure and result in errors in the reported oxygen pressures. This procedure is difficult to extend to higher pressures, partly because of equipment limitations and partly because of the difficulty in obtaining enough gas to fill the entire system to high pressure. At 1 Kb for example, the molar volume of oxygen is a few tens of cc's — comparable or smaller than the volume of solid phases.

Otto has used his method in the pressure range of 0.1 bar to 200 bars to determine univariant equilibria and thermodynamic properties in the systems Mn-O (1964, 1965), Pb-O (1966-a) and Ag-O (1966-b). Kubota (1961) and Goto and Kitamura (1962) attempted to use closed system direct manometric measurements to determine the equilibria in the $CrO_2 \rightarrow Cr_2O_3$ reaction and Figure 3 shows

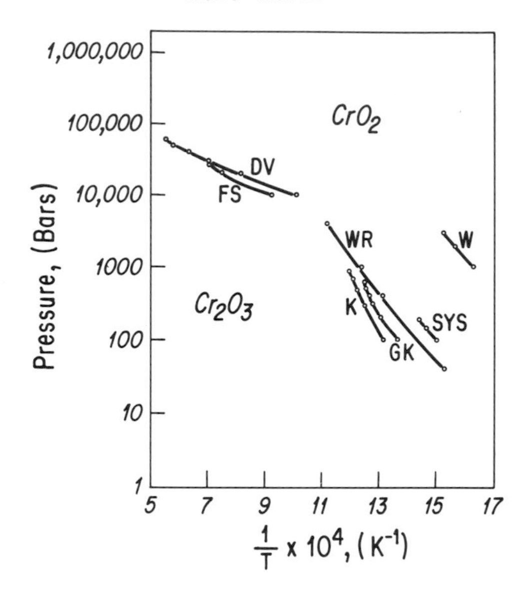

Figure 3. Comparison of univariant curves for the decomposition reaction $2CrO_2 \rightarrow Cr_2O_3 + 1/2\ O_2$ obtained by various methods. DV (DeVries, 1967) belt apparatus. FS (Fukanage and Saito, 1968) piston-cylinder high-pressure apparatus. K (Kubota, 1961) direct vapor pressure measurement on closed system. GK (Goto and Kitamura, 1962) direct gas pressure measurement on closed system. WR (White and Roy, unpublished) quench runs in open system. SYS (Sōmiya *et al.*, 1965) quench runs in open system. W (Wilhelmi, 1968) decomposition of CrO_3 in sealed gold tube in hydrothermal apparatus.

that the results are not in very good agreement with determinations made by other methods. The Cr-O system is a good illustration of the problems in this method because of the extreme hydroscopic character of CrO_3, the usual starting material. Other examples of direct tensiometric measurements are the systems Tb-O (Guth and Eyring, 1954), Pr-O (Burnham and Eyring, 1968), Ru-O (Bell and Tagami, 1963), and Cm-O (Chikalla and Eyring, 1969).

III. Use of Cold-Seal Pressure Vessels

A. Basic Equipment Design

The turbine-blade alloy pressure vessels of the Tuttle or "test tube bomb" type have been widely adapted for experimentation at high water pressures and are commercially available (see Chapters 5 and 6). With only a few modifications and restrictions on materials, these vessels can also be used at high oxygen pressures. The alloy steels of the Stellite (Haynes Specialty Steel Co.) type prove remarkably resistant to oxidation and can contain gaseous oxygen at red heat over a pressure range to 4 Kb.

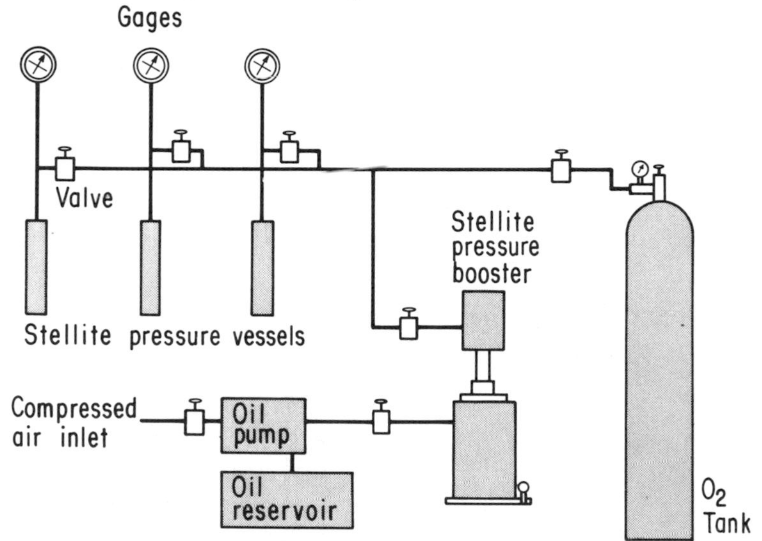

Figure 4. Apparatus for oxygen reactions to 4 Kb pressure.

The basic design is illustrated in Figure 4. Pressure vessels can be machined from Stellite bar stock with 1 inch or 1 1/4 inch o.d. with a 1/4 inch bore. Closure is effected with a cone-in-cone seal. The male cone fitting can be stainless steel with a 59° cone against a 60° seat cut in the top of the vessel (Figure 5 a.)). Pressure tubing and valves are all standard stainless steel high-pressure commercial type rated at 75,000 psi. Several types of gages have been used, all of the bourdon tube design. Stainless steel bourdon tubes are preferred although K-monel tubes have been used without obvious deterioration. Brass or high-ferrous alloy tubes are not recommended.

Pressures up to 300 bars can be obtained from a commercial oxygen cylinder using the thermal expansion of the gas to achieve higher pressure. Above 300 bars some sort of intensifier is needed (see Chapter 8). The most satisfactory of a series of designs used in the writer's laboratory has a Stellite cylinder with a 1 inch bore and a hardened steel piston with a floating upper section (Figure 5 b.)). The seal is of the Bridgman design (see Figures 4 and 5, Chapter 8), with a neoprene gasket. Silicon stopcock grease makes a satisfactory lubricant. Unlubricated gaskets tend to shear and deteriorate very badly after a few runs. Flakes of neoprene will explode at pressures of 2 Kb as they work their way into the top of the cylinder but oxidation of the gasket itself is not a serious

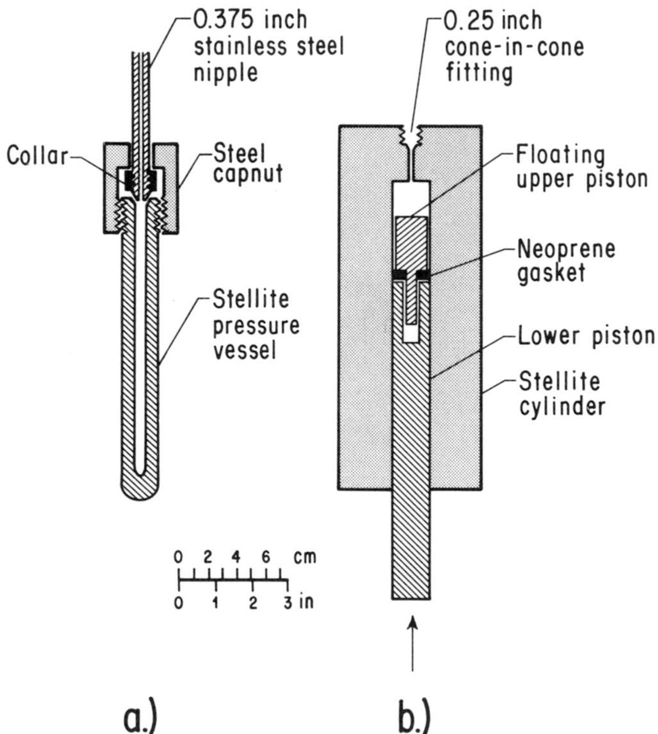

Figure 5. Details of: a.) pressure vessel closure system; and b.) intensifier cylinder.

problem. Gaskets do need frequent replacement. Force to drive the piston-in-cylinder intensifier is provided by a 20 ton hydraulic ram driven by an oil pump. Any hydraulic pump providing 3000 psi oil pressure to the ram could be used. The Sprague air-driven pump has been satisfactory.

B. Safety Hazards and Special Precautions

High oxygen pressure apparatus needs to be treated with the same respect accorded to other high-pressure equipment. In addition there are the hazards associated with combustion from the gas itself. All high-pressure fittings must be scrupulously free of grease and organic matter. High-iron alloys should be avoided. If they cannot be avoided, for example in a ferrous steel intensifier cylinder, all machined surfaces should be pre-oxidized by exposure to oxygen at tank pressures.

One instance is known in which a new Stellite pressure vessel exploded at 740 C and 4.8 Kb. The rip in the vessel wall was burned somewhat and the spray of gas and fine droplets of metal demolished the furnace. The main mass of vessel metal, however, did not ignite in spite of the freshly torn surface.

Stainless steel fittings and tubing will ignite at oxygen pressures of 4 Kb at temperatures of several hundred degrees. Stainless steel burns with explosive violence once ignited and hot gas should not be permitted in the connecting manifold. Acceptable temperature ranges appear to be up to 700 C at 4 Kb extending to 900 C at 200-300 bars (tank pressures) although the pressure vessels rapidly deteriorate. In particular there is oxidation along any flaws or hairline cracks in the original alloy. Failure takes the form of insidious leaks rather than catastrophic rupture.

C. Use in Open Systems

Direct reactions with the oxygen vapor phase can be conducted in open systems (open capsule run) with the sample in direct contact with the gas. Reaction of the sample with the bomb wall can be avoided by wrapping the specimen in gold foil envelopes. Loose powder is better than pressed pellets because of better contact of the gas with individual grains. Gold is inert to oxygen under all conditions so far investigated. [Although some oxide of gold may indeed be stable under high oxygen pressures and low temperatures, the reaction of metallic gold to form an oxide is extremely sluggish (Muller, et al., 1969).] On the other hand, platinum readily oxidizes and is not a suitable sample container at high oxygen pressures. Oxide films form directly on solid slabs of metallic platinum and these films may in turn react with the specimen introducing platinum impurities. The phase diagram of the platinum-oxygen system of Muller and Roy (1968) is shown in Figure 6. If, for some reason, platinum must be used, this diagram shows the temperature-pressure range in which the metallic phase is stable.

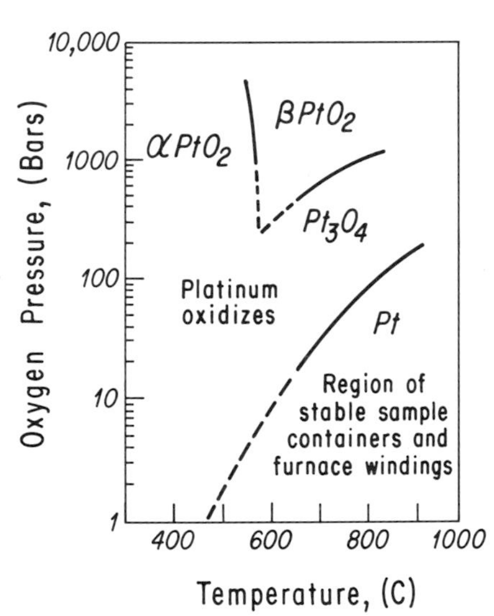

Figure 6. P-T projection for the system Pt-O showing the stability range of metallic platinum. (Modified from Muller and Roy, 1968)

Specimen sizes can be anything convenient and runs can be made with a few mg of materials. Typical run length is 24 hours although kinetics of individual reactions vary widely. Quenching can be accomplished by plunging the pressure vessel into a container of cold water. Even this extreme treatment does not seem to have detrimental effects on the steels.

D. Use in Closed Systems

High oxygen pressure experiments can be carried out with conventional hydrothermal equipment if some high oxidation state of oxide or internal oxidant can be conveniently encapsulated. For example, decomposition reactions can be followed by encapsulating the highest valence oxide in the system in a welded gold tube which is then reacted in a constant temperature, constant pressure hydrothermal run. The decomposition of the higher valence oxide provides the oxygen which, because of the collapsible gold capsules, is in pressure equilibrium with the surrounding medium. However, it is necessary to have a sufficient amount of oxygen released to fill all pore space in the capsule at the pressure of the system. If the quantity of gas is insufficient, there will be developed some unknown lower oxygen pressure in the pores, because the solid phase will continue to support the capsule walls. A second precaution is that the composition of the gas phase must be known. Unknown gaseous impurities, such as adsorbed water, will contribute to the total pressure and result in an oxygen partial pressure that is lower than the presumed value.

Closed systems of this type have been used by Wilhelmi (1968) for the system Cr-O.

E. Use with Internal Oxidants

A unique wrinkle in developing highly oxidizing conditions in a closed system with external pressure control has been developed by Shannon (1968). Shannon sealed his experimental materials, platinum or rhodium oxides, in a gold tube with water and potassium chlorate. The welded gold tubes were then placed in hydrothermal pressure vessels using water as the pressure medium at temperature to 700 C and pressures to 3 Kb. The decomposition of the $KClO_3$ provided the excess of oxygen within the limited volume of the gold tube and so reactions involving oxygen uptake could be conducted. The KCl reaction by-product did not react with any of the oxide materials in the experiment.

This technique would seem to have wide application in the preparation of higher oxide materials since the presence of the water and the halogen should greatly increase the kinetics of many oxide reactions. It is less applicable to phase equilibrium work unless one can show, in each case, that the oxidant and its reaction products are behaving as inert ingredients except for the released oxygen itself.

F. Advantages and Limitations

The main advantage of the cold-seal pressure vessel technique is that the vapor pressure is controlled independently of the particular system under study. Very small samples can be used since the pressurizing medium is supplied externally. If oxygen itself is the pressuring medium, the technique minimizes the effects of other gaseous components released from the sample and also promotes more complete reactions. Redox reactions can be approached from both oxidized and reduced starting materials, thereby providing a better criterion for equilibrium. Disadvantages include the temperature limitations of heated pressure vessels and the necessity to perform a rapid quench. Quenching can be a serious problem in systems in which oxidation or reduction reactions take place rapidly. Phase determinations are usually made by interpolating boundaries between plotted results of quench runs and thus, more data are needed to accurately locate phase boundaries in contrast to tensiometric methods which locate points on the univariant curves directly. On the other hand, the cold-seal systems are more amenable to systems with three or more components because phase assemblages can be accurately established from long duration equilibrium runs.

Application of cold-seal techniques to materials preparation are illustrated by spinels containing Ni^{3+} (Shafer, 1962) and a variety of platinum-containing compounds (Muller and Roy, 1969-a, b, 1970, Shannon, 1968). Phase equilibria have been determined in the binary systems Cr-O (Roy, 1965, Sōmiya et al., 1965,

Wilhelmi, 1968), Mn-O (Klinsberg and Roy, 1960), Pb-O (White and Roy, 1964), Pt-O and Rh-O (Muller and Roy, 1968), and Au-O (Muller *et al.*, 1969). Ternary equilibria have been measured in a few systems containing chromium (Muller *et al.*, 1968). Examples of crystal growth under high oxygen pressure in cold-seal vessels are sparse although the technique could certainly be used.

IV. Use of Internally Heated Pressure Vessels

Experiments at temperatures above 700-900 C and/or pressures above 4 Kb are usually conducted in some type of internally heated pressure vessel (see Chapter 8). The vessel wall is kept cold, often with a circulating water cooling system, and the pressure limits are those of the cold yield strength of the steel.

One of the most elaborate such vessels (Baker and Talukdar, 1968) is sketched in Figure 7. The pressure vessel was constructed of steel with heads bolted against o-ring seals. Chloroprene or acryllic rubber o-rings were satisfactory. Power leads and thermocouple (or other) electrical connections were packed through insulated phosphor bronze packings. The pressure was supplied from oxygen gas cylinders or from an intensifier system as discussed previously. The internal furnace was a resistance winding on an inner alumina tube. The inner tube was placed in the vessel and the extra space filled with powdered alumina which serves both as thermal insulation and also removes some of the free volume of the system. The open end of the furnace tube terminated in a plug of asbestos wool through which the oxygen was admitted. Nichrome was used as the heating element at temperatures up to 900 C and above that platinum was used. The Pt-O oxygen phase diagram (Figure 6) indicates the regions where platinum would be a suitable furnace winding.

An alternative design for internally heated pressure vessels was described in some detail by Ferretti *et al.* (1961). The pressure vessel was constructed of 316 stainless steel, 5 7/8 inch i.d. and 14 inches long. The vessel was vertically mounted and sealed with threaded end caps pressed against o-rings. The internal heating device was an r.f. induction coil powered through leads packed in

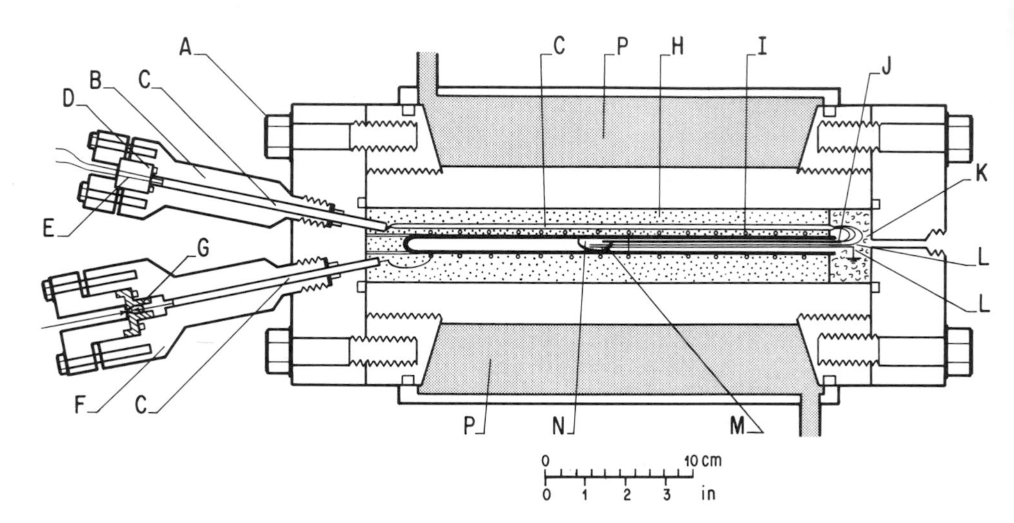

Figure 7. Baker Vessel with component parts: A. Bolts; B. Phosphor bronze electrical lead assembly; C. Alumina sleeve; D. Brass block; E. Araldite resin; F. Phosphor bronze power inputs; G. Metal-ceramic seal (Consolidated Electrodynamics type 41-901-0006); H. Alumina powder; I. Nichrome element; J. Alumina thermocouple sheath; K. Asbestos wool; L. Gold connecting wires; M. Alumina boat; N. Silver oxide; P. Cooling jacket. [From Baker and Talukdar (1968) with permission of Dr. E.H. Baker and Dr. M.J. Jones, editor, The Institution of Mining and Metallurgy.]

the end plates. This configuration allowed the apparatus to be used for crystal growth experiments. The apparatus was used at temperatures to 1600 C and at an oxygen pressure of 75 bars. Because of the large internal diameter and low strength of stainless steel, this exact design could not be used at much higher pressures.

The advantages of internally heated vessels are the essentially unrestricted range of attainable temperatures, and the flexibility in type of experiment. The Baker design has been used to conduct thermogravimetric measurements under high oxygen pressure and there is no reason that any other measurement such as differential thermal analysis (D.T.A.) or electrical properties could not be performed. Likewise the large volume of the vessel permits crystal growth experiments and it seems that much more could be done along these lines. The disadvantages include increased cost of the more complicated vessel, and a lower return of data because of the longer time needed to set up and take down experiments. If the pressures were extended to the multi-kilobar range, problems would arise in selecting furnace and packing materials that would be stable while exposed to high oxygen pressure for long periods of time.

Baker and Talukdar (1968) used their internally heated vessel to determine P-T-X phase relations in the system Ag-O including the solubility of oxygen in liquid silver. A similar sort of apparatus, not described in detail in the literature, has been used by MacChesney and coworkers to prepare a wide variety of perovskite compounds of the $SrFeO_3$ and $BaFeO_3$ families that contain iron in the tetravalent state. See MacChesney *et al.* (1966) and references contained therein for their applications of both internally heated and cold-seal vessels to high oxygen pressure usage. Van Hook (1963) also used an internally heated vessel to follow the peritectic melting of yttrium iron garnet to 100 bars pressure. The crystal growth apparatus of Ferretti *et al.* was used to grow $CoFe_2O_4$ free of ferrous iron.

To the writer's knowledge no one has extended internally heated pressure vessels to oxygen pressures above 1 Kb. Internally heated vessels are now used routinely in several laboratories to pressures of 10 Kb with argon as an inert pressure medium. Whether one could use oxygen at these pressures, or whether entirely new problems would arise due perhaps to having exceeded the flash point of some component, remains to be seen. It would seem that of the various types of high oxygen pressure equipment, the internally heated vessels are the most versatile and that much more use could be made of them.

V. Oxygen at Pressures Greater than Four Kilobars

Experimentation at very high oxygen pressures requires the use of the various techniques for achieving pressures in the tens of kilobars range. Many such devices have been built and are modifications of the opposed anvil, belt, or piston-cylinder arrangements (see Chapter 7). Any such apparatus should work about equally well at high oxygen pressure. It is not the purpose of this article to review all of the high-pressure techniques and reference should be made to other review works such as Wentorf (1962).

Very-high-pressure apparatus can be used to determine oxide decomposition reactions in much the same way that cold-seal vessels are used with sealed systems. The high oxidation state oxide is placed in the high-pressure apparatus and heated. It decomposes at some temperature to a lower oxidation state oxide with the release of oxygen gas. Since the sample volume is small and confined, the oxygen is in pressure equilibrium with the system and the measured pressure can be regarded as the oxygen pressure. This technique is subject to the same restrictions as other closed system experiments: The apparatus must not leak, there must be no extraneous component in the vapor phase, the gas must not react with other parts of the system, and the gas in the pore volume must be in pressure equilibrium with the confining pressure. The last of these is a less serious problem because high pressures exceed the crushing strength of the grains of the solid phases and pore volumes are very small.

The simplest of the high-pressure techniques, the opposed anvil apparatus, is sketched in Figure 8. Typically about 10 mg of sample is sandwiched between noble metal discs within a nickel ring on the 1/4 inch diameter anvil face. Most past experimentation has used platinum/rhodium discs, but it appears from Figure 6 that gold discs would be less subject to oxidation with resulting contamination of the sample. Pressure is applied with a hydraulic ram and the specimen pressure is calculated as the hydraulic pressure multiplied by the ratio of the ram area to the anvil area. After reaction, the temperature is quenched with an air blast, the pressure removed, and the sample extracted for analysis. Particular

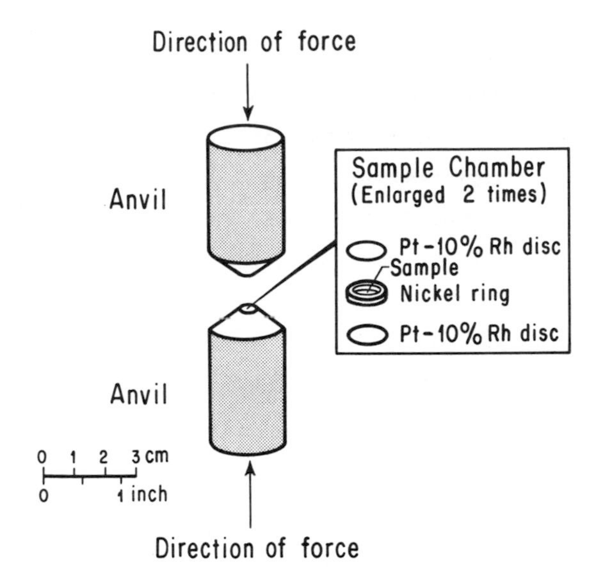

Figure 8. Opposed anvil apparatus. Only the anvils and sample assembly are shown. The hydraulic ram that applies force to the anvils and the associated pressure-controlling system has been omitted.

care must be given to an examination of the nickel ring to ascertain that it has not cracked and that there appears a clean seal between it and the platinum discs. Experimentation by the writer with lead oxides indicates that the phase boundary between PbO_2 and Pb_2O_3 is in good agreement with extrapolations made from hydrostatic experiments at lower pressures. DeVries (1967) using a belt apparatus, and Fukanage and Saito (1968) using a piston-cylinder apparatus obtained univariant curves for the decomposition of CrO_2 into Cr_2O_3 that were in reasonable agreement with lower pressure data (Figure 3.)

An interesting experiment to extend the usefulness of very high oxygen pressures would be to use Shannon's internal oxidant technique as mentioned earlier in this chapter.

VI. Oxygen Fugacity at High Pressures

Univariant curves for solid-vapor reactions determined at high oxygen pressures could be made to yield useful thermochemical information on the higher oxidation state oxides. Such calculations however, require a knowledge of the oxygen fugacity at the temperatures and pressures of the experiments. Such data are very sparse (see Chapter 9). Compressibility of oxygen to 50 C and 135 atm was measured by Michels et al. (1954) and the vapor pressure over liquid oxygen by Muijlwijk et al. (1966). The most comprehensive work is the recent paper by Weber (1970) which is unfortunately confined to temperatures below 300 K. No data for oxygen compressibility seem to be available in the temperature range of solid state reactions, 300-1500 C.

BIBLIOGRAPHY:

Baker, E.H. and Talukdar, M.I. (1968). Silver-oxygen system in the pressure range 0.2-750 atm and some measurements on silver-nitrogen and gold-oxygen equilibria. Trans. Inst. Mining Metal., 77C:128-133.

Bell, W.E. and Tagami, M. (1963). High-temperature chemistry of the ruthenium-oxygen system. J. Phys. Chem., 67:2432-2436.

Burnham, D.A. and Eyring L. (1968). Phase transformations in the praseodymium oxide-oxygen system: high-temperature x-ray diffraction studies. J. Phys. Chem., 72:4415-4424.

Chikalla, T.D. and Eyring, L. (1969). The curium-oxygen system. J. Inorg. Nucl. Chem., 31:85-93.

DeVries, R.C. (1967). Stability of CrO_2 at high pressures and temperatures in the "belt" apparatus. Mat. Res. Bull., 2:999-1008.

Ferretti, A., Wickham, D.G., and Wold, A. (1961). Induction heated pressure vessel for growing oxide single crystals. Rev. Sci. Instr., 32:566-568.

Fukunage, O. and Saito, S. (1968). Phase equilibrium in the system CrO_2-Cr_2O_3. J. Amer. Ceram. Soc., 51:362-363.

Goto, Y. and Kitamura, T. (1962). On the phase diagram of the Cr-O system at high pressure of oxygen and some properties of the compound CrO_{2+x}. J. Japan Soc. Powder Met., 9:109-113.

Guth, E.D. and Eyring L. (1954). The terbium oxides. I. Dissociation pressure measurements: x-ray and differential thermal analysis. J. Amer. Chem. Soc., 76:5242-5244.

Klingsberg, C. and Roy R. (1960). Solid-solid and solid-vapor reactions and a new phase in the system Mn-O. J. Amer. Ceram. Soc., 43:620-626.

Kubota, B. (1961). Decomposition of higher oxides of chromium under various pressures of oxygen. J. Amer. Ceram. Soc., 44:239-248.

MacChesney, J.B., Jetzt, J.J., Potter, J.F., Williams, H.F., and Sherwood, R.C. (1966). Electrical and magnetic properties of the system $SrFeO_3$-$BiFeO_3$. J. Amer. Ceram. Soc., 49:644-647.

Michels, A., Schamp, H.W., and DeGraaff, W. (1954). Compressibility isotherms of oxygen at 0°, 25°, and 50^oC and at pressures up to 135 atmospheres. Physica, 20:1209-1214.

Muijlwijk, R., Moussa, M.R., and Van Dijk, H. (1966). The vapor pressure of liquid oxygen. Physica, 32:805-822.

Muller, O., Roy, R., and White, W.B. (1968). Phase equilibria in the systems NiO-Cr_2O_3-O_2, MgO-Cr_2O_3-O_2, and CdO-Cr_2O_3-O_2 at high oxygen pressures. J. Amer. Ceram. Soc., 51:693-699.

Muller, O. and Roy R. (1968). Formation and stability of the platinum and rhodium oxides at high oxygen pressures and the structures of Pt_3O_4, β-PtO_2, and RhO_2. J. Less-Common Met., 16:129-146.

Muller, O., Newnham, R.E., and Roy R. (1969). Preliminary study of new crystalline gold oxides. J. Inorg. Nucl. Chem., 31:2966-2970.

Muller, O. and Roy, R. (1969-a). Two new ternary copper-platinum oxides. J. Less-Common Met., 19:209-214.

Muller, O. and Roy R. (1969-b). Synthesis and crystal structure of Mg_2PtO_4 and Zn_2PtO_4. Mat. Res. Bull., 4:39-44.

Muller, O. and Roy, R. (1970). Synthesis and crystal structure of $CdPtO_4$. J. Less-Common Met., 20:161-163.

Otto, E.M. (1964). Equilibrium pressures of oxygen over Mn_2O_3-Mn_3O_4 at various temperatures. J. Electrochem. Soc., 111:88-92.

Otto, E.M. (1965). Equilibrium pressures of oxygen over MnO_2-Mn_2O_3 at various temperatures. J. Electrochem. Soc., 112: 367-370.

Otto, E.M. (1966-a). Equilibrium pressures of oxygen over oxides of lead at various temperatures. J. Electrochem. Soc., 113: 525-527.

Otto, E.M. (1966-b). Equilibrium pressures of oxygen over Ag_2O-Ag at various temperatures. J. Electrochem. Soc., 113:643-645.

Roy, R. (1965). Controlled p_{O_2} including high oxygen pressure studies in several transition metal-oxygen systems. Bull. Soc. Chim. France, 1965:1065-1070.

Shafer, M.W. (1962). Preparation and properties of ferrospinels containing Ni^{3+}. J. Appl. Phys. Suppl., 33:1210-1211.

Shannon, R.D. (1968). Synthesis and properties of two new members of the rutile family RhO_2 and PtO_2. Solid State Comm., 6: 139-143.

Sōmiya, S., Yamaoka, S., and Saito, S. (1965). Phase relation between CrO_2 and Cr_2O_3 by decomposition of CrO_3 under high oxygen pressure — preliminary report. Bull. Tokyo Inst. Tech., No. 66:81-84.

Van Hook, H.J. (1963). Influence of oxygen pressure on incongruent melting temperature of yttrium iron garnet. J. Amer. Ceram. Soc., 46:248.

Weber, L.A. (1970). P-V-T, thermodynamic and related properties of oxygen from the triple point to 300 K at pressures to 33 MN/m^2. J. Res. Natl. Bur. Standards, 74A:93-129.

Wentorf, R.H., Jr. (1962). Modern very high pressure techniques. London: Butterworths.

White, W.B. and Roy, R. (1964). Phase relations in the system lead-oxygen. J. Amer. Ceram. Soc., 47:242-249.

Wilhelmi, K-A. (1968). Formation of chromium oxides in the

Cr_2O_3-CrO_3 region at elevated pressures up to 4 kilobar. Acta Chem. Scand., 22:2565-2573.

ACKNOWLEDGMENT:

Much of the high oxygen pressure research at Penn State was conducted with the support of the U.S. Army Electronics Command under contracts to Professor Rustum Roy. I am indebted to Dr. Olaf Muller and Dr. Kenneth L. Keester for sharing with me their experiences with high oxygen pressure apparatus.

CHAPTER 5

Buffering Techniques for Hydrostatic Systems at Elevated Pressures[1]

J. Stephen Huebner

I. Introduction

Many chemical reactions of general interest and of specific geologic significance evolve or consume volatiles at elevated temperatures and pressures. Experimental techniques are needed to study such reactions as functions of temperature, pressure, and either the chemical potentials of the volatile species participating in the reaction or the composition of the non-condensed phase. One approach to the problem, used successfully by Greenwood (1961, 1962, 1967), Gorden and Greenwood (1970), Metz (1966), Johannes and Metz (1968), and Holloway et al. (1968), is the analysis of the fluid phase, yielding the proportions of the species at the time of the analysis. The proportions of gas species in the fluid will change continually as the reaction progresses. The fluid composition may be determined after the experiment, or before the experiment if the composition is adjusted to compensate for changes that occurred during the run. This chapter will treat a different approach that was developed by Eugster (1957). The fugacities of gas species are fixed by equilibration of fluid with one or more solid phases. The fugacity values are maintained constant (or closely approach equilibrium) for the duration of an experiment while the phases under investigation react. For example, the mole fraction ratio, $X\ CO_2/X\ H_2O$, in the fluid phase does not vary during a run, even though these species may react with the solids to consume CO_2 and liberate H_2O as in the reaction $Mg(OH)_2 + CO_2 \rightarrow MgCO_3 + H_2O$. This method also has the advantage

---

[1] Publication authorized by the Director, U.S. Geological Survey

that a reaction can be studied in terms of a variable that cannot be measured directly by chemical analysis. For instance, a redox reaction can be studied as a function of oxygen fugacity, even though the concentration (mole fraction) of molecular oxygen may be 10^{-15}, a value below the limits of analytical detection.

A. Symbols and Abbreviations

The following table lists the symbology and notation convention used throughout this chapter:

TABLE 5.1

Symbols and Abbreviations

T	Temperature, degrees Kelvin
P	Pressure
pi	Partial pressure of species *i*
fi	Fugacity of species *i*
Xi	Mole fraction of *i* in a mixture
K	Equilibrium constant of reaction
μi	Chemical potential of *i*
O_B	Oxygen buffer
S_B	Sulfur buffer
F_B	Fluorine buffer
IW	Iron-wüstite buffer ($Fe-Fe_{1-X}O$)
IM	Iron-magnetite buffer ($Fe-Fe_3O_4$)
WM	Wüstite-magnetite buffer ($Fe_{1-X}O-Fe_3O_4$)
MH	Magnetite-hematite buffer ($Fe_3O_4-Fe_2O_3$)
FIQ	Fayalite-iron-quartz buffer ($Fe_2SiO_4-Fe-SiO_2$)
FMQ	Fayalite-magnetite-quartz buffer ($Fe_2SiO_4-Fe_3O_4-SiO_2$)
NNO	Nickel-nickel oxide buffer (Ni-NiO)
Po	Pyrrhotite ($Fe_{1-X}S$)
Py	Pyrite (FeS_2)
G	Graphite (C)
X	Charge (assemblage being investigated)

OH, CO, COH, COHS, etc. A gas phase consisting of the elemental components listed.

B. Buffers

Solution chemists consider mixtures of weak acids or bases with their corresponding salts to be buffers because these solutions resist changes in pH. Although small amounts of hydrogen ion may be added to or removed from the buffer solution, the pH of the solution tends to remain constant because of reaction. Mixtures of a metal with its most reduced oxide (Ni or NiO) or two oxides with different cation oxidation states (Fe_3O_4 and Fe_2O_3) resist changes in oxygen fugacity at constant temperature and total pressure; for this reason Eugster (1957) has termed such mixtures oxygen buffers. Addition or depletion of oxygen from the system will change the relative molecular proportions of the buffer constituents according to equations such as $Ni + 1/2\ O_2 = NiO$ or $4\ Fe_3O_4 + O_2 = 6\ Fe_2O_3$, but as long as all the phases of the buffer are present, the equilibrium oxygen fugacity is rigorously defined by the buffer.[1]

Developments of the last decade have been severalfold. Use has been made of an increasing number of buffer assemblages. Work has begun on the buffering of polycomponent gas systems, such as C-O-H, C-O-H-S, F-O-H, and H-O-S, etc., which more closely approximate the inferred composition of some geologically important fluids. Fugacity calculations of the buffered gas species can be made with increased precision because the quality of the basic thermodynamic data for the buffering solids, fluids, and fluid mixtures has improved. Several buffers have been precisely calibrated by direct physical measurement of fO_2 and fS_2. Recently Lindsley and Munoz (1969) have used an oxygen buffer at high pressure (20 Kb). (For

[1] Barton and Skinner (1967, p. 245) have applied the term "fixed-point buffer" to an invariant assemblage that defines at a constant value each chemical potential associated with the system. In contrast, "sliding-scale buffers" are solutions that will change composition as components are gained or lost from the solution. "Sliding-scale buffers" do not maintain chemical potentials at constant values.

additional high pressure buffer technique references, see Chapter 8.)

C. Application of the Phase Rule

The solid-gas buffering technique permits the fugacities of all the gas species in the system to be specified. The number of phases present ($\underline{p}$) must be related to the number of components ($\underline{c}$) and the number of restrictions ($\underline{r}$, which are predetermined intensive parameters) such that the variance ($\underline{\omega}$) of the system is zero. This relation is given by the phase rule

$$\underline{\omega} = \underline{c} + 2 - \underline{p} - \underline{r} = 0 \tag{1}$$

Temperature and pressure will be held constant; thus, $\underline{r} \geq 2$. For example, if the gas composition can be described by the system H-O, the number of components is two. One phase, gas, is present. The system is invariant (completely specified) if three intensive parameters are held constant: temperature, pressure, and the chemical potential of oxygen or hydrogen. If the gas has more than two components, invariance is achieved by increasing the number of phases, fixing more intensive parameters, or both.

The equations and thermochemical data necessary to calculate the fugacities of gas species and, if necessary, the gas composition, have been reviewed by Eugster and Wones (1962), Eugster and Skippen (1967), French (1966), and French and Eugster (1965) (see also Chapter 2). Rather than present the material again, this chapter will stress the practical aspects of the buffering technique.

II. Planning the Research

A. Planning of Experiments

It must be emphasized that before beginning experiments, the investigator should carefully plan the approach to his problem; ideally he will choose a path to a solution. Within a given system, it is quite likely that several reactions written between the phases present are not all independent; in other words, it is possible to express the equilibrium constant of one reaction in terms of the equilibrium constants of two or more other reactions.

This fact is well demonstrated by Skippen (1967) who performed an analysis of the phase relations of nine phases in the system MgO-CaO-SiO_2-CO_2-H_2O and found that of the 49 possible reactions, only five were independent. Since one of these five reactions had been previously investigated, he needed to investigate experimentally only four reactions to determine quantitatively the phase relations of the entire system of five components and nine phases.

Another approach that can aid in reducing the experimental effort is to attempt, insofar as possible, the construction of a phase diagram based upon thermochemical data (such as is tabulated by Robie and Waldbaum, 1968). Although the uncertainties in free energies of reaction and thus the positions of the reactions on the diagrams are relatively large, such diagrams will narrow the ranges of possible conditions for the first experiments. As experimentation progresses, it may then be possible to refine the thermochemical data.

Incorporation of calculated phase boundaries may obviate some experiments. For example, because of experimental difficulties, Huebner (1969) calculated (from thermochemical data) the reactions Mn_2O_3-MnO_2-Gas and $MnCO_3$-MnO_2-Gas, then incorporated them in a phase diagram describing the stability of rhodochrosite. The coefficients of reactions which involve two (or more) gas species determine the slopes of the reaction in an activity-activity type diagram (see Zen, 1961). It is then necessary to bracket only two non-colinear reactions to locate the entire grid of related reactions (in a specified system) in activity-activity space.

B. Representation of Data

Many reactions between solid assemblages and a gas can be written with alternative formulations depending upon the gas species used to balance the reaction. This change of variables actually corresponds to combining a solid-gas equilibrium with a gas equilibrium. For instance, the reaction

$$KFe_3AlSi_3O_{10}(OH)_2 + 1/2\ O_2 = KAlSi_3O_8 + Fe_3O_4 + H_2O$$

becomes, with the addition of H_2 to each side of the reaction,

$$KFe_3AlSi_3O_{10}(OH)_2 = KAlSi_3O_8 + Fe_3O_4 + H_2.$$

For purposes of plotting the data thermodynamically (variables T, P, log K) and deriving thermochemical data, use of the second equation is simpler. Hydrogen fugacity can be calculated directly from the oxygen fugacity, dissociation constant of water, pressure, and the fugacity coefficients of the gas species (see Eugster and Wones, 1962, p. 93). Data can be plotted isobarically as a function of T^{-1} and log fH_2(= log K); heat and free energy of reaction can then be determined from the plotted data.

Munoz and Eugster (1969) realized that they could express the equilibrium between wollastonite, fluorite, and quartz as either $CaSiO_3 + 2HF = CaF_2 + SiO_2 + H_2O$ or $CaSiO_3 + F_2 = CaF_2 + SiO_2 + 1/2\ O_2$. They chose to express the reaction in the first formulation because, in their experimental system and in natural systems, HF is more abundant than F_2 and H_2O is more abundant than O_2.

It is not the purpose of this chapter to review all of the kinds of diagrams on which experimental results can be portrayed. Barton and Skinner (1967, p. 280-291) and Garrels and Christ (1965) review a number of such diagram types. In general, choice of diagram depends on the variables to be plotted and the intended application.

C. Buffer Selection

Buffers should be selected with two primary considerations in mind. The buffers chosen should provide a range of conditions that is great enough to determine the reaction with the desired precision. However, the time required for the attainment of equilibrium may be a competing factor. Specific limiting factors are discussed in succeeding sections, but it is important to remember the necessity of demonstrating equilibrium. The reaction equilibrium must not only be bracketed in terms of a variable such as temperature, but the reacting assemblage must also be in equilibrium with the buffer. Otherwise a carefully bracketed reaction could represent only an apparent, and erroneous, set of intensive variables.

III. Use of Pressure Vessels

A. Hydrostatic Pressure Vessels

Pressure vessels, commonly called "cold-seal bombs," are metal chambers capable of containing a fluid pressure medium at the pressure and temperature of the experiment (see Tuttle, 1949). The seal (Figure 1) is a "cone-in-cone" type: a cone of stainless steel deforms slightly against the conical seat of the more resistant bomb metal. The threaded "heat nut" and cone are outside the

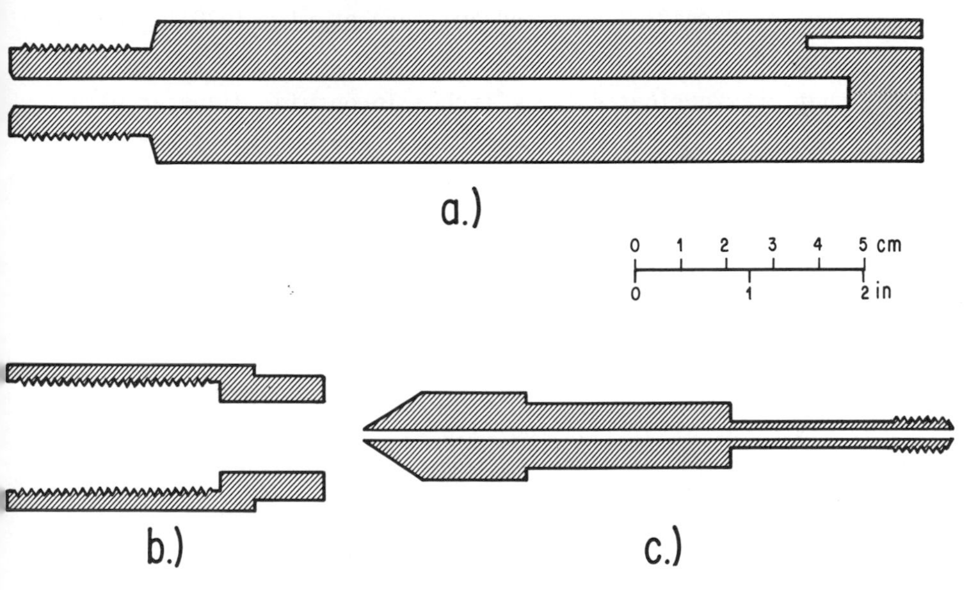

Figure 1. Cold-seal bomb. a.) Bomb showing main bore and thermocouple well. b.) Heat nut. c.) Cone and shank.

furnace, thus the name "cold-seal" bomb even though the temperature of the closure fitting may reach several hundred degrees. Samples to be studied are placed in the central bore of the bomb, and the void is filled with a loose fitting rod to prevent convection of the hydrothermal fluid and to lessen the thermal gradient (see Chapter 6 for details). A small diameter hole drilled parallel to the main bore and from the base of the bomb houses the thermocouple.

Two alloys are commonly used in the fabrication of cold-seal bombs: Stellite No. 25 (cobalt-base) and René No. 41 (nickel-base). These alloys give good service; when failure occurs during a run, it is by gradual swelling and rupture. But sudden and thus dangerous failure by longitudinal splitting has been known to occur when bombs of these or other alloys are P,T quenched in water; the author knows of a case in which a René 41 bomb split longitudinally even during an air jet quench. These situations are relatively rare but indicate the need for precautions at all times. Pertinent physical properties of these and other "superalloys" are summarized by Vaccari (1969) and Sullivan <u>et al.</u> (1970). Continuous operating limits are in the range 800-900 C and 1-2 Kb for bombs that are fabricated 1 1/4 inches in outside diameter with a 1/4 inch bore. René No. 41 bombs are supposed to withstand slightly greater temperatures for the same pressure and lifetime than do Stellite No. 25 bombs, but other factors, such as drilling technique, smoothness of the bottom and sides of the bore hole, position and diameter of the thermocouple well, and previous history of the bomb, have important effects on the performance.

The working limits of cold-seal bombs may be extended by the use of different alloys and modified designs. Williams (1966) used a molybdenum — 0.5% titanium alloy to work at 1000 bars and 1200 C. Williams (1968) also found that addition of 0.08% zirconium to the alloy permitted working limits of 1100 C and 3000 bars (the so-called TZM vessels). Luth and Tuttle (1963) have used specially heat-treated and modified René No. 41 cold-seal vessels at 10,000 bars and 750 C.

The bomb metal alloy must be compatible with the atmosphere in contact with both the inner and outer surfaces of the bomb. Stellite No. 25 and René 41 alloys develop a coating that permits use in a furnace atmosphere of air; these alloys may be used with any of the common pressure media: water, argon, methane, and carbon dioxide. (See Chapter 4 for oxygen as the pressure medium.) When a bomb is new, reducing conditions will prevail in the bore hole, but as a coating develops during use, the "bomb wall oxygen fugacity" approaches that of the nickel-nickel oxide buffer.

[Huebner (1969) cautions against the reducing effects of bomb walls when studying reactions at high oxygen fugacities.] Molybdenum alloy bombs must be used in a reducing or inert furnace atmosphere. It is important to realize that water reacts with a molybdenum pressure vessel; however, argon is a satisfactory pressure medium.

Internally heated bombs (Yoder, 1950; and Chapter 8, this volume) permit greater range of temperature and pressure (1400 C and 10 Kb) by containing the sample and furnace winding within a pressure vessel whose outer surface is cooled with water. Although some recent work has been done (Lindsley and Munoz, 1969; and Chapter 8, this volume) internally heated vessels have not been used much in conjunction with the buffering technique because gold or silver capsules, used to contain the buffer, melt at relatively low temperatures (1064 C at 1 atm for gold, 961.9 C for silver). Because the pressure medium is in contact with the windings, it must have a high resistance and not precipitate a conductive solid such as graphite. Water-pumped argon is a satisfactory fluid.

B. Furnaces

Furnaces for use with cold-seal vessels may be either tubular or split. The bomb must be positioned such that the sample is in the hot spot of the furnace, allowing the minimum thermal gradient over the sample. Because the bomb itself conducts heat away from the hot spot, the position of the hot spot shifts when the bomb is placed in the furnace (see also Chapter 6). To keep the "cold-seal" outside the furnace and the hot spot at the sample, it may be necessary to increase the density (turns per inch) of the furnace winding at the front (toward the heat nut and cold-seal) of the furnace. One successful configuration used by the author for a horizontally mounted tubular furnace uses 0.040 inch kanthal wire helically wound at 7 turns per inch for six inches followed by 12 turns per inch for 3 inches on an alundum tube 1 1/2 inch i.d. x 2 inch o.d.; the total resistance of the element is 21 ohms. Various commercial "split" furnace models with longitudinal windings have also proven to be satisfactory. Careful measurements by the author and others in this laboratory using a cold-seal bomb with filler rod in a tubular furnace have revealed that the temperature gradient over a 1 1/4 inch long charge can be held to 1 C or less. The length of the bomb is an important factor in placing the bomb in the hot spot, but increasing the length of the bombs does not greatly decrease the thermal gradient in the sample position.

C. Pressure Media, Pumping, and Measurement

The most commonly used pressure medium is water because it can be raised to the desired pressure with relatively inexpensive equipment. Other fluids are used when necessary: carbon dioxide and methane in schemes in which the pressure medium participates directly in the buffering reaction, and argon in H_2-Ar mixtures and as a nonconductive medium in internally heated bombs.

Water is easily raised to pressure with a reciprocally driven piston-cylinder pump powered by air pressure, an electric motor, or by hand, and is stored at pressure in a small reservoir. The low density gases Ar, CH_4, and H_2 are pumped most quickly with a two (or more) stage metal diaphragm "gas booster" pump. However, a more economical system consists of filling a reservoir with gas, and then compressing it by pumping oil against a floating or traveling piston. The chief disadvantages are slow speed, the ever present possibility of oil leaking past the O-rings that separate piston and reservoir walls, the awkwardness of switching from one gas to another, and the hazard of containing a larger than necessary volume of gas at pressure. CO_2 is best pumped in a separator with floating piston. Note that a tank equipped with a siphon will supply liquid CO_2 to the separator. The use of a reciprocally driven piston pump is inefficient because the carbon dioxide changes from a high density (liquid) to a low density (gas) phase as it is drawn into the pumping chamber; then the piston's compressive stroke merely changes the gas back to liquid. The net passage of CO_2 through the piston pump is small.

Perhaps the least expensive method of pressurizing fluid is to admit it to the bomb, close the valve to the bomb, and heat the assembly. As the confined fluid is heated, its pressure rises. Fluid is then released until the pressure drops to the desired value. Pressure can be raised only by increasing temperature, however, and each bomb should be equipped with a pressure gage to monitor pressure. Great caution is necessary: because certain dense fluids such as water are relatively incompressible, exceedingly great pressure can be developed on heating a closed system, causing failure of the bomb or plumbing system. At high pressure even tiny water jets cause severe damage to living tissues, and are reported to cause great pain. There is also always the possibility that pieces of ruptured tubing will be thrown across the room as shrapnel.

Pressure is most easily measured with a bourdon tube gage. Pressures to 75,000 psi can be routinely measured with a precision exceeding one percent. Most gases are compatible with standard bourdon tube materials; however, hydrogen and methane may cause embrittlement of stainless steel tubes. Special tube compositions are obtainable, or the gas can be isolated from the gage with a mercury trap (see Shaw, 1967, Figure 1) or a double trap with gas in contact with mercury which is in turn in contact with oil in the gage. Pressures can also be measured with a manganin cell (change of manganin wire resistance with pressure) or with a small pressure transducer (change of magnetic properties with pressure), both of which are available commercially. (See Chapter 8, Appendix 8-A.1)

D. Temperature Measurement

Temperature measurement in cold-seal bombs is accomplished using standard thermocouples and measurement circuits. Chromel-alumel couples are stable over the temperature range for cold-seal bombs and give a larger output (0-40 mv) than platinum-platinum rhodium couples. The thermocouples may be either sheathed or bare; however, sheathed thermocouples survive high temperatures many times longer than unsheathed couples, and are less likely to change composition by alloying with impurities. (In Chapter 9,

Presnall points out that impurity of sheathing ceramic can also be a problem and cause signal drift.) Thermocouple extension wire ("compensation" wire which at low temperature has the same thermal-e.m.f. properties as the thermocouple) is used to connect the thermocouple to a cold (reference) junction and potentiometer with null indicator. E.m.f. readings are converted to temperatures using tables such as those tabulated by Shenker *et al.* (1955). [The adoption of the International Practical Temperature Scale of 1968 by the Comité International des Poids et Mesures (Barber, 1969) necessitates a minor revision of the tables.]

The entire thermocouple measuring circuit should be calibrated against known melting points. This is most commonly done using pure NaCl and metals (see Roeser and Lonberger, 1958) sealed in silica-glass tubes that can be placed in the position of the charge in the bomb well. Because the melting point can be easily bracketed to ±1-2 C, the potentiometer should be capable of measuring voltages with a precision that exceeds the voltage change for one degree (0.04 mv per degree with chromel-alumel couples). The total uncertainty in the run temperature routinely can be held to <±5 C; this estimate is taken as the sum of ±<2 C temperature calibration, ±<2 C temperature variation during the run, and ±1 C thermal gradient.

The entire temperature history of the run including runup and quench, may be recorded; standard multipoint recorders permit recording the temperature of as many as 24 bombs sequentially at periodic intervals. A carefully calibrated recorder strip chart may prove sufficient for actual temperature determinations (in lieu of a potentiometer and null indicator). But even when not used for this purpose, the recorder is very useful in indicating changes in temperature during a run, and approaches to a preselected temperature during runup.

E. Runup and Quench

The method of temperature and pressure quench chosen depends upon the rates of possible reactions in the system being investigated. Normally it is desirable to preserve on cooling the phase assemblage present during a run, that is, to prevent retrograde or back reaction. If the bomb is cooled as a closed system, the P-T path of the charge tends to follow a constant volume condition, _i.e._, a vapor isochore (actually, a mean or average isochore because temperature is not uniform throughout the bomb bore). When the liquid + vapor curve is intersected, the P-T path follows the liquid + vapor curve. Depending upon rates of cooling desired, a bomb may be cooled slowly merely by removing it from the furnace, or more quickly by removing it and placing it in a jet of air. A faster quench is achieved by placing the bomb in a water bath — a more hazardous procedure because the bomb could split, releasing its contents explosively, and perhaps throwing out shrapnel. If fast quenching is desired, the bomb should be suitably shielded during the quench. Recently Wellman (1970) has described a very rapid quenching method in which the charge capsule slides from the hot to the cool ends of a long cold-seal bomb. Incidentally, in rare cases it may be desirable to pressure quench by releasing the bomb fluid — but beware, because the fluid is hot and under pressure. Recognition and interpretation of quench phenomena, and determination of a satisfactory quenching procedure, is an essential part of every investigation (see, for instance, Wones, 1967).

The considerations used in determining the quench method are also applicable to the heating cycle; however, even by preheating the furnace to a temperature that is in excess of the desired temperatures (thereby letting the bomb cool the furnace down to the desired temperature), it is difficult to bring a bomb from room temperature to, say, 700 C in less than 20-30 minutes. Commonly, runup times are about an hour.

The most commonly used runup procedure is to pressurize the bomb and charge at room temperature, and then increase the temperature at nearly constant pressure. During heating, fluid in the bomb is permitted to expand into the pressure line and into a pressure reservoir (commercially available) kept at room temperature. The reservoir should be large compared with the bomb being heated. If a water pressure medium expands into a two-liter reservoir from a standard bomb (Figure 1), the pressure will rise only about one percent as the bomb is heated to 700 C.[1]

F. Safety Precautions

Use of standard hydrostatic apparatus presents no special hazards, provided the equipment is maintained in good condition, used under conservative working conditions, and handled with reasonable care. A good rule is to think carefully about all the steps in a complete operation before beginning it.

Bombs run at 2 Kb are commonly unshielded. For use at pressures in excess of 4 Kb, shielding should be mandatory; the shielding considerations discussed by Holloway (in Chapter 8) are applicable.

Several guidelines to safe practice are listed below:

(1) Avoid, if possible, quenching with water. Never use coolants that may evolve toxic vapors. Use shielding with all rapid quenching methods.

(2) Never point the hot end of a bomb toward someone; be careful of the orientation of a thermocouple which could be projected during a bomb failure.

(3) Do not bend capillaries or pressure tubing to a small radius of curvature; this is particularly important at soldered junctions.

(4) Use a valve to isolate each bomb from the fluid reservoir except on runup or when checking pressure.

(5) Never apply excessive force to any fitting because it may place metal parts under excessive stress and cause failure immediately or at some later date when least expected. In particular, do not strip the head-nut threads, and do not over-tighten the plugs or caps on pressure reservoirs. The author knows of a case in which an improperly tightened reservoir split longitudinally

---

[1] Fyfe and Godwin (1962) discuss a situation in which the temperature was raised before the pressure.

when pressurized.

(6) Do not store more than the necessary amount of pressurized fluid; place reservoirs behind shields or in an isolated location.

(7) Never face or place hands in front of the open ends of pressure tubes.

(8) Consider a bomb to be hot unless you know it is cool. Don't place hot bombs where they might be picked up by someone else.

IV. The Sample Containers

A. Sample Containment

The charge (sample) and buffer are contained within malleable inert metal tubes during a run. Commonly used metals with their inner and outer diameters are listed in Table 5.2. Tubing is

TABLE 5.2

Sizes of Metal Tubing Commonly Used to Contain Sample and Charge

Metal	Inner Diameter, mm	Outer Diameter, mm
Gold	4.0	4.6
	2.4	3.0
Silver	4.0	4.4
	2.4	3.0
	1.4	1.8
Platinum	3.5	4.0
	2.6	3.0
Silver-palladium alloy	3.8	4.5
	2.4	3.0

obtained in a convenient length (1 foot) and is seamless. The metal should be "pure" although the degree of purity will, of course, depend on the nature of the contaminants in the metal and the specific application of the tubing. The tubing may be cut to the desired length with a jeweler's saw, but it is much easier and neater to score the outer surface of the tube (by rolling it across a flat surface with a small applied force from a sharp knife blade), then break it apart with fingers. Any grease or other contaminant may be removed by one or more of the following processes: (1) wiping with tissue moistened in acetone, (2) heating in gas flame (gold melts at 1064 C; silver will oxidize), (3) treatment in either warm HCl or HNO_3 solution.

The choice of tubing depends upon the specific use of the tube. Platinum or silver-palladium alloys ($Ag_{60}Pd_{40}$ to $Ag_{80}Pd_{20}$) are permeable to hydrogen. Gold and silver are nearly impervious to the passage of hydrogen across the tube wall. The melting point is a consideration [silver melts at 962 C, gold at 1064 C, platinum and silver palladium alloy (Hansen and Anderko, 1958) at higher temperatures]. However, all metals crystallize and may fail at temperatures below their melting points. The metal must be compatible with the materials in which it is in contact; silver-palladium forms alloys with nickel, changing the activity of nickel in the Ni-NiO buffer reaction and giving erroneous fugacities. Some sulfurous gases attack silver and silver-palladium. Iron dissolves in platinum and gold; at low oxygen fugacity the effect is so serious with platinum that the charge composition can change appreciably. Silver-palladium alloy, which dissolves little iron (Muan, 1962, 1963), can then be used on all buffers but Ni-NiO. If it is necessary to use the Ni-NiO buffer with an iron-bearing charge, a welded platinum tube with a crimped, very small inner silver liner is used (Eugster and Wones, 1962, Figure 1.)

B. Hydrogen Diffusion Rates

The diffusion rates of gases through metals have been discussed by Barrer (1951). The temperature dependence of the diffusivity of hydrogen through platinum at low pressures is given by Richardson *et al.* (1904); the success of Eugster's buffer technique indicates that platinum functions well at elevated pressures, and Shaw (1967) has discusses some of the limitations on osmotic equilibrium.

Hydrogen diffusion in palladium has been studied by Davis (1954), Makrides and Jewett (1966), Rubin (1966), and Sieverts and Hagen (1935). Because palladium is not easily fabricated into shapes, it is alloyed with silver to make it more manageable. Sieverts and Hagen show that the solubility of hydrogen in Ag-Pd alloys has a maximum at Ag_{20-40} at moderate temperatures (<418 C); see also Hickman (1969). No H_2 diffusion rate measurements are known to the author for the compositions $Ag_{80}Pd_{20}$ to $Ag_{60}Pd_{40}$, but again the successful use of these alloys in buffering capsules (Wones and Eugster, 1965; Lindsley, 1962, p. 101) and by Shaw (1967) indicates their usefulness.

There have been no systematic studies of relative hydrogen diffusion rates between platinum and Ag-Pd alloy capsules. The data presented by Barrer (1951, Table 42, Figure 52) indicate only that at high temperatures, platinum is the better metal for hydrogen diffusion, whereas at low temperatures palladium is preferable. Shaw (1967), using data in Barrer's Table 42, concluded that palladium was the preferable metal below 1000 C. Shaw (1967, and oral communication) also notes that $Ag_{70}Pd_{30}$ alloy gave a markedly faster approach to osmotic equilibrium than did platinum at 700 C.

C. Tubing Closure

Tubing closure is accomplished by either crimping or welding the ends shut. Welding is necessary when the contents are to be isolated from the pressure medium except for a net gain or loss from the system by diffusion through the tube walls. Tubing contents are enclosed by crimping when the free passage of all volatiles into and out of the capsule is desired.

1. Crimping

Tubes are crimped with small, smooth-jawed duckbilled pliers to give a flat crimp or with a machinists three-jawed chuck (complete closure of the chuck is necessary to crimp small diameter tubing) to give a triangular crimp. A triangular crimp is preferable because it enables a 4 mm tube to pass smoothly into a 1/4 inch bore, whereas a flat crimp must be bent to avoid seizing the bomb walls.

2. Welding

"Welding" of a crimped area that has been folded back over

itself can be accomplished by heating with a small gas torch, but most investigators use a small arc welder (and are careful to use welder's goggles). The circuit (Figure 2) requires few components. One electrode is the crimped capsule clamped in a small vise (the undersides of the jaws are milled out) to conduct heat away from the tube; the second electrode is a graphite (1/4 inch spectroscopic) or tungsten (1/16 inch) rod. The arc is made by touching the tip of the electrode to the gold tubing, quickly breaking the physical contact between the electrode and tube, and then drawing the electrode and the arc across the end of the crimp. If necessary, the process is repeated until the weld is smooth and shiny. A good weld is usually indicated when the arc moves smoothly and sometimes emits a characteristic high frequency sound, which is indeed music to the ears. During a poor weld, the arc hops across the surface and may sputter, throwing out bits of metal. (To avoid this, it is essential that the inner surface of the crimped portion of the tube be cleaned before crimping.) All welds should be checked under a microscope — a 30X binocular with good illumination will reveal incomplete welds and "pinholes" through which gases and dissolved species may pass during a run. Testing for escape of gas by heating on a hot plate (100-150 C) for 15-30 minutes and then checking for weight loss is a useful technique.

The various metals require different arc intensities. The variable resistance (R_1 in Figure 2) and the arc intensity vary inversely; required arc intensities increase in the following order (for tubing of a given diameter and wall thickness): platinum, gold, silver, and silver-palladium alloy. The effectiveness of the arc can also be changed by sharpening the electrode (thus concentrating the arc's energy over a smaller area) or by raising the tube in the vise (giving less rapid heat conduction from weld).

If the metal capsule to be welded is loaded with volatiles, it is advisable to cool the lower portion of the tube with water,

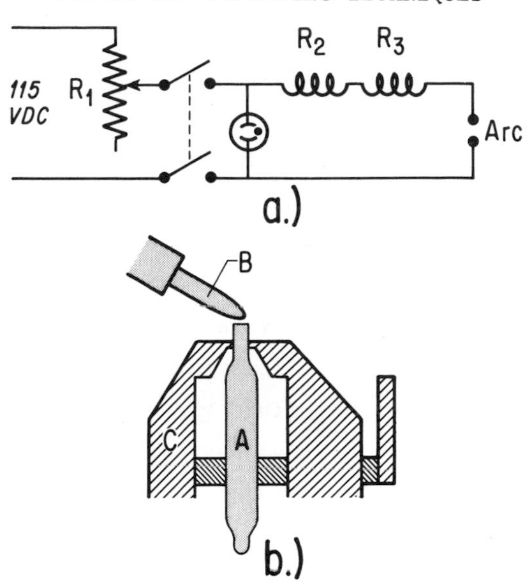

Figure 2. Welder. a.) Schematic diagram of welding circuit. R_1 is 27Ω, 5.5-17 amp. Normal range 8-27Ω. R_2 is 10Ω, 75 watt resistor wirewound for 2 1/4 inches on 1 1/8 inch ceramic tube. R_3 is the same as R_2 except 20Ω. b.) Sketch of welder. A - Capsule to be welded, B - Electrode, and C - Vise.

using either a small beaker, or a water-soaked tissue paper wrapped about the metal capsule.

Because silver and its alloys dissolve oxygen, it is necessary to weld these metals with an electrode holder that directs a stream of nitrogen or argon at the electrode tip (such "inert gas welders" are available commercially). Silver-palladium alloy is very easily welded in this manner. Welded silver-palladium commonly has a dull or even tarnished finish that does not rub off, and the weld may be slightly pitted; a possible explanation is the formation of palladium nitride (see Savitskii et al., 1969, p. 32). Most people find silver to be very difficult to weld, and use gold instead.

It is again emphasized that before welding, the inner lip of the tube must be cleaned of all foreign particles. If the capsule is loaded with too much powder, particles may be forced into the area to be welded by the crimping process. If the weld is uneven and contains large pits, it is often easier to cut back some of the tube (using sharp side-cutting or end-cutting pliers) than to continue welding over the pit. Also, the electrode should be clean and sharp. The arc should be drawn smoothly across the end of the capsule; the suggested rate is about 1/2 inch per second, but the speed depends to some extent on the technique of the individual user. The author finds that silver is more easily welded using a tungsten rather than a graphite rod electrode.

D. Loading and Opening the Capsules

The placing of reactants into capsules requires no elaborate equipment: an assortment of small spatulas, brushes, syringe (50μl) with needle, glassine paper, and perhaps a tiny funnel plus a block with holes to support the tubes in an upright position. For the handling of magnetic or corrosive chemicals, the tools should be constructed of nonmagnetic or inert materials. Use of deliquescent chemicals may require working in an enclosed atmosphere of dried gas.

It is important to measure each quantity of the constituents added in making up the capsule. If this is not done, the capsule may rupture during the run because of too much material and the development of excessive pressure. Also it is necessary to add enough material to produce the phases needed and the proportions must be known to perform calculations relating the composition of the products to the reactants. Finally, the capsule weight gives an important check for loss or addition of material during a run. The author and others in this laboratory prepare capsules and record weights in the following order:

(1) Empty tube, welded or crimped at one end.

(2) Tube with water, or other source of volatile (if volatile is to be added).

(3) Crimped tube with volatile, solids and, if present, a smaller enclosed capsule.

(4) Welded capsule. Weight No. 4 should be smaller than No. 3 by less than 1 mg. A greater weight loss indicates escape of volatile from the capsule, or the loss of an unusually large amount of metal during the weld. These alternatives may later introduce an ambiguity in the interpretation of the run. If it is necessary to cut off any metal, Weight No. 3 should be repeated before welding again.

(5) After the run, the capsule is removed, cleaned, and dried before weighing. The weight change during a run should be small relative to the total weight; it is commonly <1 mg for a 2 g capsule, but the weight loss will be larger if much hydrogen diffused out of the capsule. If the weighing is done carefully, the passage of small amounts of hydrogen into or out of the capsule can be detected by weight changes.

(6) If leakage of gases is suspected, a charge can be checked by heating it slightly above 100 C to force vapor out (being careful to avoid explosion), then cooling and weighing the capsule again. If the weights are the same, vapor was retained.

Capsules are opened with an assortment of small tools, including small side-cutting, duckbill, and long-nosed pliers, steel needles, spatulas, small brushes, and a hot plate to dry the charge. The escape of gas, odor, moistness, color, crystallinity, and homogeneity or heterogeneity should be noted. At this time, examination with a binocular microscope may be desirable. In particular, note the presence or absence of water in the buffer or charge; if an aqueous phase was to have been present during a run, look for evidence of it in the run products. A moist powder will cause glassine paper to buckle; water can be driven off with heat and detected by weight loss, or it can be distilled from the run products placed at the bottom of a glass tube and detected by condensation on the cooler walls higher in the tube.

V. Buffers: Review of Specific Buffering Methods

The following paragraphs will review the specific buffering schemes and give references pertaining to the development and use of these buffers. They are arranged according to the components in the fluid to be buffered. The buffer notation is that introduced by Eugster and Skippen (1967) and is adhered to for consistency and simplicity. The parentheses "( )" indicate a welded capsule semipermeable to hydrogen; the symbols within the parentheses indicate the contents of the capsule; "| |" similarly refers to a crimped tube through which fluid can pass freely. Other symbols are explained in Table 5.1.

A. $\underline{O_B,OH(X,OH)}$

This class includes the original hydrothermal buffers developed by Eugster (1957). Two or more solid phases react to maintain (buffer) the oxygen fugacity of an aqueous fluid at a constant value in an outer welded gold or silver capsule (Figure 3 a.)).

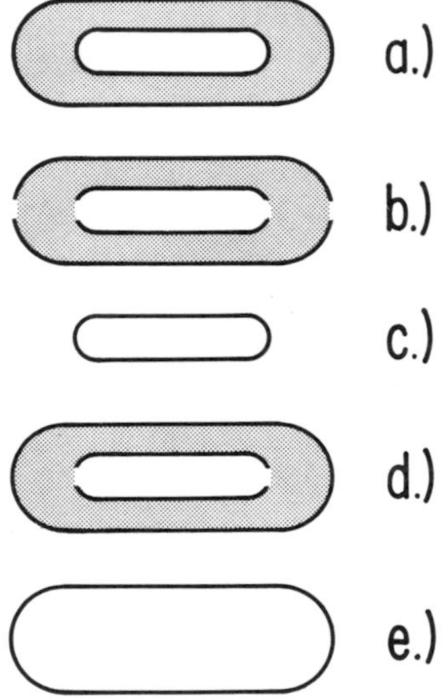

Figure 3. Capsule configurations for buffers. Outer buffer is shaded. The charge (plus internal buffer, if present) is white. a.) Hydrogen buffer isolated from charge; outer and inner capsules welded. b.) Charge open to buffer. Both capsules crimped. c.) Pressure medium is hydrogen buffer. Welded capsule. d.) Charge open to buffer; sealed outer capsule. e.) Internal buffers only. Contents separated from pressure medium.

At known temperature, pressure, and fO_2, fH_2 can be calculated (see Eugster and Wones, 1962). Hydrogen diffuses through a welded platinum or silver-palladium alloy inner capsule and equilibrates so that the fugacities (or chemical potentials) of the hydrogen are the same in the inner and outer compartments. Actually this is a hydrogen buffer because the reaction controls the hydrogen fugacity in the inner capsule. Hydrogen interacts with the aqueous phase in the inner tube, fixing fO_2 and fH_2O in the vicinity of the charge. Reactions which involve O_2, H_2 or H_2O can be investigated with this buffering system. Even if a reaction is of the simple dehydration type, this buffering system is useful to maintain the oxidation state of an element, or to permit fH_2O to be varied independently of pressure (see Shaw, 1967, p. 534-535).

The oxygen buffers (O_B) used with this and other buffering schemes are listed in Table 5.3 a and 5.3 b which includes equations for log fO_2, and references to the derivation of the equation and the use of the buffer. The relative positions of the buffers may be seen from diagrams of log fO_2 versus temperature; e.g., Huebner (1969), French and Eugster (1965), and Eugster and Wones (1962).

As a general "rule of thumb," 5-10 mg of H_2O is loaded into the inner, charge capsule and 15-50 mg in the outer, buffer capsule. Less water should be used at high temperatures or low pressures, than at low temperatures or high pressures, to avoid rupturing the capsules. (See also Chapters 11 and 12)

Some general comments on the use of these buffers follow. Most investigators use gold outer capsules because gold is easy to weld. However, hydrogen does slowly diffuse through gold; in cases in which the fH_2 is much higher (FIQ, IM, WM, IW) or much lower (MH) than in the pressure medium, hydrogen passes out of or into the outer gold buffer capsule, eventually oxidizing or reducing one of the solid buffer phases, consuming the water, or reacting to form so much water that the gold capsule bursts.

TABLE

Equations for Buffer

Buffer	Name	Type	Equation for Log fi
NNO	Ni-NiO	fO_2	$9.36 - \frac{24930}{T}$
IW	$Fe-Fe_{1-X}O$	fO_2	$6.57 - \frac{27215}{T}$
IM	$Fe-Fe_3O_4$	fO_2	$8.99 - \frac{29260}{T}$
WM	$Fe_{1-X}OFe_3O_4$	fO_2	$13.12 - \frac{32730}{T}$
MH	$Fe_3O_4-Fe_2O_3$	fO_2	$13.966 - \frac{24634}{T}$
IQF	$Fe-SiO_2-Fe_2SiO_4$	fO_2	$7.51 - \frac{29382}{T}$
FMQ	$Fe_2SiO_4-Fe_3O_4-SiO_2$	fO_2	$9.00 - \frac{25738}{T}$
	$Mn_{1-X}O-Mn_3O_4$	fO_2	$13.38 - \frac{25680}{T}$
	$Mn_3O_4-Mn_2O_3$	fO_2	$7.34 - \frac{9265}{T}$
	$Mn_2O_3-MnO_2$	fO_2	$11.14 - \frac{8810}{T}$
	$Cu-Cu_2O$	fO_2	$6.34 - \frac{17093}{T}$
Graphite	$G-CO_2-CO-O_2$	fO_2, fCO_2	$-0.044 - \frac{20586}{T}$
Methane	$G-CH_4-H_2$	fH_2	
PoPy	$Fe_{1-X}S-FeS_2$	fS_2	$19.687 - \frac{21197}{T} + \frac{2007500}{T^2}$

5.3 a)

Curves in Atm and K

References	Pressure Correction	Reference	References to Use
10	$+ 0.046(\frac{P-1}{T})$	10	4
4,1	$+ 0.055(\frac{P-1}{T})$	4	4
4,11	$+ 0.061(\frac{P-1}{T})$	4	15
4,1	$+ 0.083(\frac{P-1}{T})$	4	4,2
7	$+ 0.019(\frac{P-1}{T})$	4	4,2,5
4	$+ 0.050(\frac{P-1}{T})$	4	4
16	$+ 0.092(\frac{P-1}{T})$	4	4,2
10	$+ 0.081(\frac{P-1}{T})$	10	9
10	$+ 0.005(\frac{P-1}{T})$	10	9
10	$+ 0.019(\frac{P-1}{T})$	10	9
11	$+ 0.098(\frac{P-1}{T})$	17	8,3
6	$+ \log P - 0.028(\frac{P-1}{T})$	6	6,5
13			13,12
Calc. from 15	$0.3 - 0.5\ Kb^{-1}$	14	

TABLE 5.3 b)

(Footnotes)

1 Darken and Gurry (1945)
2 Ernst (1960)
3 Ernst (1962, p. 691, 722)
4 Eugster and Wones (1962)
5 French (1970)
6 French and Eugster (1965)
7 Haas, J.L. (Personal communication, 1968)
8 Huebner (1967)
9 Huebner (1969)
10 Huebner and Sato (1970)
11 Norton (1955)
12 Rutherford (1969)
13 Skippen (1967)
14 Toulmin and Barton (1964)
15 Wise and Eugster (1964)
16 Wones and Gilbert (1969)
17 This paper

Obviously, in any of these cases the buffering capacity is lost, and the run gives no quantitative information on the reaction.[1] Because silver tubing is less permeable to hydrogen than is gold, it slows the diffusion rate, and extends the lifetime of the buffer.

---

[1]Some investigators choose proportions of buffer constituents such that one of the solid buffer phases is completely oxidized before all the water is lost to dissociation and loss of hydrogen. This is because detection of solids by optical examination or x ray diffraction is easier than detecting a small quantity of water. Furthermore, because it is possible to estimate the amount of hydrogen that diffused out of a capsule by measuring the weight loss during a run, one can determine (with good probability) the condition of a buffer such as IW or WM within an unopened capsule.

Another procedure is to replace the buffer from time to time as required by the run duration. It may even be necessary to reload a buffer contained in silver, although at less frequent intervals. The lifetime of a buffer depends primarily upon its buffering capacity (amounts of exchangeable oxygen and hydrogen), the temperature, the relative oxidation states and amounts of the inner charge (compared to that of the buffer), the thickness of the gold or silver tubing, and the hydrogen fugacity of the external pressure medium.

The iron-wüstite assemblage lasts only a few hours at 600-700 C in gold capsules contained in used Stellite or René No. 41 bombs because hydrogen is lost to the aqueous pressure medium until either all the iron is oxidized to wüstite and thence magnetite, or all the water (in the buffer capsule) is consumed.[1] This illustrates the importance of identifying water and the solid phases after a run. The Ni-NiO buffer assemblage lasts a very long time because its hydrogen fugacity is close to that of the external pressure medium (which interacts with the bomb walls). The MH buffer is short lived because hydrogen enters the buffer capsule from the external pressure medium and reduces the hematite, thereby forming magnetite and H_2O. The MH buffer will last 2-3 days at 750 C in gold capsules when water is the external pressure medium (the use of argon or carbon dioxide as a pressure medium should prolong the life of the MH buffer). The FMQ buffer is only slightly more reducing than the pressure medium, but its lifetime is generally short because its buffering capacity is small: FMQ has little exchangeable oxygen per unit weight (or volume) of buffer. A reasonable lifetime at 700 C is 4-15 days in gold tubing, depending upon the thickness of the tubing walls and the fH_2 of the pressure medium.

The kinetics of the buffering process have not been systema-

---

[1] Hamilton _et al._ (1964) used hydrogen-argon mixtures to prolong the life of wüstite-magnetite and fayalite-magnetite-quartz buffers.

tically investigated. Equilibration seems to be attained more quickly at high hydrogen fugacity values than at low values, suggesting that the rate of hydrogen diffusion through platinum or silver palladium is the rate limiting step. Buffered experiments, in which hydrogen does not diffuse through metal, will equilibrate at less than 400 C (Eugster and Wones, 1962). With hydrogen diffusion through metal, the more reducing hydrogen buffers will function at temperatures below 500 C (Skippen, 1967; Huebner and Eugster, 1969; Wise and Eugster, 1964; Eugster and Wones, 1962), and perhaps as low as 400 C. The magnetite-hematite buffer produces such a low hydrogen fugacity that equilibration across platinum membranes is achieved only at temperatures above about 600 C (Eugster and Wones, 1962).

The manganosite-hausmannite buffer, which lies between the NNO and MH buffers (Huebner and Sato, 1970) does not equilibrate readily in hydrothermal systems involving hydrogen diffusion through platinum (Huebner, unpublished data; Ernst, 1966, p. 51; Lindsley, 1963, p. 61). A possible explanation of this fact is that the manganese oxide "poisons" the surface of the metal, inhibiting the diffusion reaction at the metal surface (see Makrides and Jewett, 1966).

B. CO,G|X,CO|

The graphite buffer (French and Eugster, 1965) uses the reaction between carbon dioxide fluid pressure medium and graphite to buffer the fugacities of the gas species present, CO_2, CO, and O_2. This buffer has been used to study the decomposition of siderite (French, 1970) and rhodochrosite (Huebner, 1969). The graphite is spectroscopic grade and the carbon dioxide is commercially available "bone dry" grade. Carbon in other forms is to be avoided because the thermodynamic equations of state are unknown.

At oxygen fugacity values smaller than those of the graphite buffer, the system is "condensed" (see French and Eugster, 1965, p. 1534) if at equilibrium. Carbon dioxide and carbon monoxide decompose to graphite plus oxygen. The only gas species of the system C-O present are oxygen and carbon whose partial pressures are exceedingly small — much less than the total pressure of the CO_2 + CO gas phase in equilibrium with graphite. The region of T-P-fO_2 space at fO_2 values below those of the graphite buffer is therefore experimentally inaccessible in the system C-O; to work in this region, another (inert gas) component must be added to the system.

There is some question whether the (C-O) gas + graphite buffer assemblage always gives the stable equilibrium values of gas fugacities. French (1970) and French and Rosenberg (1965) report lower temperatures for the equilibrium assemblage siderite-magnetite-graphite-gas than do Weidner and Tuttle (1965). Because $\left(\frac{\partial fO_2}{\partial T}\right)_P$ for this assemblage is negative, a possible explanation is the metastable reaction of graphite + gas (C-O) at fO_2 values below those of the graphite buffer (that is, metastable persistence of CO_2 and CO at fO_2 values below the graphite buffer) in Weidner and Tuttle's work. However, Huebner (1969, p. 466) gives evidence that equilibrium did prevail at 1-2 Kb and 703-778 C during a study of the assemblage rhodochrosite-manganosite-gas.

The capsule configuration (Figure 3 b.)) for the graphite buffer is the simplest and easiest to prepare of the buffers. A large tube (commonly 4.4 mm o.d. x 4.0 mm i.d. x 25 mm long) is crimped at one end filled about 1/3 full of graphite powder; the smaller, inner tube is crimped but unsealed and is placed inside, and more graphite is added to fill the larger capsule nearly to the top. The longer tube then is crimped shut. It is important that the inner capsule be positioned near the middle of the outer capsule so that gas from the bomb must pass through (and equilibrate with) the graphite buffer before entering the inner or charge capsule. Otherwise, the fO_2 inside the charge will be intermediate between that of the pressure medium (and bomb walls) and the graphite buffer.

Silver tubing is most commonly used with the graphite buffer because it is inert, inexpensive, and malleable; none of the special properties of gold, platinum, or silver-palladium alloy are needed. Silver foil can be used, although it is less convenient and the foil packets may become lodged inside the bomb.

It is possible to use a welded outer capsule (Figure 3 d.)) if a source of CO_2 gas is sealed within the capsule; French and Rosenberg (1965) report the use of siderite, $FeCO_3$. Water can then be used as the external pressure medium.

The anticipated decomposition product of the charge can be mixed with the graphite buffer, as done by French and Eugster (1965). In this case, siderite was the initial charge, and magnetite, the decomposition product, was mixed with the buffer. After the run, the direction of reaction was determined;[1] either siderite decomposed to form magnetite, or magnetite reacted to form the carbonate. Similarly, Huebner (1969) used a rhodochrosite ($MnCO_3$) charge and mixed manganosite ($Mn_{1-X}O$) with the graphite buffer. Reaction of oxide to form carbonate was indicated by the formation of carbonate in the center of the buffer. Carbonate formation near the crimped ends of the outer capsule was not conclusive because the gas may not have equilibrated with the buffer. It must be emphasized that the magnetite or manganosite does not participate in the buffering equilibrium graphite + gas. The oxide could as well have been placed in a separate charge capsule.

C. $\underline{O_B, CO| C, CO|}$

The dissociation of carbon dioxide according to the reaction

---

[1] It is very satisfying to be able to determine the direction of reaction in almost every experiment. If a run contains the assemblages present on both sides of an equilibrium, the direction of reaction can always be determined as long as there was reaction. However, if the run initially contains only one assemblage, reaction will be observed (and direction of reaction determined) only if that assemblage was unstable and the reaction proceeded at such a rate that detectable amounts of the reaction products were formed.

$CO_2 \rightleftarrows CO + 1/2\ O_2$ can be used to control the fugacities of all gas species in the system C-O in equilibrium with an oxygen buffer. In this case there is no graphite; all carbon is present in the gas phase. The oxygen fugacity values are greater than those of the graphite buffer (at the same P and T), and only those oxygen buffers which fulfill this condition can be used. At 500-2000 bars, French (1970) used the MH buffer in this way, and Huebner (1969) similarly used the NNO buffer and three manganese oxide buffers.

Capsule configuration (Figure 3 b.)) and preparation is similar to that used for the graphite buffer. It is especially important that the inner charge capsule be isolated from the crimped ends of the outer capsule to permit equilibration of the gas and buffer. Huebner (1969) presents evidence for an oxygen fugacity gradient in buffers having equilibrium oxygen fugacity values that are much greater than the bomb walls and pressure medium.

It is sometimes possible to choose as buffer the decomposition products of the inner charge. Both reactants and products are present initially. By examining the charge and a sample of the buffer taken from near the center inner charge capsule (away from the crimped ends of the outer capsule, where gas may not have equilibrated with buffer), the direction of reaction can be determined. Furthermore, the possibility of contamination by foreign ions entering the charge is eliminated.

Reactions involving a C-O gas + solids proceed slowly. For example, at low temperatures (300-400 C) and high oxygen fugacities ($>10^{-10}$ atm), reaction to crystallize or decompose rhodochrosite in visible amounts takes many weeks or months (Huebner, 1969). Several factors determine the rate of reaction, among which is the net transport rate of oxygen. By contrast, in an O-H system gas, oxidation or reduction can be brought about by the movement of hydrogen into or out of the system and the resultant equilibration of H_2O, H_2 and O_2:

$$H_2O \rightleftarrows H_2 + 1/2\ O_2.$$

In such aqueous buffering systems, the partial pressure of hydrogen is one to several hundred bars, and the hydrogen molecule is small, easily diffusing through the gas. The relatively large amount of hydrogen and its rapid diffusion rate favors a fast net transport rate of hydrogen (and thus oxygen by subsequent homogeneous gas reaction). But in a C-O gas at "high" oxygen fugacities and low temperatures, both the pO_2 and pCO are very small (10^{-10} atm is a typical value) and do not favor appreciable net transport. In addition, the diffusion rates of CO and O_2 are slow because these molecules are relatively large (compared to hydrogen).

D. CH, G(X, OH)

The equilibrium between methane and graphite was recognized as a possible buffering assemblage by Eugster and Skippen (1967) and has been used to determine reaction equilibria by Skippen (1967) and Rutherford (1969). At constant temperature and pressure, the equilibrium between a methane pressure medium[1] and graphite, $C + 2\ H_2 \rightleftarrows CH_4$, controls the hydrogen fugacity in much the same way that the equilibrium between carbon dioxide and graphite controls oxygen fugacity. One or two sealed platinum or silver-palladium alloy capsules (Figure 3 c.)) are placed with some powdered graphite in a bomb, and graphite rod is used as the filler rod. Hydrogen equilibrates with the contents of the charge such that fH_2 in the charge is equal to the fH_2 defined by the methane-graphite assemblage. Hydrogen fugacity values are tabulated as a function of T and P by Skippen (1967, Table 2, p. 21) and Eugster and Skippen (1967, Table 1, p. 499). If the charge capsule contains an aqueous phase, fH_2O and fO_2 are buffered; tabulated values are given by Skippen (1967, Table 9, p. 73).

The range of hydrogen fugacities provided by the methane buffer extends from slightly higher than that of FMQ,OH at 450 C to that of WM,OH at 825 C (see Rutherford, 1969, Figure 3; Eugster and Skippen, 1967, Table 1). This buffer provides needed hydrogen

[1] At equilibrium, the gas is dominantly CH_4 and H_2.

fugacities in an area that is important because equivalent oxidation potentials occur in nature. Furthermore, this buffer has a distinct advantage over the conventional oxygen buffers because it is virtually inexhaustible. There is no passage of hydrogen between the buffer capsule and the bomb because the bomb bore is the buffer capsule. Loss of hydrogen results in decomposition of methane to form graphite and more hydrogen; added hydrogen consumes graphite to form methane. Changes in total pressure are minimized by intermittently opening the bomb to a reservoir which is maintained at constant pressure. Hydrogen slowly diffuses through the bomb walls,[1] as is evidenced by the slight pressure drop with time and by the fact that methane-filled bombs only tarnish instead of developing an oxidized patina on the outer surface. Thus, the only limit on buffering capacity is the amount of graphite that can be precipitated in the bomb bore. Although graphite has been observed to precipitate on the charge capsules, the bomb bore has never become filled.

E. O_B,OH,(XG,COH) and CH,G(XG,COH)

If a third component, carbon, and a second phase, graphite are added to a conventional hydrothermal buffer such as O_B,OH(X,OH), control of all gas species in a C-O-H system gas is possible. The system was originally investigated for its numerous biochemically important species. Dayhoff et al. (1964) and Suess (1962) calculated the equilibrium distribution of a number of important species as a function of composition at various conditions. French (1966) calculated the equilibrium fugacities of the significant gas species (H_2,H_2O,CH_4,CO, and CO_2) in a C-O-H gas in equilibrium with graphite as a function of T,P and fO_2 values defined by some conventional oxygen buffers. For each oxygen buffer, there

[1]The effect of hydrogen on the strength of the bomb material has not been investigated in these studies; however, bombs used to contain methane do not behave differently from bombs used with water.

is a temperature above which $pCO_2 + pCO > P$, and equilibrium is not possible. The more oxidizing the buffer, the smaller is the value of the maximum temperature. At 2000 bars pressure, maximum permissible temperatures for several buffers O_BG,COH are: WM, 940 C; FMQ, 590 C; NNO, 425 C; MH, 90 C (see French, 1966, Figure 11); thus, the usefulness of a buffering system involving an oxidizing buffer in equilibrium (and contact) with graphite and a COH gas is restricted.

1. O_B,OH(X,G,COH)

By placing the hydrothermal oxygen buffer in an outer capsule and the C-O-H gas and graphite within an inner capsule semipermeable to hydrogen, Eugster and Skippen (1967) devised a practical buffering system that can be used over a wide range of conditions. The hydrogen fugacity of the C-O-H gas is equal to that of the outer buffer O_B,OH, but the oxygen fugacities are not the same. Oxygen fugacity in the outer capsule is equal to that of the oxygen buffer (or, in the case of the methane-graphite buffer, zero), but the oxygen fugacity of the C-O-H gas is near that of the graphite buffer in the system C-O. However, fCO_2 and fH_2O vary widely, and of course fH_2 varies according to the outer hydrothermal buffer chosen. This buffering system is ideal for the study of reactions in which both CO_2 and H_2O participate, such as $Mg_3Si_4O_{10}(OH)_2 + 3\ CaCO_3 + 2\ SiO_2 = 3\ CaMgSi_2O_6 + 3\ CO_2 + H_2O$ (Eugster and Skippen, 1967). In another application, Huebner and Eugster (1969) found that the reaction $MnCO_3 + SiO_2 = MnSiO_3 + CO_2$ would not proceed in an atmosphere of CO_2 + CO, but would proceed quickly in the $CO_2 + H_2O$ gas mixture given by the buffer NNO,OH(XG,COH).

Run preparation for the O_B,OH(XG,COH) configuration (Figure 3 a.)) is similar to that of the conventional hydrothermal buffers, O_B,OH(X,OH), except that graphite and a source of COH gas must be placed in the inner capsule. Water is not a satisfactory source because it reacts very slowly with graphite to form carbon-bearing gaseous species. More suitable is the use of oxalic acid, $(COOH)_2 \cdot 2H_2O$, which decomposes at about 150 C on runup and yields a gas whose initial C/O ratio is 0.33. During buffer equilibration, the gas composition is adjusted by the diffusion of hydrogen into or out of the inner capsule, and by the precipitation (or consumption) of graphite. Oxygen cannot pass out of the capsule or be precipitated as a buffer component. Because the reaction of graphite with gas to form gaseous carbon-bearing species is slow, benzoic acid, C_6H_5COOH, with a C/O ratio of 3.5 can be added to the oxalic acid to give a C/O ratio exceeding 0.33 in the initial gas. Huebner (1967) investigated the reaction of $MnCO_3$ to $Mn_{1-X}O$ in a C-O-H gas. When using oxalic acid as the sole source of an atmosphere that is dominantly CO_2, he observed that the carbonate partially decomposed as the bomb was brought to a temperature within the $MnCO_3$ stability field. The addition of a small amount of benzoic acid to the oxalic acid alleviated the problem, presumably by increasing the C/O ratio of the initially formed gas, providing enough CO_2 to prevent decomposition during the early stages of the run.

A mixture of 80% (wt.) oxalic acid and 20% (wt.) benzoic acid, giving a C/O atomic ratio of 0.58, works well with C-O-H gases whose C/O ratios range up to 0.5 (nearly pure CO_2). When using an inner capsule that is 10-15 mm long, 3-5 mg of organic compounds should be added. Of course, H_2O-rich gas mixtures can be obtained by adding water to the inner charge.

Hydrogen is the only component permitted to pass between the O-H gas of the outer capsule and the C-O-H gas of the inner capsule. Leakage of any carbonaceous species into the outer capsule will effect fH_2, possibly drastically, creating the buffer $O_B C, COH| XG, COH|$. The following criteria, to be observed when opening the capsules after the run, indicate that the inner capsule did not leak:

(1) No escape of gas when outer capsule is opened because an aqueous phase condenses as liquid on the quench. (It is possible that some H_2 can persist metastably when quenching the reducing buffers such as IM, but gas should not escape from the HM, NNO, and

FMQ buffers.)

(2) There must be no appreciable weight loss of the inner capsule during the run or with gentle heating (100 C) on hotplate or over match flame.

(3) There should be an audible loss of pressurized gas from the inner capsule (unless the equilibrium gas composition was largely water that condensed during the quench).

F. O_BG,COH(XG,COH) and O_BG,COH|XG,COH|

These buffering configurations (Figures 3 a.) and 3 d.)) provide equivalent gas fugacities whether or not the inner capsule is welded or merely crimped; a welded inner capsule will prevent contamination of the charge by the oxygen buffer (and vice-versa). Skippen (1967, p. 59-64) tabulates log fugacity values of the important gas species as a function of various T and P. The ratio fCO_2/fH_2O varies widely as a function of T and fO_2. Because the oxygen fugacity is that of the oxygen buffer, a large range in oxygen fugacities is possible, the particular value of fO_2 depending upon the choice of buffer, T, and P. However, as noted previously, the low temperature at which P CO + P CO_2 exceeds P limits the use of this buffer when the O_B is MH, NNO, or FMQ.

G. O_BG,COH(X,OH)

This buffer uses the O_B + graphite + C-O-H gas equilibrium to buffer the hydrogen fugacity of an O-H gas (Figure 3 a.)). Because carbonaceous species are present, fH_2 is depressed in the C-O-H gas relative to the conventional hydrothermal buffer O_B,OH(X,OH) using an O-H gas. The buffer FIQG,COH(X,OH) provides hydrogen fugacities (and thus oxygen fugacities within the inner capsule) midway in the range of the O_B,OH(X,OH) buffers. The assemblage FIQG,COH should last longer than FIQ,OH because fH_2 is lower and will not diffuse as quickly through the outer capsule wall. Use of O_BG,COH buffers where O_B is FMQ, NNO, or MH is limited to low temperatures because, as discussed previously, P CO_2 + P CO rises quickly with increasing temperature to exceed the total pressure.

Skippen (1967) reports the satisfactory use of approximately 15 mg of organic acids to generate the C-O-H gas phase in the outer capsule.

H. Buffers for Fluorine Species

Fluorine is an essential constituent of some common minerals and proxies for hydroxyl in other common minerals, suggesting that at least some rock forming fluids can be modeled by the system O-H-F. In his work on lepidolites, Munoz (1966) suggested fixing T, P, fH_2, and fO_2 of an O-H-F atmosphere by using a conventional hydrogen buffer in an outer capsule and an oxygen buffer in contact with the O-H-F gas of the inner or charge capsule: O_B,OH(XO_B,OHF). Fluorine could be added to the inner capsule as hydrofluoric acid and sealed in by welding, a hazardous procedure. To avoid this problem, Munoz and Eugster (1969) experimented with new buffer configurations (Figure 3 a.)), including O_B,OH(XF_B,OHF) and O_BG,COH-(XF_B,OHF), in which the fF_2 is controlled internally by assemblages of solids such as wollastonite + fluorite + quartz, or anorthite + fluorite + sillimanite + quartz. Because the fluorine-bearing gas species are not abundant, the solid fluorine buffer can act as a source of fluorine for the gas, obviating the need for addition of HF. Carbon can be added to the charge atmosphere and graphite to the charge, giving O_B,OH(XF_BG,COHF) or O_BG,COH(XF_BG,COHF), and depressing the value of fHF. Fugacities of the important species O_2,F_2,H_2,H_2O, and HF, or O_2,F_2,H_2,H_2O, HF,CO_2CO, and CH_4 at 2 Kb are tabulated by Munoz and Eugster (1969, Table 2, p. 948-949). The authors do not suggest the constituents to be added to give a gas, but it would be reasonable to add water in the case of an O-H-F gas and organic acids (plus water if necessary) for the C-O-H-F gases. The amounts should be similar to the amounts added to buffers discussed previously.

I. Buffers Controlling the Fugacities of Sulfur Species

Like the oxygen fugacity buffers, the pair of iron sulfides pyrrhotite + pyrite will specify the sulfur fugacity value at P and T (Toulmin and Barton, 1964, have experimentally determined the fS_2 of this buffer from 743-325 C). Various other simple sulfide pairs, or metal + metal sulfide pairs could also be used as sulfur buffers, but to date this has not been done. (See Chapter 11 for additional information)

More important is the incorporation of the sulfur buffer into an aqueous buffering system. The gas system H-O-S is completely specified if P,T, and two chemical potentials (or chemical potential ratios) are fixed. If elemental sulfur is present, only one μi need be specified. Two buffering configurations can result. In the simpler configuration all gas fugacities are defined internally in a single capsule. The second configuration uses a membrane permeable to hydrogen. (See Chapter 12 for additional information)

1. S_BO_BX,HOS

This buffering configuration (Figure 3 e.)) uses only a single capsule, although the buffers and charge can be isolated by S_BO_B, HOS|X,HOS| (Figure 3 d.)). The sulfur and oxygen fugacities can be controlled by completely separate buffers; however, it is more convenient to use an assemblage to define a fugacity and a fugacity ratio: MHPy, MPoPy, IWPo, IMPo, or FMQPo. The use of such assemblages to study a silicate sulfide reaction such as

$$KFe_3AlSi_3O_{10}(OH)_2 + 3\ S_2 = KAlSi_3O_8 + 3\ FeS_2 + H_2O + 3/2\ O_2$$

was recognized by Toulmin (see Roedder, 1965, p. 1392), who with coworkers made preliminary runs on the assemblages

(biotite + sanidine) + (MPoPy-Gas) or (MHPy-Gas).

Eugster and Skippen (1967) graphically present the fugacity values of important species in several of the H-O-S buffering systems.

2. O_B,OH(S_BX,HOS) and O_BG,COH(S_BX,HOS)

The oxygen buffer can be isolated from the charge (Figure 3 a.)) by use of a platinum inner capsule and hydrogen permeation. The sulfur buffer PoPy must be located inside the charge capsule. Fugacity values differ from the previous buffer for controlling an H-O-S gas. The only published calculations are those of Eugster and Skippen (1967; Figure 9, p. 515) who plotted log K as a function of T at, presumably, 2 Kb; K is the equilibrium constant of the biotite-sanidine-pyrite-gas equilibrium; log K = log fH_2O + 3/2 log fO_2 - 3 log fS_2.

3. O_B,OH(S_BGX,CHOS) and O_BG,COH(S_BGX,CHOS)

Fugacities of all species in a C-H-O-S system gas can be specified by a system that is analogous to the previous H-O-S buffers; by adding one phase and one component, the invariance of the system is maintained. The fugacities of the gas species have been calculated by Eugster (personal communication), but have not been published. However, Speidel and Heald (1967) present partial pressures of 14 C-H-O-S system gas species as a function of T, the volume percent of SO_2 in the gas, and the CO_2/SO_2 ratio.

4. O_B,OH($O_B S_B$X,CHOS) and O_BG,COH($O_B S_B$X,CHOS)

Graphite need not be added to a C-H-O-S gas if sufficient intensive parameters are specified. This system buffers fO_2 and fS_2 within the inner capsule, and fH_2 through the platinum membrane (Figure 3 a.)). At P and T, the system is invariant. No fugacity calculations have been published.

J. <u>Buffering in Other Systems</u>

Any phase of fixed composition in equilibrium with fluid will define and buffer the thermodynamic activity of that phase in the fluid. In the well known KCL-saturated calomel electrode cell, the activity of KCl in a solution is maintained constant by the presence of pure, solid KCl in equilibrium with the solution. Likewise, the activity of any species is buffered if that species is present both in the fluid, and as a pure solid in equilibrium with the fluid. For example, if a fluid is in equilibrium with pure quartz, the activity of SiO_2 is fixed; furthermore, if the standard state of silica is pure quartz at T and P, the SiO_2 activity is unity at that P and T.

Several possible silica buffers have been recognized by Carmichael <u>et al.</u> (1970), who review reactions in which free silica is a component and a participating phase. Reactions such as 2 $MgSiO_3$= Mg_2SiO_4 + SiO_2 and $NaAlSi_3O_8$ = $NaAlSiO_4$+ 2 SiO_2 also define a constant silica activity at P,T. (Note that silica need not be present as a phase to participate in the reaction.) Use of such silica buffers, in combination with oxygen buffers, would be a technique for controlling both fO_2 and a_{SiO_2} when investigating, for instance, the reactions between the manganese ore mineral braunite, $(Mn,Si)MnO_3$, and either $Mn_3O_4 + SiO_2$ or $Mn_2O_3 + SiO_2$.

It is theoretically possible to buffer the alkali activity a_{Na_2O} or a_{K_2O}, or activity ratio a_{Na_2O}/a_{K_2O}, of a gas in equilibrium with a solid assemblage. Korzhinskii (1959) indicates several such reactions:

$$2\ KMg_3AlSi_3O_{10}(OH)_2 + 7\ CaMg(CO_3)_2 =$$
$$6\ Mg_2SiO_4 + 7\ CaCO_3 + MgAl_2O_4 + 2\ H_2O + 7\ CO_2 + K_2O$$

$$5\ CaCO_3 + 2\ Mg_2SiO_4 + MgAl_2O_4 + 5\ CO_2 + Na_2O =$$
$$2\ NaAlSiO_4 + 5\ CaMg(CO_3)_2$$

$$2\ KMg_3AlSi_3O_{10}(OH)_2 + 7\ CaCO_3 + MgAl_2O_4 + 2\ Na_2O + 7\ CO_2 =$$
$$7\ CaMg(CO_3)_2 + 4\ NaAlSiO_4 + K_2O + 2\ SiO_2 + 2\ H_2O$$

Where necessary, the activities of H_2O and CO_2 can be specified by one of the system C-O-H buffers. The reaction between solid solutions of nepheline and leucite can be expressed as

$$KAlSi_2O_6 + 1/2\ Na_2O = NaAlSiO_4 + SiO_2 + 1/2\ K_2O.$$

At T and P, a_{K_2O}/a_{Na_2O} will be buffered (silica activity is fixed by the presence of a silica phase or by a silica buffer).

Other assemblages exist which will, in theory at least, buffer the activities of Al_2O_3, MgO, $Fe_{1-X}O$, etc.[1] Few such assemblages have been used as buffers, in large part due to a lack of calibrated numerical data for calculating the activity or fugacity values of components defined by these buffers.

---

[1]Obviously the oxygen buffer, for instance, also buffers the chemical potentials of other components. The assemblage Fe_3O_4-Fe_2O_3 buffers not only the oxygen potential, but also Fe_3O_4, Fe_2O_3, and even $Fe_{1-X}O$ (through an equilibrium such as 6 "FeO" + O_2 = 2 Fe_3O_4).

K. Buffering in Gas Systems of Many Components

Three component buffering systems are derived from two component systems by either adding one component and one phase, or adding one component and defining (fixing) another intensive parameter. Four-component gas buffers are derived similarly (or by combining simpler buffers). In principle, this process can be extended to five- or six-component buffers such as:

$$O_BOH(F_BS_BG,COHFS) \text{ or } O_B,OH(F_BS_BO_BG,COHFSCl).$$

By incorporating a chlorine buffer (personal communication, Frantz and Eugster) it is theoretically possible to buffer a seven-component gas such as C-O-H-F-S-N-Cl.

The practicality of these very elaborate buffers is questionable. The solids that are in equilibrium must be chosen such that they do not react to consume an internal buffer, or form a solid solution whose thermochemical properties are so poorly known that the fugacities (and fugacity ratios) of the system cannot be computed. For instance, combining the oxygen buffer MHG with the sulfur buffer PoG will result in reaction between hematite and pyrrhotite to form magnetite and pyrite, resulting in the buffer assemblage MPoPy or MHPy. A second limitation is that of buffering capacity. Only a small amount of buffer can be placed in a capsule, yet that buffer must have sufficient buffering capacity to define the fugacities of the gas species for a significant time (i.e., time for the charge to react). An assemblage ceases to function as a buffer after one or more of its phases is consumed.

L. Non-Buffering Assemblages That Act as Indicators of Chemical Potentials

Minerals and mineral assemblages which are univariant at constant T and P do not buffer a system in the sense herein described, but they may be capable of indicating chemical potentials or chemical potential ratios.[1] Examples are wüstite (Darken and

[1] Such assemblages have been termed "sliding-scale buffers" by Barton and Skinner (1967, p. 245).

Gurry, 1945), for which the atomic ratio Fe^{2+}/Fe^{3+} (or the chemical potential ratio $\mu Fe^{2+}/\mu Fe^{3+}$ varies as a function of fO_2; pyrrhotite (Toulmin and Barton, 1964), for which Fe^{2+}/Fe^{3+} varies with fS_2; and the assemblage biotite-magnetite-sanidine-gas (Wones and Eugster, 1965), in which the Fe^{2+}/Fe^{3+} of the assemblage (or the Fe^{2+}/Mg^{2+} of the biotite) varies as a function of fH_2.

Because of the difficulty of calibration, extensive application of such indicators has not been made. Much of the existing data used in deriving an equation of state for a solid solution was obtained at higher temperatures than those of interest here or attainable in experimental apparatus of the type described here. However, several investigators have used carefully studied solid solutions as fugacity meters.

The electrum-tarnish method (Barton and Toulmin, 1964) permits the estimation of fS_2 by noting the decomposition or development of sulfide tarnish on a series of silver-gold alloys of known compositions. The electrum is physically isolated from the solid charge, but gas is permitted to equilibrate with both charge and electrum. Although application has been made only in silica glass tubes at relatively low pressures, the electrum tarnish method could be used at higher pressures providing that the thermodynamic effect of pressure were known (see Barton and Toulmin, 1964, p. 640) and that the electrum and tarnish were not otherwise changed in composition. An advantage of the method is the relatively wide range of T and fS_2 over which it is operative.

The use of pyrrhotite to indicate sulfur fugacity was suggested by Toulmin and Barton (1964). Pyrrhotite composition, $Fe_{1-X}S$ varies as a known function of fS_2, and pyrrhotite composition can be obtained by x ray powder diffraction measurements. The pyrrhotite sulfur barometer has been used by Barton (1969) and Barton and Toulmin (1966). Huebner and Toulmin (unpublished results) also used the method to determine the fS_2 of the assemblage biotite-pyrrhotite-magnetite-sanidine-gas at 600-800 C and 2000 bars. Huebner noted that when a bomb containing this assemblage was T,P quenched with an air jet, the pyrrhotite composition was not uniform [based on the breadth of the (102) x ray diffraction peak], whereas the pyrrhotite was homogeneous (sharp peak) if the bomb was quenched in a bucket of water. Evidently the pyrrhotite composition can change rapidly, over the times characteristic of the cooling of hydrothermal bombs.

By permitting a charge to equilibrate with a relatively large quantity of pyrrhotite, Barton and Toulmin (1966) maintained the sulfur fugacity at an approximately constant value. However, the pyrrhotite composition, and hence the sulfur fugacity, changes as sulfur is consumed or evolved by the reaction within the charge. The sulfur fugacity is not buffered in the sense that the system is invariant.

The use of a solid solution in the outer, buffer compartment of $O_BOH(X,OH)$ has been proposed by Carapezza (1969). The oxygen fugacity of the assemblage (Ni,Mg)O-Ni-Gas is known as a function of temperature and oxide composition, which can be measured by x ray powder diffraction techniques. The oxygen and hydrogen fugacities will vary during an experiment due to exchange of hydrogen with both the charge and the pressure medium surrounding the capsule.

Shaw (1963, 1967) has described an experimental technique that permits the hydrogen fugacity to be maintained at any desired value; fH_2 need not be restricted to the values provided by solid buffers. A desired hydrogen fugacity is imposed on a H_2-argon mixture pressure medium in the bomb by equilibration through a platinum (or silver-palladium alloy) membrane. The charge to be studied is contained in a capsule permeable to hydrogen. The hydrogen fugacity within the charge capsule is equal to the fH_2 of the hydrogen-argon pressure medium, which is equal to the fH_2 of pure hydrogen (of known pressure) within the large membrane. Shaw discusses the calculations for converting from hydrogen pressure, measured outside the bomb, to hydrogen fugacity within the charge.

The fO_2 of the FMQ oxygen buffer has been determined at 600-800 C with this technique (Wones and Gilbert, 1969). Because of hydrogen pressure variations due to diffusion into the charge capsule, diffusion through the bomb walls, and pressure variations in the hydrogen reservoir caused by the variations in room temperature, it cannot be assumed that at any given time the fH_2 of the charge can be simply determined by measuring the hydrogen pressure with a gage. Complete bracketing of a reaction consists of not only determining the temperature limits, but also the fH_2 limits, for reaction reversal (made possible by Shaw's technique). (Additional examples are also discussed in Chapter 8)

VI. Run Interpretation and the Problem of Equilibrium

A. Interpretation of Runs

The charge and buffer should be examined by macroscopic, microscopic, and x ray powder diffraction techniques. Solid phases should be thoroughly defined with regard to their composition and structural state (see, for instance: Stewart and Ribbe, 1969). The importance of a microscopic examination cannot be overemphasized. Small quantities of crystalline phase (commonly less than five percent) that cannot be detected by x ray powder diffraction of a mixture, can almost always be observed optically. Non-crystalline phases (glasses) do not diffract x rays. X ray diffraction will not reveal textural evidence of disequilibrium. Many papers attest to the important observations obtained with a microscope (see Schairer and Bowen, 1955; Schairer, 1942). There is a tendency with present technological innovations to forget the microscope in favor of the x ray machine, but examination of the run products by <u>both</u> techniques avoids some embarrassing misinterpretations. At the completion of the examination of a series of runs, the investigator can often tell what reactions occurred during runup (see Wones, 1967, p. 2249; Rutherford, 1969, p. 390; Huebner, 1967, p. 196) and quench (Fyfe and Godwin, 1962; Nolan, 1966), and can distinguish these effects from reactions that took place during the time the bomb was at P and T.

B. Criteria of Equilibrium

The determination of equilibrium is very elusive; there is no "equilibriometer"[1] to measure the attainment of equilibrium. Most experimental results identify some degree of disequilibrium. When an assemblage is formed that shows no evidence of reaction, there is no way of telling by this observation alone whether or not the assemblage represents equilibrium. When we observe an assemblage that shows evidence of reaction (disequilibrium), we must determine in which direction the reaction is going (to equilibrium). In some cases the rate of reaction can be used as evidence, but only a few investigators (notably Greenwood, 1963) have studied the rate constant of a reaction using hydrothermal techniques at elevated pressures, and none have performed a rate study of a buffered reaction.

Various authors have proposed criteria of equilibrium and disequilibrium; some of these criteria are summarized in the following list:

(1) At equilibrium, the variance of the system, as given by the phase rule, is $\omega \geq 0$. A negative variance indicates that too many phases are present for true chemical equilibrium. (See Yoder, 1952, p. 588.) If the compositional tie (conjugation) lines between pairs of phases cross, these phases must be out of equilibrium because the variance of the system will be negative.

(2) The presence of a mineral subsystem that is known to be at disequilibrium, from previous studies or from textural evidence, is sufficient to demonstrate disequilibrium in the parent system (Zen, 1963, p. 940).

(3) Textural evidence for disequilibrium includes compositional zoning of phases (heterogeneous grains), reaction rims, and the "replacement" textures familiar to ore microscopists. Recrystallization of phases suggests but does not prove approach to equilibrium (Zen, 1963, p. 939).

---

[1] Barton and Skinner (1967, p. 247).

(4) If an experiment has reached equilibrium, continuing the experiment for a longer period of time will produce no further change (Roedder, 1951, p. 89); furthermore, repeating the experiment under the same conditions will duplicate the result.

(5) The conditions of equilibrium can be estimated by approaching them from different directions. This is the method used by almost everyone, often called "reversing the experiments." It must be determined which phases are being formed (the more stable assemblage) and which phases are being consumed; see Greenwood (1963) and Ernst (1966) for good examples. Then some condition (usually T) is changed, and the experiment is repeated at otherwise similar conditions, and the result noted. Eventually the equilibrium is located by "bracketing," that is, by noting the direction in which equilibrium lies, and approaching the equilibrium as closely as possible from different directions. (This is also discussed in Chapter 7.)

(6) The phase diagram based on the data must be internally consistent. The individual elements of the phase diagram must relate to one another in a manner that is correct, that is, fulfills the topological criteria for thermodynamic consistency, as outlined by Schreinemakers (1915-1925), Morey and Williamson (1918), Korzhinskii (1959), and others.

C. Equilibrium in Buffered Capsules

It is the responsibility of the individual investigator to design his experiments such that the reactions being studied can be demonstrably reversed. However, because the essence of the buffering method is to provide constant values of certain intensive parameters, these parameters cannot be varied independently of P and T. It is impossible with the buffering techniques described here to bracket an equilibrium in terms of fH_2 only. Five facts suggest that this impossibility does not seriously limit the usefulness of the data, that equilibrium is in fact achieved:

(1) The data obtained by the buffering method are internally consistent and agree with data determined by other methods.

(2) The technique described by Shaw (1967) demonstrates that H_2 equilibration is easily achieved across the membrane that separates the H_2 reservoir and the H_2-Ar mixture in the bomb. The buffer method uses a much smaller volume (charge capsule), so the approach to equilibrium should be faster.

(3) Buffers demonstrably react, as evidenced by hydrogen diffusion through a gold or silver capsule into the pressure medium, or reaction rims about buffer grains (Huebner, 1969, Figure 2). Diffusion from the buffer into the charge capsule (Pt or Ag-Pd alloy) is faster than diffusion into the pressure medium, so the charge should more nearly reflect the conditions of the buffer rather than the pressure medium.

(4) Several investigators have shown that the solid solutions in certain oxidation-reduction reactions will converge upon a constant composition (for a given buffer, temperature, and pressure), regardless of whether the initial assemblage is more oxidized or reduced than the final assemblage. This bracketing of equilibrium in terms of composition has been demonstrated with Fe-Ti oxides (Lindsley, 1963) and biotites (Wones and Eugster, 1965).

(5) Equilibration among gas species is generally faster than equilibration between solids.

In cases in which there is doubt that the buffering system defines equilibrium values of gas species, the equilibrium conditions can be checked using Shaw's technique (1967), a fugacity indicator such as pyrrhotite (Toulmin and Barton, 1964), or a solid solution buffer (Carapezza, 1969).

BIBLIOGRAPHY:

Barber, C.R. (1969). The international practical temperature scale of 1968. Metrologia, 5:35-44.

Barrer, R.M. (1951). Diffusion in and through solids. Cambridge: University Press.

Barton, P.B., Jr. (1969). Thermochemical study of the system

Fe-As-S. Geochim. Cosmochim. Acta, 33:841-857.

Barton, P.B., Jr. and Skinner, B.J. (1967). Sulfide mineral stabilities, in Geochemistry of hydrothermal ore deposits, edited by H.L. Barnes. New York: Holt, Rinehart and Winston, Inc.

Barton, P.B., Jr. and Toulmin, P., III (1964). The electrum-tarnish method for the determination of the fugacity of sulfur in laboratory sulfide systems. Geochim. Cosmochim. Acta, 28:619-640.

Barton, P.B., Jr. and Toulmin, P., III (1966). Phase relations involving sphalerite in the Fe-Zn-S system. Econ. Geol., 61:815-849.

Carapezza, M. (1969). A method for continuously variable control of the fugacity of a gas in hydrothermal synthesis at high pressure. Geochem. International, 6:819-823.

Carmichael, I.S.E., Nicolls, J., and Smith, A.L. (1970). Silica activity in igneous rocks. Amer. Mineral., 55:246-263.

Darken, L.S. and Gurry, R.W. (1945). The system iron-oxygen. I. The wüstite field and related equilibria. J. Amer. Chem. Soc., 67:1398-1412.

Davis, W.D. (1954). Diffusion of gases through metals. I. Diffusion of hydrogen through palladium. Knolls Atomic Power Laboratory Report No. KAPL-1227:6-39. (Prepared for the U.S. Atomic Energy Comm.)

Dayhoff, M.O., Lippincott, E.R., and Eck, R.V. (1964). Thermodynamic equilibria in prebiological atmospheres. Science, 146:1461-1464.

Ernst, W.G. (1960). The stability relations of magnesioriebeckite. Geochim. Cosmochim. Acta, 19:10-40.

Ernst, W.G. (1962). Synthesis, stability relations, and occurrence of riebeckite and riebeckite-arfvedsonite solid solutions. J. Geol., 70:689-736.

Ernst, W.G. (1966). Synthesis and stability relations of ferrotremolite. Amer. J. Sci., 264:37-65.

Eugster, H.P. (1957). Heterogeneous reactions involving oxidation and reduction at high pressures and temperatures. J. Chem. Phys., 26:1760-1761.

Eugster, H.P. and Skippen, G.B. (1967). Igneous and metamorphic reactions involving gas equilibria, in Researches in geochemistry, Vol. 2, P.H. Abelson, editor. New York: John Wiley and Sons, Inc.

Eugster, H.P. and Wones, D.R. (1962). Stability relations of the ferruginous biotite, annite. J. Petrology, 3:82-125.

French, B.M. (1966). Some geological implications of equilibrium between graphite and a C-H-O gas at high temperatures and pressures. Rev. Geophys., 4:223-253.

French, B.M. (1970). Stability relations of siderite ($FeCO_3$), determined in controlled-fO_2 atmospheres. Greenbelt, Md., Goddard Space Flight Center Report X-644-70-102.

French, B.M. and Eugster, H.P. (1965). Experimental control of oxygen fugacities by graphite-gas equilibriums. J. Geophys. Res., 70:1529-1539.

French, B.M. and Rosenberg, P. (1965). Siderite ($FeCO_3$): Thermal decomposition in equilibrium with graphite. Science, 147: 1283-1284.

Fyfe, W.S. and Godwin, L.H. (1962). Further studies on the approach to equilibrium in the simple hydrate systems, $MgO-H_2O$ and $Al_2O_3-H_2O$. Amer. J. Sci., 260:289-293.

Garrels, R.M. and Christ, C.L. (1965). Solutions, minerals, and equilibria. New York: Harper and Row.

Gilbert, M.C. (1966). Synthesis and stability relations of the hornblende ferropargasite. Amer. J. Sci., 264:698-742.

Gordon, T.M. and Greenwood, H.J. (1970). The reaction: dolomite + quartz + water = talc + calcite + carbon dioxide. Amer. J. Sci., 268:225-242.

Greenwood, H.J. (1961). System $NaAlSi_2O_6-H_2O$-argon. J. Geophys. Res., 66:3923-3946.

Greenwood, H.J. (1962). Metamorphic reactions involving two volatile components. Carnegie Inst. Washington Year Book 61: 82-85.

Greenwood, H.J. (1963). The synthesis and stability of anthophyllite. J. Petrology, 4:317-351.

Greenwood, H.J. (1967). Mineral equilibria in the system MgO-SiO_2-H_2O-CO_2, in Researches in geochemistry, Vol. 2, P.H. Abelson, editor. New York: John Wiley and Sons, Inc.

Hamilton, D.L., Burnham, C.W., and Osborn, E.F. (1964). The solubility of water and effects of oxygen fugacity and water content on crystallization in mafic magmas. J. Petrology, 5:21-39.

Hansen, M. and Anderko, K. (1958). Constitution of binary alloys. New York: McGraw Hill.

Hickman, R.G. (1969). Diffusion and permeation of deuterium in palladium-silver at high temperature and pressure. J. Less-Common Metals, 19:369-383.

Holloway, J.R., Burnham, C.W., and Hillhollen, G.L. (1968). Generation of H_2O-CO_2 mixtures for use in hydrothermal experimentation. J. Geophys. Res., 73:6598-6600.

Huebner, J.S. (1969). The stability of rhodochrosite in the system manganese-carbon-oxygen. Amer. Mineral., 54: 457-481.

Huebner, J.S. (1967). Stability relations of minerals in the system Mn-Si-C-O. Ph.D. thesis, The Johns Hopkins University, Baltimore, Md.

Huebner, J.S. and Eugster, H.P. (1969). Rhodochrosite decarbonation in the system MnO-SiO_2-CO_2 (Abstract). Geol. Soc. Amer. Spec. Paper, 121:144-145.

Huebner, J.S. and Sato, M. (1970). The oxygen fugacity-temperature relationships of manganese and nickel oxide buffers. Amer. Mineral., 55:934-952.

Johannes, W. and Metz, P. (1968). Experimentelle bestimmungen von gleichgewichtsbeziehungen im system MgO-CO_2-H_2O. Neues

Jahrb. Mineral. Montash., 1/2:15-26.

Korzhinskii, D.S. (1959). Physicochemical basis of the analysis of the paragenesis of minerals. New York: Consultants Bureau, Inc.

Lindsley, D.H. (1962). Investigations in the system $FeO-Fe_2O_3-TiO_2$. Carnegie Inst. Washington Year Book, 61:100-106.

Lindsley, D.H. (1963). Fe-Ti oxides in rocks as thermometers and oxygen barometers: Equilibrium relations of coexisting pairs of Fe-Ti oxides. Carnegie Inst. Washington Year Book, 62: 60-66.

Lindsley, D.H. and Munoz, J.L. (1969). Subsolidus relations along the join hedenbergite-ferrosilite. Amer. J. Sci., (Schairer Vol.) 269-A:295-324.

Luth, W.C. and Tuttle, O.F. (1963). Externally heated cold-seal pressure vessels for use to 10,000 bars and 750°C. Amer. Mineral., 48:1401-1403.

Makrides, A.C. and Jewett, D.N. (1966). Diffusion of hydrogen through palladium. Engelhard Ind. Tech. Bull., 7:51-54.

Metz, P. (1966). Untersuchung eines heterogenen bivarianten gleichgewichts mit CO_2 und H_2O als fluider phase bei hohen drucken. Berichte der Bunsengesellschaft für physikalische Chemie (früher Zeitschrift für Elektrochemie), 70:1043-1045.

Morey, G.W. and Williamson, E.D. (1918). Pressure-temperature curves in univariant systems. J. Amer. Chem. Soc., 40: 59-84.

Muan, A. (1962). The miscibility gap in the system Ag-Fe-Pd at 1000°, 1100°, and 1200°C. Trans. AIME, 224:1080-1081.

Muan, A. (1963). Silver-palladium alloys as crucible material in studies of low-melting iron silicates. Amer. Ceram. Soc. Bull., 42:344-347.

Munoz, J.L. (1966). Synthesis and stability of lepidolites. Ph.D. thesis, The Johns Hopkins University, Baltimore, Md.

Munoz, J.L. and Eugster, H.P. (1969). Experimental control of fluorine reactions in hydrothermal systems. Amer. Mineral.,

54:943-959.

Nolan, J. (1966). Melting-relations in the system $NaAlSi_3O_8$-$NaAlSiO_4$-$NaFeSi_2O_6$-$CaMgSi_2O_6$-H_2O, and their bearing on the genesis of alkaline undersaturated rocks. Quart. J. Geol. Soc., (London) 122:119-157.

Norton, F.J. (1955). Dissociation pressures of iron and copper oxides. General Electric Res. Lab. Rept. 55-Rl-1248.

Richardson, O.W., Nicol, J., and Parnell, T. (1904). The diffusion of hydrogen through hot palladium. Phil. Mag., 43:1-29.

Robie, R.A. and Waldbaum, D.R. (1968). Thermodynamic properties of minerals and related substances at 298.15°K (25.0°C) and one atmosphere (1.013 bars) pressure and at higher temperatures. U.S. Geol. Survey Bull., 1259.

Roedder, E. (1965). Report on S.E.G. symposium on the chemistry of the ore-forming fluids, August-September, 1964. Econ. Geol., 60:1380-1403.

Roedder, E.W. (1951). The system K_2O-MgO-SiO_2. Amer. J. Sci., 249:81-130; 224-248.

Roeser, W.F. and Lonberger, S.T. (1958). Methods of testing thermocouples and thermocouple materials. U.S. Natl. Bur. Standards Circ., 590.

Rubin, L.R. (1966). Permeation of deuterium and hydrogen through palladium and 75 palladium-25 silver at elevated temperatures and pressures. Engelhard Ind. Tech. Bull., 7:55-62.

Rutherford, M.J. (1969). An experimental determination of iron biotite-alkali feldspar equilibria. J. Petrol., 10:381-408.

Savitskii, E.M., Polyakova, V.P., and Tylkina, M.A. (1969). Palladium alloys. Primary Sources Publishers, translated by R.E. Hammond. Original title: Splavy palladiya (Moscow, 1967).

Schairer, J.F. (1942). The system CaO-FeO-Al_2O_3-SiO_2: I. Results of quenching experiments on five joins. J. Amer. Ceram. Soc., 25:241-274.

Schairer, J.F. and Bowen, N.L. (1955). The system K_2O-Al_2O_3-SiO_2.

Amer. J. Sci., 253:681-746.

Schreinemakers, F.A.H. (1915-1925). In-, mono-, and divariant equilibria. Koninkl. Akad. Wetenschappen te Amsterdam Proc., English ed., Vol. 18-28 (29 separate articles in the series).

Shaw, H.R. (1963). Hydrogen-water vapor mixtures; control of hydrothermal atmospheres by hydrogen osmosis. Science, 139:1220-1222.

Shaw, H.R. (1967). Hydrogen osmosis in hydrothermal experiments, in Researches in geochemistry, Vol. 2, P.H. Abelson, editor. New York: John Wiley and Sons, Inc.

Shenker, H., Lauritzen, J.I., Jr., Corruccini, J., and Lonberger, S.T. (1955). Reference tables for thermocouples. U.S. Natl. Bur. Standards Circ., 561.

Sieverts, A. and Hagen, H. (1935). Der elektrische widerstand wasserstoffbeladener drähte aus legierunger des palladiums mit silber und mit gold. Z. Physik. Chem., 174:247-261.

Skippen, G.B. (1967). An experimental study of the metamorphism of siliceous carbonate rocks. Ph.D. thesis, The Johns Hopkins University, Baltimore, Md.

Speidel, D.H. and Heald, E.F. (1967). Gaseous equilibria in portions of the system C-H-O-S. Bull. Earth and Mineral Sciences Experiment Station, No. 83.

Stewart, D.B. and Ribbe, P.H. (1969). Structural explanation for variation in cell parameters of alkali feldspar with Al/Si ordering. Amer. J. Sci., 267-A:444-462.

Suess, H.E. (1962). Thermodynamic data on the formation of solid carbon and organic compounds in primitive planetary atmospheres. J. Geophys. Res., 67:2029-2034.

Sullivan, C.P., Donachie, M.J., Jr., and Morral, F.R. (1970). Cobalt-base superalloys — 1970. Brussels: Center d'Information du Cobalt.

Toulmin, P., III and Barton, P.B., Jr. (1964). A thermodynamic study of pyrite and pyrrhotite. Geochim. Cosmochim. Acta, 28:641-671

Tuttle, O.F. (1949). Two pressure vessels for silicate-water studies. Geol. Soc. Amer. Bull., 60:1727-1729.

Vaccari, J.A. (1969). Today's superalloys: good; tomorrow's: even better. Materials Engineering, 5:21-31.

Weidner, J.R. and Tuttle, O.F. (1965). Stability of siderite, $FeCO_3$ (abstract). Geol. Soc. America, Special Paper, 82:220.

Wellman, T.R. (1970). The stability of sodalite in a synthetic syenite plus aqueous chloride fluid system. J. Petrology, 11:49-71.

Williams, D.W. (1966). Externally heated cold-seal pressure vessels for use to 1200°C at 1000 bars. Mineral. Mag., 35: 1003-1012.

Williams, D.W. (1968). Improved cold-seal pressure vessels to operate to 1100°C at 3 kilobars. Amer. Mineral., 53:1765-1769.

Wise, W.S. and Eugster, H.P. (1964). Celadonite: synthesis, thermal stability and occurrence. Amer. Mineral., 49: 1031-1083.

Wones, D.R. (1967). A low pressure investigation of the stability of phlogopite. Geochim. Cosmochim. Acta, 31:2248-2253.

Wones, D.R. and Eugster, H.P. (1965). Stability of biotite: experiment, theory, and application. Amer. Mineral., 50: 1228-1272.

Wones, D.R. and Gilbert, M.C. (1969). The fayalite-magnetite-quartz assemblage between 600° and 800°C. Amer. J. Sci., (Schairer Vol.) 267-A:480-488.

Yoder, H.S., Jr. (1950). High-low quartz inversion up to 10,000 bars. Trans. Amer. Geophys. Un., 31:827-835.

Yoder, H.S., Jr. (1952). The $MgO-Al_2O_3-SiO_2-H_2O$ system and the related metamorphic facies. Amer. J. Sci., (Bowen Vol.): 569-627.

Zen, E-an (1961). The zeolite facies: an interpretation. Amer. J. Sci., 259:401-409.

Zen, E-an (1963). Components, phases, and criteria of chemical equilibrium in rocks. Amer. J. Sci., 261:929-942.

ACKNOWLEDGMENT:

Numerous individuals contributed to the development of the buffering technique, but unfortunately many of the "tricks of the trade" cannot be identified with the original inventor. However, I would particularly like to acknowledge Prof. H.P.Eugster, who introduced me to the techniques when I was a graduate student at The Johns Hopkins University. I also benefited greatly from discussions with B.M. French, G.B. Skippen, and D.R. Wones. D.H. Lindsley and H.R. Shaw carefully reviewed the manuscript. To all these individuals, many thanks, with the understanding that I alone am responsible for any inadequacies in the chapter.

CHAPTER 6

Temperature Calibration in Cold-Seal Pressure Vessels

A.L. Boettcher

D.M. Kerrick

ABSTRACT

Experiments at P_{H_2O} = 1 Kb with a 1 inch o.d. hydrothermal vessel showed that a filler rod is an important factor in measurement of internal temperatures with an external thermocouple. Without a filler rod, significant temperature differences exist between the external and internal thermocouple readings with the long axis of the vessel positioned horizontally, and in the vertical position with the end of the vessel containing the closure nut oriented up. Steady-state convection of water within the vessel removes sufficient heat that the internal thermocouple is maintained at lower temperatures than the external thermocouple. The temperature difference is markedly reduced in runs in these orientations with a filler rod in the vessel. Furthermore, the presence of a filler rod reduces thermal gradients along the axis of the vessel. Little difference exists between the external and internal thermocouple readings with the axis of the vessel vertical and with the closure nut down. Similar results were obtained with a 1 1/4 inch o.d. vessel.

In the same orientation of the vessel, temperature differences between external and internal thermocouples are much less with an argon pressure medium than with water. With most of the vessel filled with a snug-fitting filler rod, and with the sample at the hot spot along the length of the furnace tube, external thermocouple measurements of sample temperatures can be accurate to within 5 C.

I. Introduction

A. Statement of the Problem

Since about 1949, most hydrothermal research at pressures up to 4 Kb has been conducted in relatively simple cold-seal vessels of the type originally described by Tuttle (1949) and shown in Figure 1. (See Chapters 4 and 5 also) The utility of this well

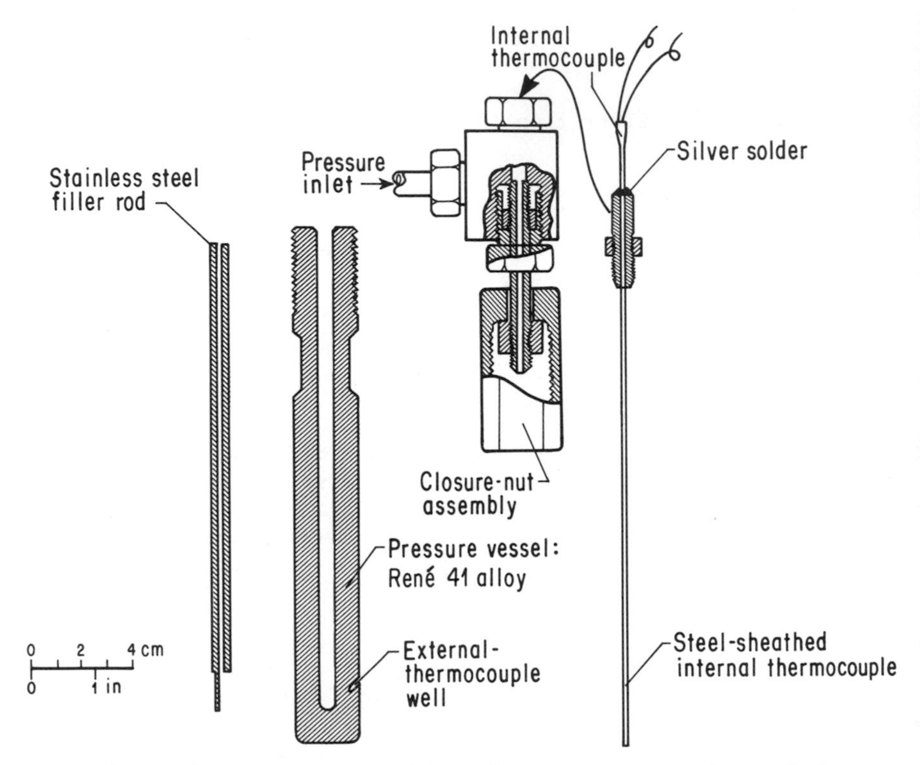

Figure 1. Scale drawing of cold-seal pressure vessel, including filler rod, closure nut, and thermocouple assemblies. Assembled unit is shown in Figure 2.

known device was extended by Luth and Tuttle (1963) when they modified the design to permit the attainment of pressures up to about 10 Kb. All of the cold-seal vessels in use up to about 1966 were limited to temperatures below about 900 C, and considerably less at the highest pressures for long runs. The advent of new materials, particularly molybdenum alloys (Williams, 1966, 1968) has extended the temperature capabilities of cold-seal vessels to about 1150 C at pressures below 4 Kb (the so-called TZM vessels), and the prospects of new refractory alloys to extend this range are encouraging.

Consequently, it appears safe to conclude that externally-heated cold-seal pressure vessels will continue as the most important device for producing hydrothermal pressures below 10 Kb. Compared to internally heated vessels described in Chapters 4 and 8, and also discussed by Yoder (1950) and Burnham et al. (1969), cold-seal vessels are safe, inexpensive, conceptually simple, and operationally routine. Some externally heated pressure vessels, such as Morey bombs (Morey, 1953), have the disadvantage that the entire pressure vessel, including closure nut and seal, are subjected to the high temperatures in the furnace. This produces a high incidence of failure resulting from leakage, and it commonly leads to galling of the threads of the vessel and closure nut. These problems are largely eliminated in the cold-seal vessel because the closure-nut assembly is outside the furnace (Figure 2). However, this advantage is won at a cost — the "cold"

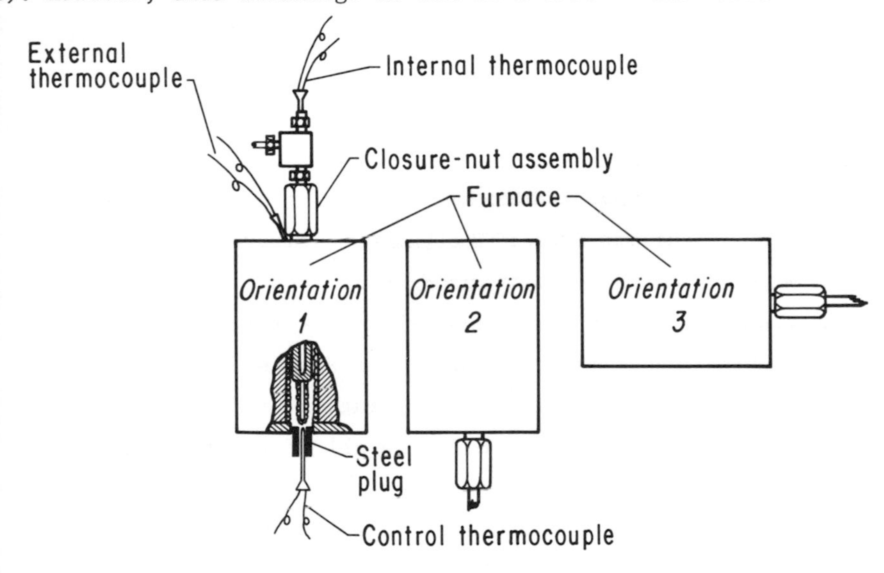

Figure 2. Scale drawing of cold-seal pressure vessels and furnaces in the orientations considered in this study.

exposed closure-nut assembly and long configuration of the pressure vessel result in large temperature gradients along the length of the vessel, resulting from conduction in the vessel and convection in the pressure-transmitting fluid medium.

It is the purpose of this study to evaluate errors in temperature measurement resulting from thermal gradients in cold-seal vessels under numerous operational conditions, and with various types of equipment and appurtenances.

B. Previous Work

There is no previously published systematic study of temperature measurement in externally heated pressure vessels of the type under consideration here, but we are aware that there have been systematic and comprehensive studies in other laboratories. The results of this study will aid investigators when they design and build high-pressure apparatus, and it is hoped that they will contribute toward a meaningful interlaboratory comparison of experimental data.

An important paper on temperature calibration in quench furnaces has recently been published by Biggar and O'Hara (1969). The interested reader is referred to their results.

II. Experimental Details

A. Experimental Equipment

For most of the experiments, René No. 41 cold-seal pressure vessels of the size and type illustrated in Figure 1 were used. As some workers now use or have used vessels with 1 1/4 inch outside diameter, some studies described in a following section used these larger vessels fabricated from Haynes No. 25 to evaluate the influence of size on temperature profile. Where so stated, a stainless steel rod filled all of the interior well of the vessel except for a 5/8 inch length at the hot end that normally is occupied by the sample. This filler rod was held in its proper position by a 3/32 inch diameter, 5/8 inch long extension of the rod. The filler rods had a 1/16 inch axial hole to accommodate a thermocouple inside the pressure vessel.

Most of the experiments used split-wound Lindberg Hevi-Duty furnaces, 9 inches long, with 7 inch long exposed heating elements, an o.d. of 5 1/2 inches and an i.d. of 1 5/16 inches. For experiments on the 1 1/4 inch o.d. pressure vessel, we used a furnace of unknown manufacture that had a single, concealed element, an o.d. of 9 inches, i.d. of 1 1/2 inches, and a length of 12 inches. These combinations of pressure vessels and furnaces are thought to reproduce the gamut of experimental conditions employed by most investigators.

Temperatures were measured with steel-sheathed chromel-alumel thermocouples 1/16 inch o.d. manufactured by Thermo Electric ("Ceramocouples"). The relative precision of all thermocouples was tested by recording the e.m.f. when their recording tips were in juxtaposition in a furnace. This showed that temperature differences between the thermocouples were less than 2 C in the temperature ranges investigated.

Temperatures were controlled with Thermo Electric 400-Series stepless proportional controllers. The control thermocouple was placed in an axial 1/16 inch diameter hole in a stainless steel plug (Figure 2). Measurement of temperature was obtained with a Texas Instrument model 35531 multipoint recorder with a 12 inch wide chart that provided a continuous record of temperatures. Accuracy of the measurement of temperature of the thermocouple is judged at ±3 C, but precision was better than this as determined by periodic checks with a potentiometer.

B. Experimental Procedures

As shown in Figure 2, internal and external temperatures were measured with the vessel in three orientations. In each orientation, the thermal gradient along the length of the furnace was determined by first placing the bomb at the position of maximum immersion[1] in the furnace, and then taking measurements at successive 1/4 inch intervals out to 2 inches from the maximum immersion position. Following this, these readings were repeated by measurements at 1/2 inch intervals inward to the maximum immersion position; in all cases, repeated temperature measurements

[1] Because the closure nut is larger than the interior of the furnace tube, the bomb can be immersed a maximum of 6 3/4 inches in the furnace.

were within 3 C of the original readings. After moving the vessel to each position, temperatures were recorded after allowing adequate time for thermal equilibration. Equilibrium temperatures were readily determined by examination of the continuous chart print-out of the temperature recorder. Although the vessel was left at each position for a minimum of 1 1/2 hours, it was observed in all cases that temperature stabilized within about 20 minutes of moving the vessel to a new position. With the 1 inch o.d. vessel, a stainless steel sleeve around the vessel positioned the bomb so that its axis coincided with that of the furnace tube. Runs were carried out at 300-400 C and 600-700 C, with P_{H_2O} = 1 Kb. At each of the three vessel orientations, measurements were made with and without a filler rod.

III. Experimental Results

A. Experiments with 1 Inch o.d. Vessel

Preliminary experiments with a 1 inch o.d. vessel were carried out with a brick plug in the end of the furnace opposite the vessel. In these runs the hot spot could not be reached in certain orientations of the vessel. Therefore, in all experiments recorded here, a stainless steel plug was used as a "thermal-sink" to shift the hot spot such that it could be reached with the vessel in all orientations.

1. Runs without Filler Rod

The results of these experiments are shown in Figures 3-6. The most significant results are the large differences in temperatures between external and internal thermocouple reading in orientation 3 (Figure 5); the temperature difference is less in orientation 1 (Figure 3) and least in orientation 2 (Figure 4). In the 600-700 C range, for example, the internal thermocouple reading was about 20-30 C lower than that of the external thermocouple in the horizontal orientation of the vessel (Figure 5). With similar external thermocouple temperatures, and with the tip of the external thermocouple at the hot spot, temperature difference between external and internal thermocouples is greatest

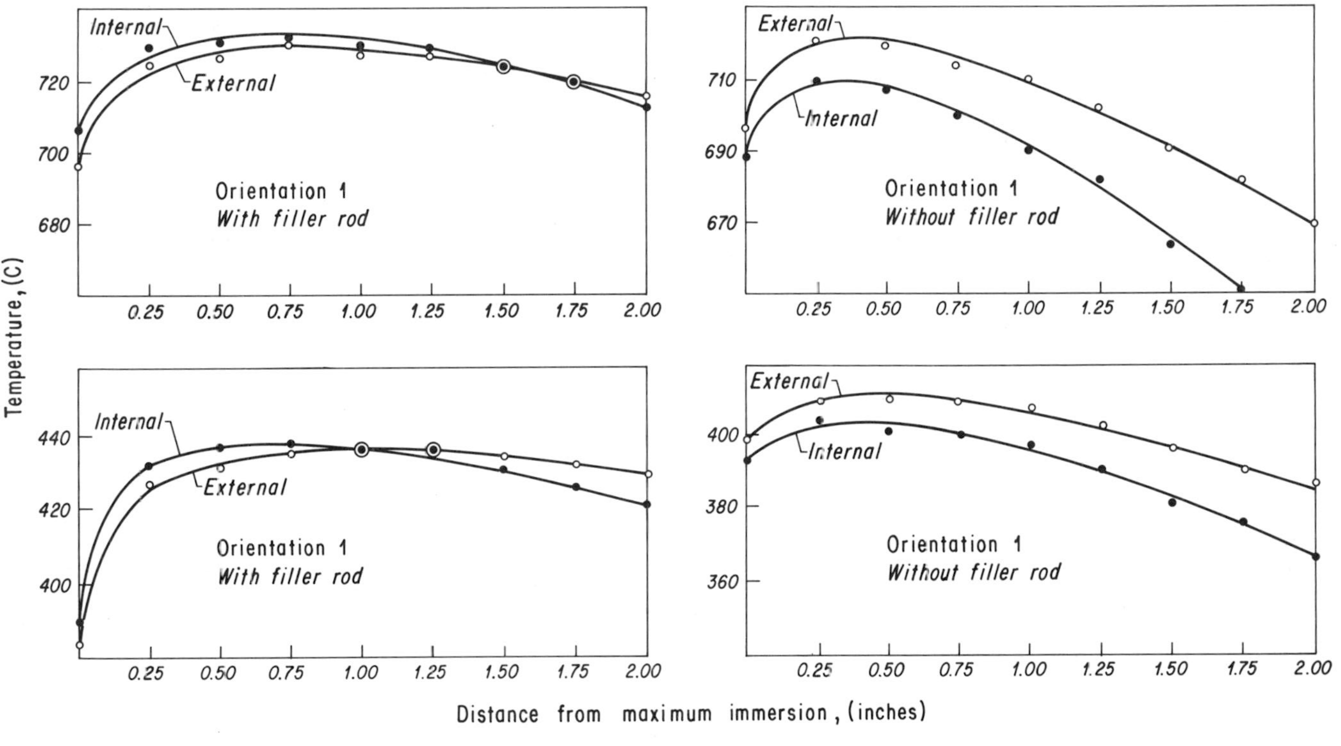

Figure 3. Temperature measurements with 1 inch o.d. vessel in orientation 1 (Figure 2) using water as pressure medium. Position of "maximum immersion" explained in text. External thermocouple readings given by open circles; solid circles represent internal thermocouple measurements.

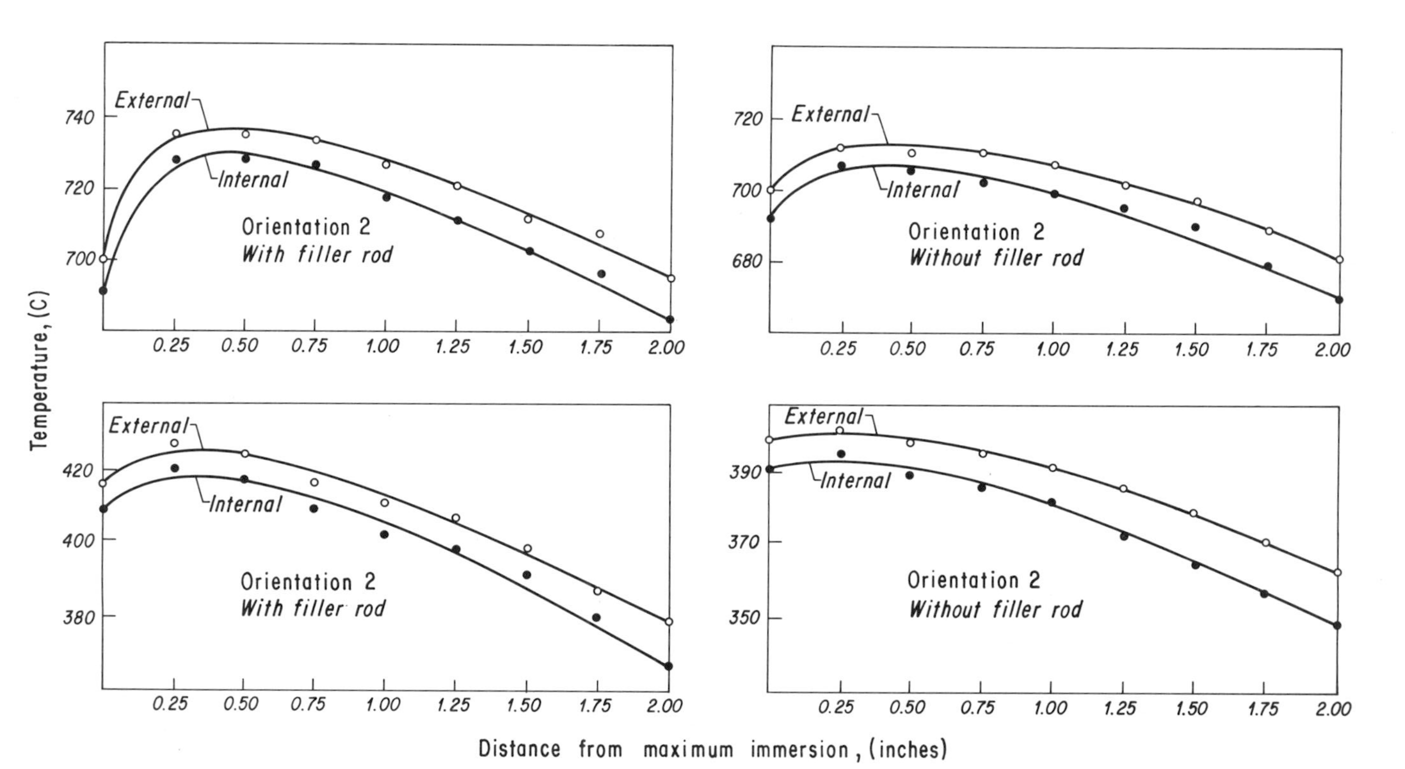

Figure 4. Temperature measurements with 1 inch o.d. vessel in orientation 2 (Figure 2) using water as pressure medium. Symbols as in Figure 3.

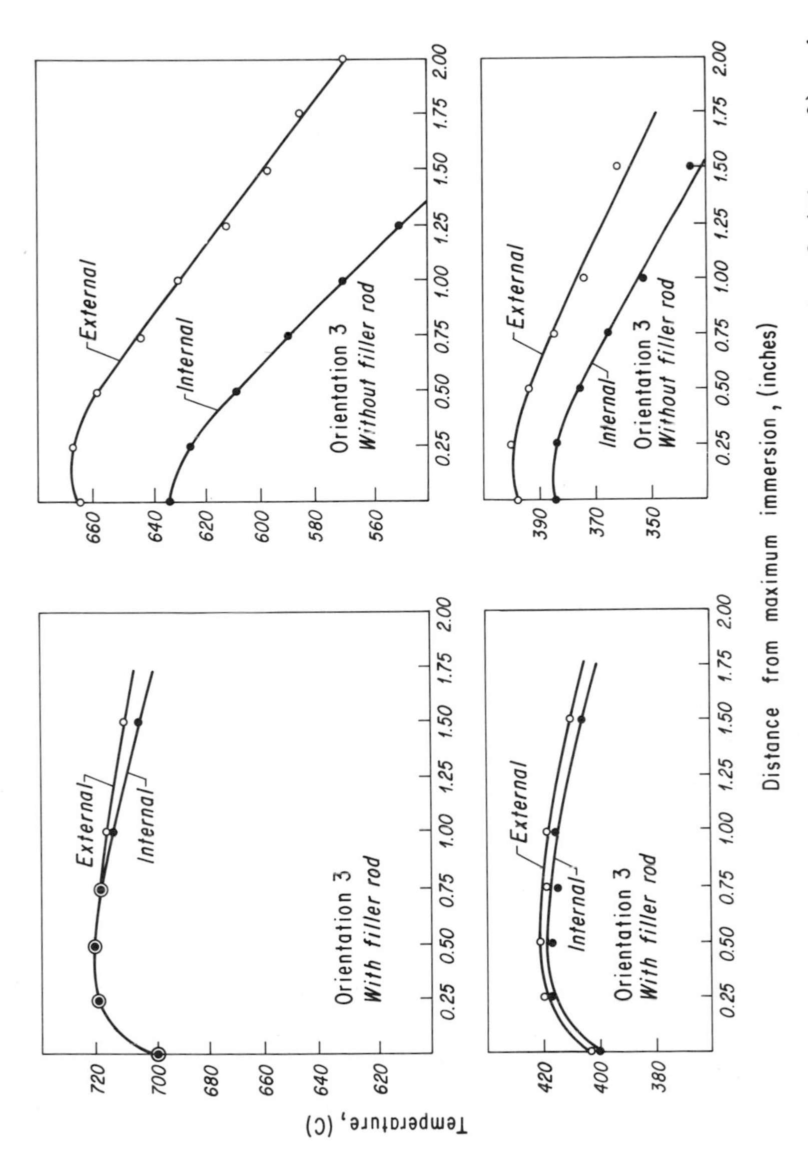

Figure 5. Temperature measurements with 1 inch o.d. vessel in orientation 3 (Figure 2) using water as pressure medium. Symbols as in Figure 3.

in bomb orientation 3, intermediate in orientation 1, and least in orientation 2; in all orientations this difference becomes larger with increasing distance of the bomb from the hot spot.

The effect of pressure on the internal thermocouple was tested with the bomb at maximum immersion, with the bomb in orientation 2, and with the external thermocouple recording at 697 C. No differences were observed in the internal thermocouple readings at pressures of 4 Kb, 3 Kb, 2 Kb, 1 Kb, and 500 bars. (See Chapter 7 for pressure correction of thermocouples in piston-cylinder apparatus.)

2. Runs with Filler Rod

In bomb orientations 1 and 3, smaller differences exist between the temperatures of external and internal thermocouples with filler rods than in the runs without filler rods. The differences observed in orientation 2 (Figure 4) between external and internal temperatures are nearly the same either with or without filler rods. In orientations 1 and 3 (Figures 3 and 5, respectively), the thermal gradients along the axis of the furnace are less with the filler rod than without the filler rod.

B. Runs with Argon Pressure Medium

Temperature measurements were also carried out with a 1 inch o.d. vessel in orientation 3 using an argon pressure medium (Figure 6). Without a filler rod, the temperature differences between external and internal thermocouples are much less with argon than in the runs in the same orientation using water as the pressure medium (compare Figures 5 and 6). As with the runs with water, a filler rod reduces the differences between external and internal thermocouple readings, and it produces flatter thermal gradients along the axis of the vessel. With a filler rod, the internal temperature was slightly higher than that of the external thermocouple between the position of maximum immersion and the hot spot. Here, it is probable that the internal thermocouple was maintained at a higher temperature by argon conducting heat away from the hot spot. From the standpoint of temperature

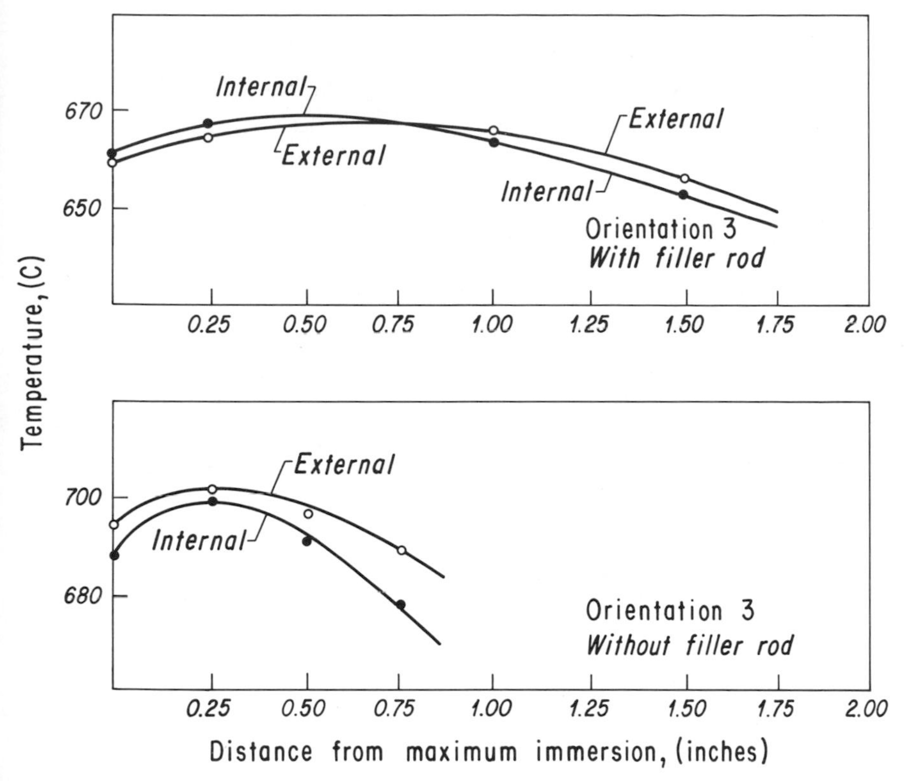

Figure 6. Temperature measurements with 1 inch o.d. vessels in orientation 3 (Figure 2) using argon pressure medium. Symbols as in Figure 3.

measurement, argon is a better pressure medium than water; however, argon has the disadvantage of leaking more readily than water.

C. Runs with 1 1/4 Inch o.d. Vessel

As shown in Figure 7, data for a water pressure medium with a 1 1/4 inch o.d. vessel without a filler rod are similar to the results of experiments for a water pressure medium with a 1 inch o.d. vessel. (Compare Figure 7 with Figures 3, 4, and 5.) Again, the temperature difference between outside and inside thermocouples is largest in orientation 3, intermediate in orientation 1, and least in orientation 2.

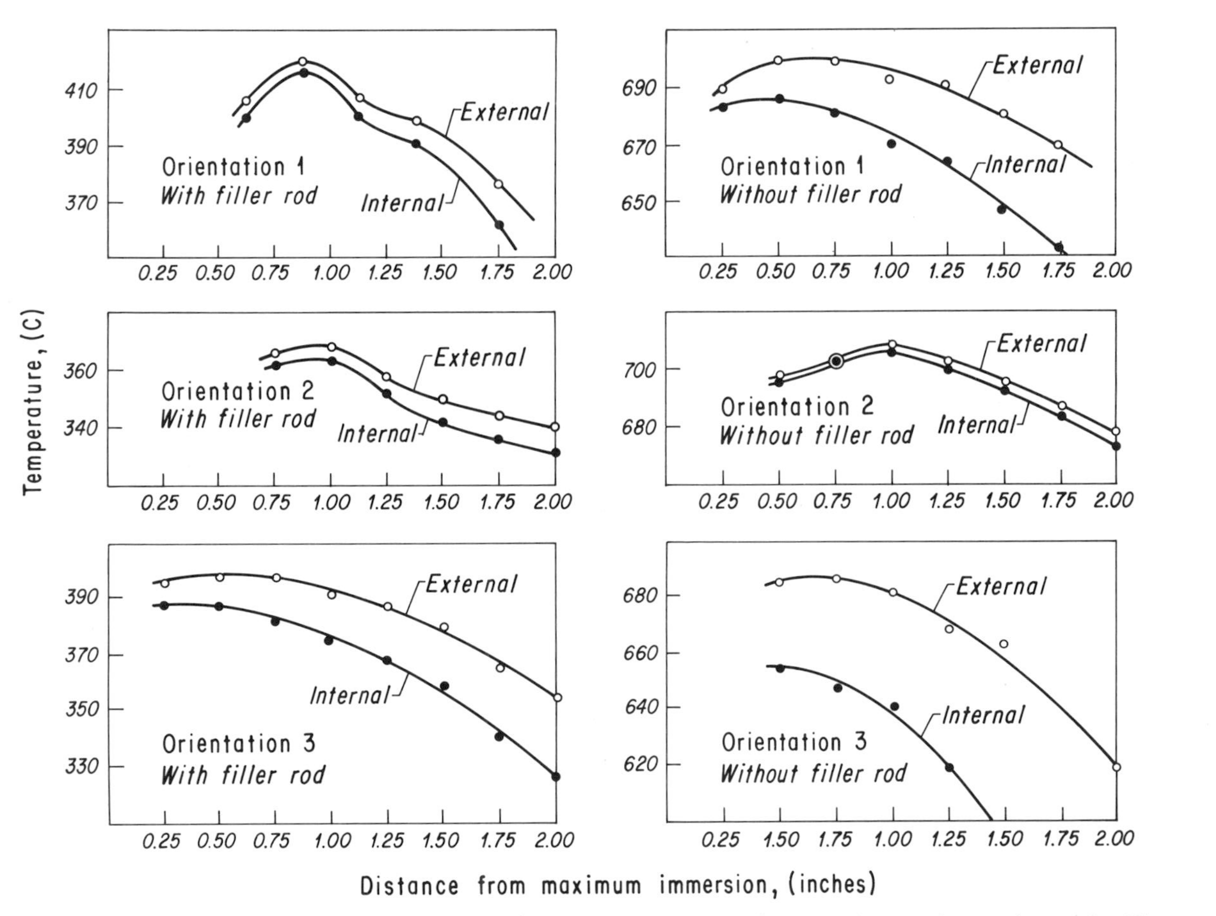

Figure 7. Temperature measurements with 1 1/4 inch o.d. vessel in orientations shown in Figure 2 (water pressure medium). Symbols as in Figure 3.

IV. Discussion

Data from the 1 inch vessel allows comparison of runs with and without filler rods in all three orientations; consequently, most of the following discussion deals with experiments using the 1 inch vessel. The results of these experiments suggest that in bomb orientations 1 and 3, convection of water has an important influence on the magnitude of the thermal gradients that are both normal and parallel to the bomb axis. Steady-state convection is produced by water circulating between the hot and cold ends of the vessel such that at any point along the length of the bomb, the temperature at the center of the vessel is below that of the concentric exterior of the bomb. Comparing orientations 1 and 3 at a given temperature range, the part of the vessel not immersed in the furnace is hottest in orientation 1 resulting from upward conduction of heat through the vessel. Heating the exposed part of the vessel in orientation 1 reduces the temperature difference between the hot and cold ends of the vessel, and, compared to orientation 3, there is a decrease in the rate of heat removal from the hot end of the vessel by convecting water. The data suggest that in orientation 2, convection of water between the hot and cold ends of the vessel is offset by the gravitative effect of hot water remaining in the upper or hotter end of the vessel; consequently, in this orientation there is no significant difference between measurements with and without the filler rod. Comparing runs in the same temperature range in orientations 1 and 3, the thermal gradient along the axis of the vessel is less with a filler rod than without a filler rod. This is expected because a filler rod reduces the rate of heat conduction away from the hot end of the vessel, and it leads to a flatter thermal gradient along the axis of the bomb.

The curves shown may not be true representations of the thermal gradients along the length of the vessel at all immersion positions because the mass of metal immersed in the furnace differs at different points along the abscissa of these graphs; the amount of heat conducted and convected to the cold end of the vessel will differ from one immersion position to the next.

The crossing of curves for outside and inside thermocouple readings in orientation 1 (Figure 3) with the filler rod inserted shows that, below the hot spot, the external thermocouple will be at a lower temperature than the internal thermocouple. Here, the internal thermocouple is probably kept hotter than the external thermocouple by water convecting heat downward away from the hot spot.

In all curves shown in Figure 3, a steep thermal gradient is noted between the position of maximum immersion and that with the vessel 1/4 inch out from maximum immersion. This reflects the large heat conductivity of the metal plug in the end of the furnace opposite the vessel.

In orientations 1 and 3 (Figures 3 and 5, respectively) without the filler rod, the difference between outside and inside temperatures increases with increasing temperature. This is expected, as convection of water within the vessel becomes stronger with increasing temperature.

V. Conclusions

The results of this study show that with vessels containing no filler rod in orientations 1 and 3, significant errors occur in measuring the sample temperature with external thermocouples. The combined effects of thermal gradients normal and parallel to the axis of the vessel can lead to errors in temperature measurement greater than 40 C. This factor may account for some discrepancies in previously determined experimental investigations of phase equilibria. For accurate estimates of sample temperatures using external thermocouples, it is necessary to fill most of the vessel with a snug-fitting filler rod and to place the sample at the hot spot along the length of the furnace tube. With these modest precautions, temperatures can in most cases be measured to well within 5 C. As well as improving measurement of the sample temperature, filler rods are an added safety feature because they reduce the volume of water within the vessel and thus decrease the intensity of any explosive rupture of the vessel.

BIBLIOGRAPHY:

Biggar, G.M. and O'Hara, M.J. (1969). Temperature control and calibration in quench furnaces and some new temperature measurements in the system CaO-MgO-Al_2O_5-SiO_2. Mineral. Mag., 37:1-15.

Burnham, C. Wayne, Holloway, J.R., and Davis, N.F. (1969). The specific volume of water in the range 1000 to 8900 bars, 20° to 900^oC. Amer. J. Sci., 267-A:70-95.

Luth, W.C. and Tuttle, O.F. (1963). Externally heated cold-seal pressure vessels for use to 10,000 bars and 750^oC. Amer. Mineral., 48:1401-1403.

Morey, G.W. (1953). Hydrothermal synthesis. J. Amer. Ceram. Soc., 36:279-285.

Tuttle, O.F. (1949). Two pressure vessels for silicate-water studies. Geol. Soc. Amer. Bull., 60:1727-1729.

Williams, D.W. (1966). Externally heated cold-seal pressure vessels for use to 1200^oC at 1000 bars. Mineral. Mag., 35:1003-1012.

Williams, D.W. (1968). Improved cold-seal pressure vessels to operate to 1100^oC at 3 kilobars. Amer. Mineral., 53:1765-1769.

Yoder, H.S., Jr. (1950). High-low quartz inversion up to 10,000 bars. Amer. Geophys. Union Trans., 31:827-835.

ACKNOWLEDGMENTS:

This research was supported by National Science Foundation Grants NSF GA-1364 and NSF GA-12737 and by the Department of Geochemistry and Mineralogy. We are grateful to Dr. Victor Wall for assistance with the experiments using argon and to Scientific Systems, Inc., State College, Pa., for loan of some equipment.

CHAPTER 7

Pressure Calibration in Piston-Cylinder Apparatus at High Temperature

P.M. Bell

D.W. Williams

I. Introduction

Experiments in physical chemistry usually require pressure control in order to contain liquid and vapor phases. Pressure is, of course, a thermodynamic variable, and commonly must be set at values greater than one atmosphere for the requirements of a particular study. Unfortunately, this type of research has historically been plagued by problems of calibration because it has not been possible to make accurate temperature and pressure measurements inside the massively enclosed chambers required for the experiments. Many of the causes of error are now recognized, so that most of the uncertainties of measurement can be evaluated.

Technical difficulties arise if pressures greater than about 7 Kb at 800 C or higher are necessary. Metal failure is frequent in this range. Few of the liquids used to transmit pressure are stable at high temperature, and the great amount of energy stored by gases at pressures greater than 10 Kb makes their confinement an exceedingly difficult task. Problems with the use of weak solids for pressure-transmitting media exist as well, but are of a different nature, most being related to errors in calibration of pressure.

Solid materials such as talc, pyrophyllite, boron nitride, silicate glasses, graphite, and sodium chloride have been exploited extensively for most high-pressure experiments, but many of the properties that make them ideal for obtaining extreme conditions cause nonuniform pressure distribution. Usually a high-pressure hydraulic ram thrusts a tungsten-carbide piston into a supported carbide cylinder. The piston compresses a furnace assembly, and pressure is transmitted to a sample located in the interior of the furnace. The hydraulic ram pressure is measured directly by reading a Bourdon gage, but the pressure actually delivered to the sample charge is calculated and usually not well known.

In detail, the high pressure cell is fabricated from several materials as shown in Figure 1. The cell consists essentially of the following parts (numbers keyed to the figure): sample assembly (11, 12, 13), thermocouple (3, 4, 10), electrical resistance furnace (9), pressure medium (14, 15), gasketing or swaging assembly to prevent extrusion of electrical leads (1, 2, 5). The various parts are machined in quantity, making experimentation rapid and

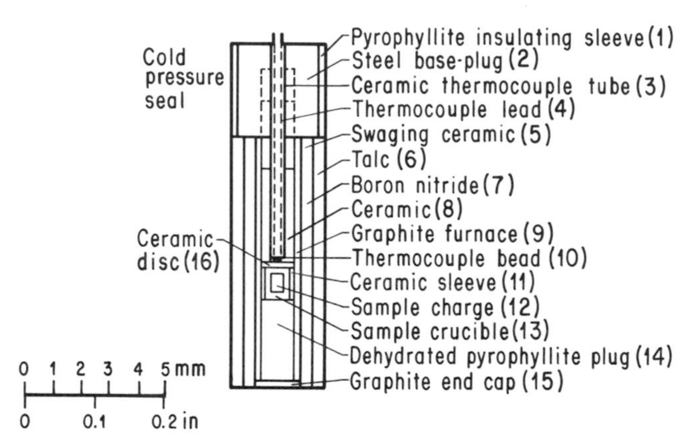

Figure 1. High-pressure cell (after Mao and Bell, 1971).

efficient. Pressures and temperatures can be conveniently applied and controlled in this cell with the use of a single-stage, piston-cylinder press. Convenience and moderate accuracy of this cell are the main reasons for its widespread use since it was first introduced by Boyd and England (1960). Various modifications and substitutions of materials have been made to improve performance of the cell for specialized experiments and to establish better temperature and pressure distribution. These modifications usually involve the introduction of multiple electrical leads into the cell, parts of which may consist of glass.

Piston-cylinder apparatus with solid-pressure transmitting media have proved most suitable for routine experimentation. At present, highly accurate means of calibrating pressure and temperature are at best time-consuming and difficult, and at worst impossible. Experience has led to the development of relatively simple procedures whereby measurements can be standardized (not calibrated) and the limits of uncertainty estimated.

II. Sources of Error

A. Pressure

Pressure is the force per unit area, so if the force is held constant, pressure is increased simply by reducing the area. Experimental pressures higher than 2 Kb are normally generated by means of the "intensifier" principle, where oil is pumped against a large-area piston, which in turn drives a small-area piston against the sample assembly. The first source of error arises in the frictional drag of the large (hydraulic) piston, which can vary anywhere from 0 to as much as 20% of the applied force. Then there is the frictional drag of the high-pressure piston along the walls of the carbide cylinder containing the sample assembly. Drag on the high-pressure piston is caused not only by the contact friction of the piston and cylinder but also by extrusion of the high-pressure cell. Additional friction is caused by strengths of the materials of the cell. Evaluation of the various frictions becomes complicated as the cell is heated. Such factors as creep and phase change in the pressure media must be considered, and it must be noted that the direction of piston travel — whether in-stroke or out-stroke — will define the sign of the frictional forces.

If the sign of the error changes with piston stroke direction, the "double-value" of friction can be evaluated at a given pressure and temperature by making observations of a univariant reaction under "in-stroke" and "out-stroke" conditions, and by subtracting the pressure measured for the transition during the former from the pressure measured during the latter. There are additional sources of error, however, such as "pistoning" or "antipistoning," where the sign does not change with piston stroke. In the usual case of pistoning the relatively incompressible components of the pressure media are supported by relatively compressible parts of the cell. The resulting error is usually not detected because the sign of the error does not change with the direction of piston stroke. The phenomenon can be evaluated only by trial and error with various pressure-transmitting materials.

Assuming piston and cylinder dimensions are well known, errors in the calculated ("load") pressure due to friction are in the range 0 to $\pm$ 6 Kb. In many cases the friction may not be symmetrical. In-stroke friction often is not equal and opposite to out-stroke friction (Boyd et al., 1967), but it is possible to design a cell for maximum symmetry (Williams and Kennedy, 1969), where the true value of friction can be obtained simply by halving the double value.

Errors caused by "pistoning" are in the range 0-3 Kb at about 700 C (Richardson et al., 1968) and have not been demonstrated to exist above 1200 C. This type of error can usually be detected only by comparison of values measured with gas-media apparatus.

There are many well known fixed points available for pressure calibration, such as the transitions of bismuth, but these are only suitable for room-temperature experiments. At high temperatures the assembly behaves very differently, so the usual calibration points cannot be used. Melting curves can be detected by differential thermal analysis (D.T.A.) and therefore have some application in pressure-temperature (P-T) calibration, but in nearly all known examples the P-T slopes tend to be relatively flat (2-15 C/Kb), making their measurement insensitive to pressure. Subsolidus reactions involving silicates often are favorably sloped in P and T for calibration purposes but are seldom known accurately because of kinetics.

B. Temperature

For years Pt/Pt10%Rh thermocouples have been used with the rationales that they behave well under most conditions, they are accepted as an international standard, and if some unknown pressure effect should be discovered a correction factor could be applied. This was a reasonable, but faulty, approach. The problem turns out to be more complex; at the present time serious pressure effects are known which are manifested in several ways. Specifically, the application of pressure to a thermocouple does not generate an e.m.f. but does change the characteristics of the metals so that the function of e.m.f./degree is changed. Studies of thermocouples under pressure are currently being conducted in several laboratories (e.g., Getting and Kennedy, 1970, Mao and Bell, 1971).

Modern explanations of the thermoelectric effect still require significant assumptions in quantum theory, some of which do not hold when one attempts to calculate pressure effects on the work functions of the metals. Barring acceptance of the "black box" approach when thermocouples are employed, it is useful to have a rough idea of their behavior. Simply, a potential difference (e.m.f.) exists between the hot and cold ends of any metal or alloy wire. This e.m.f. can be measured conventionally only by introducing another wire which, if made of the same metal, will cancel the e.m.f. of the first wire. Wires of different composition generate different e.m.f.'s at the same temperature. The small e.m.f. difference is measured, and its change with temperature is the function listed in thermocouple tables. The simultaneous application of pressure on both legs of the thermocouple changes the characteristics of both metals. The thermocouple e.m.f., being the small resultant difference between the two relatively large e.m.f.'s of the individual wires, is not so severely affected as when pressure is applied to one leg only. In the solid-media apparatus most of the error is caused by pressure applied unevenly to both legs of the thermocouple. Strong materials surrounding the wires support pressure differences, and the error can be increased or decreased, depending on the sign.

The pressure effect causes an error of about 0.4% of the temperature per 10 Kb with a Pt/Pt10%Rh thermocouple (Getting and Kennedy, 1970). The additional error ("random" error, Mao and Bell, 1971) from differential stresses in the individual thermocouple wires amounts to about ±0.25% of the measured temperature per 10 Kb. (See Chapter 6 for the influence of hydrostatic pressure up to 4 Kb on chromel-alumel thermocouples.)

Recently a thermocouple of W3%Re/W25%Re has gained popularity. It causes problems of basic calibration (average calibration error ±0.5%) and has a short life under oxidizing conditions, but it has certain advantages over Pt/Pt10%Rh. The basic pressure effect is comparable, but the "random" error is less because the e.m.f.'s of the individual legs are lower. The e.m.f./degree function for this tungsten-rhenium thermocouple is greater than that of Pt/Pt-10%Rh and can be used at considerably higher temperatures (2500 C for W-Re; 1750 C for Pt-Rh). The W-Re alloy appears to be considerably less sensitive to chemical contamination from various sources in the high-pressure cell, and there is a chance that most experimenters will change over to it in the future.

If results are to be compared from different types of equipment it is essential to make a correction for the effect of pressure on the thermocouples used, since P-T gradients in the assemblies will be different in each case, leading to different corrections. It is best to publish both the uncorrected experimental data and the corrected temperatures, so that a new correction can be applied should new work show that the corrections used at present are in error.

III. Routine Pressure-Temperature Calibration

Independent methods, such as thermal noise measurement for temperature under pressure and x ray diffraction determinations of equations of state, are known but have not been developed for routine P-T calibration of the solid-media, piston-cylinder apparatus. Currently the best hope of obtaining routine calibration is by determining a known equilibrium boundary with a slope sensitive to both temperature and pressure and making comparisons with published values. Absolute determinations of such a boundary will be improved as apparatus with hydrostatic pressure media becomes more widely available.

Static methods, where slow reactions are determined with a long series of experiments, are appropriate for quenching. Dynamic methods, such as D.T.A., are employed for rapidly reversible reactions. These methods are described separately in the following sections, with discussions of the most suitable reactions. In principle, the objective is to simultaneously achieve calibration of pressure and temperature.

A. Static Methods

1. Experimental

Methods include determining known phase transitions by quenching. Some solid-solid phase reactions, either polymorphic or chemical, are conducive to the quenching method, so one can simply determine a univariant curve to obtain relative corrections.

In practice the in-stroke-out-stroke friction of the particular pressure vessel, piston, and type of furnace assembly is determined. It is essential to determine P and T simultaneously, so a curve of appropriate slope must be chosen. Curves with slopes in the range 30-80 C/Kb are usually recommended. Three such univariant curves are kyanite $\rightleftarrows$ sillimanite, albite $\rightleftarrows$ jadeite + quartz, and quartz $\rightleftarrows$ coesite. Absolute determinations of these curves have not been made for all ranges of P and T, but enough measurements of the three curves have been made in solid-media, piston-cylinder experiments for satisfactory application as relative or secondary standards.

a. Kyanite $\rightleftarrows$ Sillimanite

Kyanite-sillimanite equilibrium has been determined by overlapping experiments of gas and solid-media apparatus by Richardson et al. (1968). Figure 2 shows a diagram explaining

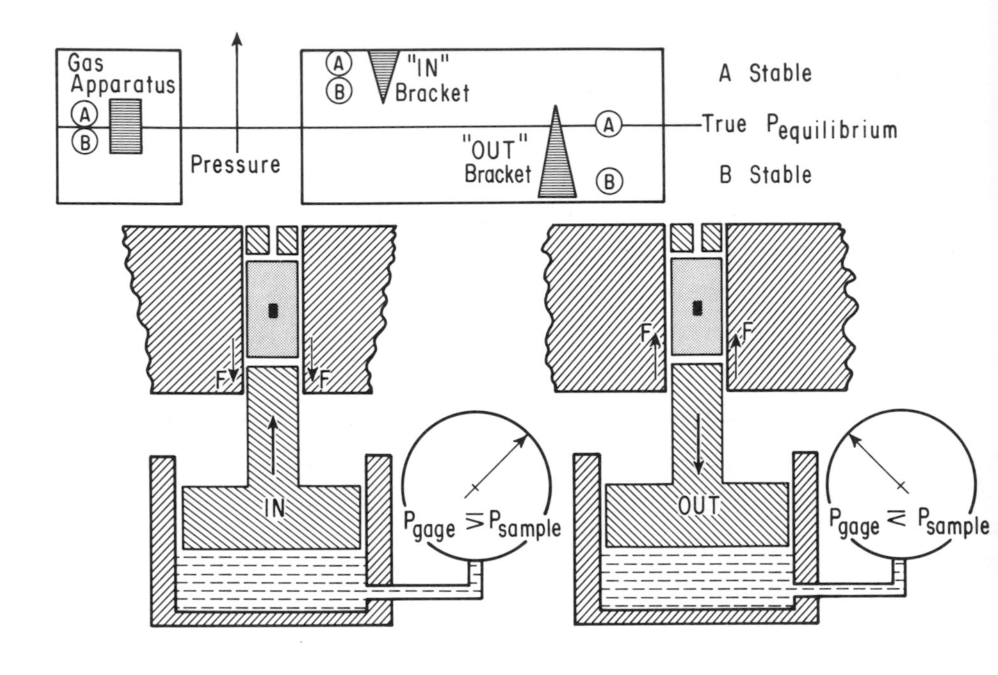

Figure 2. "In-stroke-out-stroke" brackets (after Richardson et al., 1968).

in-stroke and out-stroke brackets for the solid-media apparatus. Several brackets are plotted for the range 800-1500 C, 10.5-28.5 Kb in Figure 3 with overlapping results obtained with gas apparatus. Figure 3 also shows a shaded zone of the experimental uncertainties, which can be conveniently used for reference in calibration experiments. Probably no more thorough example of an absolute and relative comparison is available at the present time.

As an extension of kyanite-sillimanite one can also use the metastable extension of the kyanite-andalusite curve (Richardson et al., 1969). There is no absolute determination of this curve, but interlaboratory agreement in the range 700-850 C, 6-8 Kb, is excellent, hereby providing a relative standard.

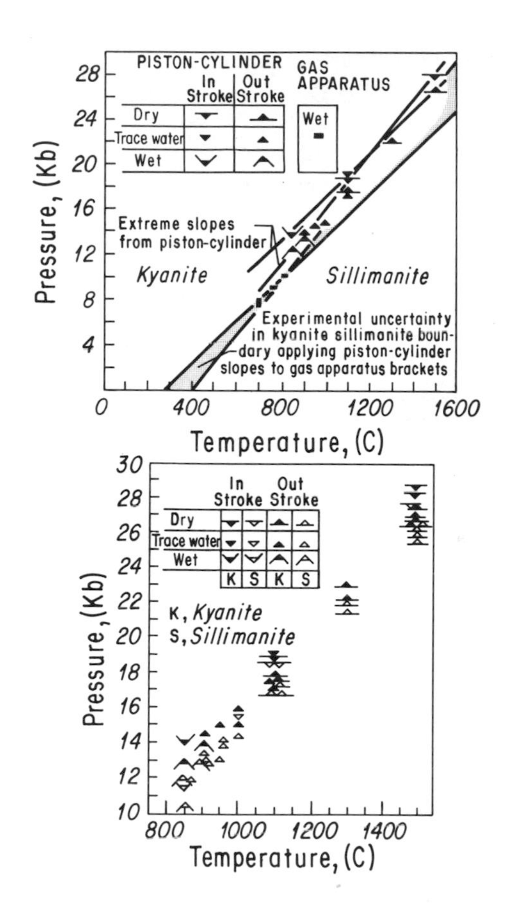

Figure 3. Uncertainties in determinations of the kyanite-sillimanite transition (after Richardson et al., 1968).

b. Albite $\rightleftarrows$ Jadeite + Quartz

After extensive interlaboratory efforts, determination of the reaction albite $\rightleftarrows$ jadeite + quartz is useful for calibration, particularly at 600 C, 14.5-18.5 Kb. The wide spread of these results (Figure 4) emphasizes the need for absolute calibration.

c. Quartz $\rightleftarrows$ Coesite

The quartz-coesite transition has been well studied in the solid-media apparatus. The most recent and comprehensive experi-

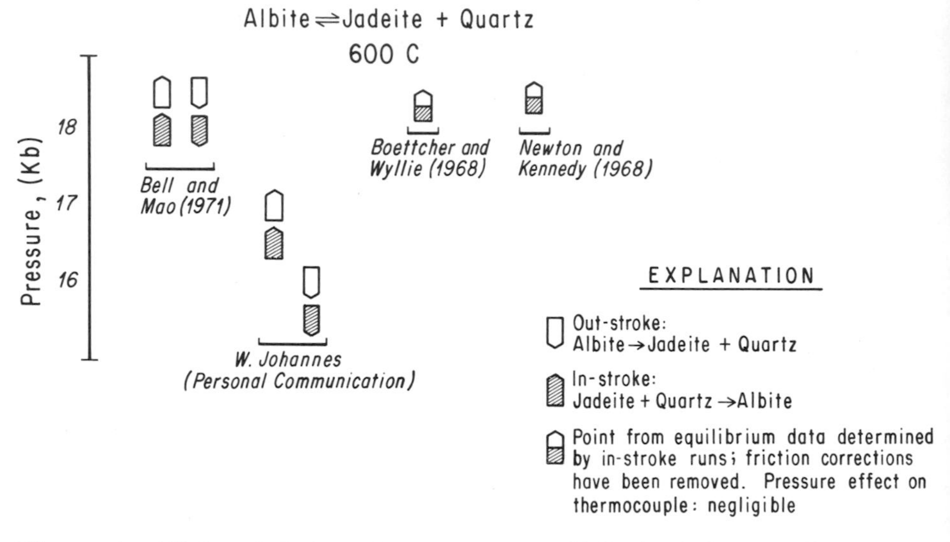

Figure 4. Albite-jadeite plus quartz equilibria (after Bell and Mao, 1971).

ments were carried out by Boyd et al. (1967). Absolute determinations are not available, but the transition is particularly useful as a relative standard in the high P-T range at about 1400 C, 37-38 Kb.

In the examples used here, the suggested procedure is simply making quench experiments in the desired range to locate a point on the univariant curve. The starting materials are usually powdered mixtures of equal volumes of reactants and products sealed in Pt capsules, with water added in some experiments. Demonstration of a reaction over a narrow P-T bracket requires a combination of (1) applying the correct conditions, (2) observing the product assemblage to grow at the expense of the reactant assemblage and vice versa, and (3) making certain of substantial completion of the reaction (usually by powder x ray diffraction).

Reactions are observed to proceed in both directions [i.e., (A) + B → A; A + (B) → B] on each side of the equilibrium curve. In the first set of experiments the apparatus must be operated under in-stroke conditions, preferably by monitoring actual piston motion. The second set is simply a repeat, except with out-stroke conditions. In both sets, special note should be made of the experimental path followed. Excursions through the high- and low-pressure sides of the univariant curve should be made, and care should be taken to avoid the problem of forming metastable products during the initial application of pressure and temperature and during the quench.

Somewhere between the set of two brackets is the relative transition point, and it is recommended procedure to determine at least three points along the equilibrium curve. The resulting values of the double value of friction give the experimental uncertainty and the location of the calibration curve.

B. Dynamic Methods

The normal method of dynamic pressure calibration of solid-media apparatus at room temperature involves monitoring some rapidly reversible (usually polymorphic) phase change, whose transition pressure has been previously accurately determined, as pressure is cycled up and down. In a similar fashion the exact transition pressure of a phase change may be determined in piston-cylinder apparatus, once frictional losses are accurately known. Careful work on polymorphic transitions in high purity materials in piston-cylinder apparatus, where special care has been taken to minimize and allow for friction, can yield very accurate results (compare Kennedy and La Mori, 1962, and Heydemann, 1967, for the bismuth I-II transition). The most accurately known fixed points on the high-pressure scale at pressures beyond the range of gas apparatus have been determined in this way in piston-cylinder apparatus.

The parameters most commonly used to detect phase transitions for calibration purposes at room temperature are change in volume (Kennedy and La Mori, 1962) or in electrical resistivity (Haygarth et al. 1967), although other features such as optical properties, x ray spacing, and thermal conductivity have been extensively used in other types of apparatus. The value of this kind of calibration experiment is that by continuous monitoring the transition can be observed as the pressure is changed at different rates, and a series of experiments with different pressure media, different lubricants, etc., can help to establish the various components of "friction" present. The best results are obtained when the transition is completed over a narrow pressure interval and is accompanied by a large discontinuity in volume or electrical resistivity.

The same technique can be applied to pressure calibration at high temperature, provided a transition can be found with a substantial slope (preferably 30 C/Kb) that can be readily monitored. Changes in sample volume are difficult to detect at elevated temperature, since much of the assembly is composed of furnace and insulation material. It is possible to follow some of the room-temperature fixed points on the high-pressure scale to moderate temperatures by monitoring electrical resistance (Haygarth _et al._, 1969), but this becomes very difficult at high temperature. For work at elevated temperatures the latent heat of a phase transformation is most usually monitored, being detected by a D.T.A. technique. This is a very sensitive way to detect melting phenomena, where an endothermic (melting) and an exothermic (freezing) signal can often be observed as a sample is heated and cooled.

1. Procedure

A typical assembly for high-temperature D.T.A. is similar to that shown in Figure 1. A four-bore thermocouple tube is used, containing two thermocouples, one with the hot junction as close as possible to the sample, the other set back a little (Figure 5). Normally a somewhat larger sample is used (40 mg) than for a quenching experiment, to give a good D.T.A. signal, but it must not be so large as to lie across a temperature gradient or a D.T.A. signal "smeared out" over a considerable temperature range will be obtained.

It is essential to use a pressure medium of low shear strength, so that small changes in piston position are rapidly translated to real changes in sample pressure. Pressure is held constant for

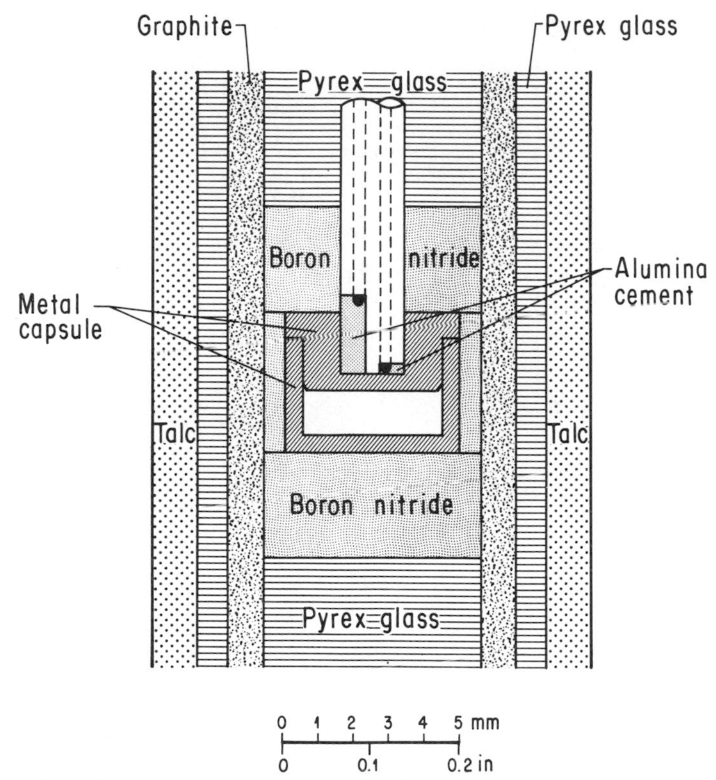

Figure 5. Detail of sample assembly for high-pressure melting studies by D.T.A.

periods of much less than an hour in D.T.A. work, compared with the many hours often available for a quenching experiment to stabilize. Thus materials such as talc, boron nitride, sodium chloride, or Pyrex glass arc used around the sample and not stronger ceramic materials. The assembly shown in Figure 5 has been used for melting work on pure silicate minerals to temperatures as high as 2000 C. At temperatures where the glass begins to partially melt and so form a good hydrostatic pressure medium, the boron nitride jacket provides positioning and insulation for the metal sample capsule inside the furnace. The Pyrex glass outside the furnace also softens at run temperature and tends to prevent water released from the talc from getting to the sample.

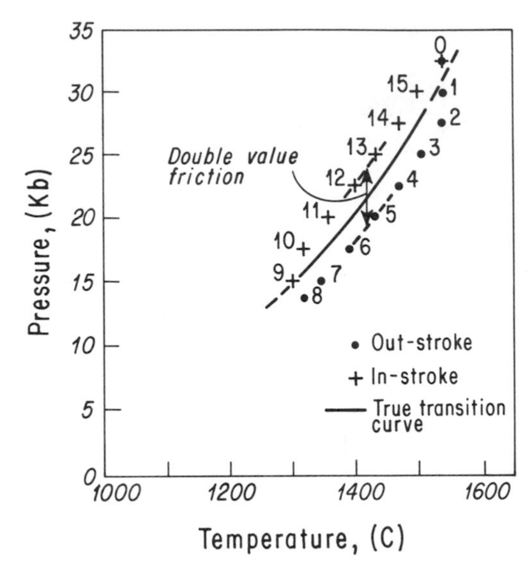

Figure 6. P-T plot (diagrammatic) of melting detected by D.T.A. for an out-stroke and in-stroke cycle. True melting curve lies at the center of the hysteresis loop of the results. Width of hysteresis loop is the double value of friction. (Numbers refer to points at which results were obtained.)

The technique for detecting a high-temperature phase transition such as melting can be summarized as follows (Figures 6 and 7 show a diagrammatic representation of the melting signal versus pressure and piston displacement versus pressure, respectively):

(1) The sample is contained in an inert capsule whose temperature is monitored by the lower thermocouple (Figure 5) on a strip chart recorder, and the difference in temperature between the two thermocouples is similarly displayed, usually on a second channel of high sensitivity (1-mv, full-scale deflection).

(2) The sample is brought to the maximum pressure to be used in the experiment, while being heated to just below the melting point.

(3) The temperature is then cycled to detect melting and freezing. This yields the first results; pressure and tempera-

ture increase at the same time, however, and we do not know whether in-stroke or out-stroke conditions prevailed for this point (0 in Figures 6 and 7).

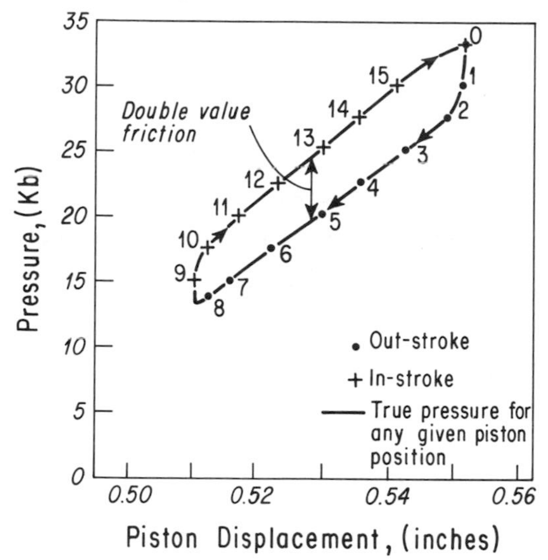

Figure 7. P versus piston displacement plot for the same experiment as Figure 6. Again true pressure lies at the center of the hysterisis loop.

(4) Pressure is then reduced, and a new determination of melting and freezing is obtained at each of several out-stroke conditions (1-8 in Figures 6 and 7). It can be seen from Figure 7 that point 1 is not a true out-stroke result, since piston movement has barely begun, and this is reflected in a melting temperature only marginally different from points 0 and 2. After each reduction in pressure it will be noticed that initial melting temperatures are higher than those for later results at the same pressure, owing to the time taken for the pressure medium to adjust to the new pressure. For accurate work it is necessary to make several temperature cycles at each pressure until successive determinations show no change. The minimum temperature oscillation to detect melting and freezing is used, so as to minimize the effects of thermal expansion of the cell on the pressure readings.

(5) The process is then repeated for piston-in conditions (points 9-15). Again Figure 7 clearly shows that point 9 is not a true in-stroke result, and it is seen to plot below the other points in Figure 6.

Having arrived back at the starting point 0, one should repeat some of the piston-out results to check that "drift" due to sample or thermocouple contamination has not occurred. Further experimental cycles can then be carried out at higher or lower pressure until the assembly fails, usually by breaking of the thermocouple wires. Further details of D.T.A. technique in piston-cylinder apparatus are given by Cohen <u>et al.</u>,(1966-a).

When the results from out-stroke and in-stroke conditions are plotted as in Figure 6, a "hysteresis loop" is obtained whose width is equal to the double value of friction in this experiment. The true transition pressure lies at the center of this hysteresis loop, provided that "pistoning" has been avoided, and that friction al losses are symmetrical on in-stroke and out-stroke conditions. A useful check is to repeat experiments with different pressure media. If the same result is obtained, one can be confident about most of the above assumptions. In the same way the pressure on the sample can be estimated from the center of the hysteresis loop in Figure 7. This type of plot is very useful to determine just when piston movement has been reversed.

Samples that show sluggish melting or supercooling, giving melting signals spread over a wide temperature interval and freezing signals up to hundreds of degrees below the true melting point, may be treated in the following way. The sample is held in the vicinity of the melting point long enough to come to equilibrium, and then rapidly cooled. The presence or absence of a freezing signal is taken to indicate "melting" or "no melting" at the hold temperature. This technique is then repeated at successive temperatures at a series of in- and out-stroke conditions, and the results are plotted as before. Further details of this technique are given by Williams and Kennedy (1969).

2. Precision and Accuracy of D.T.A. Method

It is possible to reproducibly obtain a precision of 1-2 C for sharp melting and freezing signals at a given nominal (ram) pressure, which can be read to 0.1%. The experimental data can thus be plotted as in Figure 6 with great precision, but the accuracy of the final result depends on several other factors:

(1) The accuracy of the standard curve must be considered if a comparison experiment is being done.

(2) If pressure is being determined from "ram" pressure by allowing for frictional losses, it may be necessary to allow for piston swelling, the thickness of material extruded around the piston, etc., as well as factors such as "pistoning" and assymetry in the hysteresis loop. These factors are all discussed by Haygarth _et al._ (1967).

(3) Errors in temperature measurement are important. Individual pairs of thermocouple wires can be calibrated to an accuracy of about 1 C, but larger errors are involved in making the correction to the observed temperature reading, to allow for the effect of pressure on the thermocouple e.m.f. Getting and Kennedy (1970) fully discuss the corrections to be applied to chromel/alumel and Pt/Pt10%Rh thermocouples and the uncertainties involved.

(4) The slope of the calibration curve is also significant. With a slope of 10 C/Kb, and a curve located to ± 2 C, pressure is fixed to ± 0.2 Kb, but a value of ± 0.5 Kb is probably more realistic.

Since all the curves that can be used for high-temperature, high-pressure calibration have been determined themselves in piston-cylinder apparatus, it is customary not to use such a curve to calibrate a piston-cylinder apparatus but to calculate pressure directly from the "ram" pressure by allowing for the friction given by a hysteresis loop as in Figures 6 and 7. For calibration of apparatus such as the "belt" (see for example Bell and England, 1967), however, reference to an accurate standard curve is necessary, since there are large uncertainties involved here in force/area calculations.

3. Some Suggested Calibration Curves

There has been little work done to date to provide reliable standard curves that can be used for dynamic experiments for high-temperature, high-pressure calibration. The following curves have all been fairly accurately detected by D.T.A. and should prove useful for interlaboratory comparison of equipment and technique.

a. High-Low Quartz Inversion

Following the earlier work of Yoder (1950) in gas apparatus, Cohen and Klement (1967) established the transition from 6-35 Kb. Initial slope is 26 C/Kb, with a curvature of less than 0.4 C/Kb. Their values are given in Table 7.1.

b. Melting of Sodium Chloride

The most recent D.T.A. work of Akella *et al.* (1969) reported an initial slope of 21.2 C/Kb, with considerable curvature at pressures up to 65 Kb.

c. Melting of Gold (or Silver)

Cohen *et al.* (1966-b) reported initial slopes of 6.5 C/Kb for gold and 6.0 C/Kb for silver. Silver has a linear melting curve to 40 Kb; gold has a slight curvature. In spite of these relatively low slopes, their linearity and the ready availability of gold and silver in high purity make them good candidates for pressure calibration work.

d. Melting of Pure Silicate Minerals

Following earlier D.T.A. work in gas apparatus by Yoder (1952) and quenching work in piston-cylinder apparatus by Boyd and England (1963), Williams and Kennedy (1969) have reported the melting of diopside to 50 Kb, using the modified "hold and quench" D.T.A. technique already described. The initial slope is 14.6 C/Kb, but there is considerable curvature at high pressure. Details are given in Table 7.1.

Descriptions of sample containers for these materials and the best techniques for each transition are given in the papers cited.

TABLE 7.1

Phase Transition Data for Pressure Calibration at High Temperature by D.T.A.

Transition	Pressure (Kb)	Transition Temperature (C) (as measured)	(corrected*)
Low-high quartz	0	573	573
(Cohen and Klement,	10	815	819
1967)	20	1045	1053
	30	1261	1272
	35	1361	1374
Melting of NaCl	0	800.5	800.5
(Akella *et al.*, 1969)	10	1000	1004
	20	1158	1165
	30	1273	1283
	40	1362	1377
	50	1440	1458
	60	1504	1527
Melting of gold	0	1063	1063
(Cohen *et al.*, 1966-b)	10	1124	1128
	20	1182	1190
	30	1233	1244
	40	1283	1297
Melting of diopside	0	1392	1392
(Williams and Kennedy,	10	1527	1531
1969)	20	1637	1650
	30	1735	1752
	40	1822	1850
	50	1904	1945

*Pt/Pt10%Rh thermocouples, corrected according to the data of Getting and Kennedy (1970) except for diopside where W3%Re/W25%Re thermocouples were used, corrected by intercomparison experiments with Pt/Pt10%Rh.

BIBLIOGRAPHY:

Akella, J., Vaidya, S.N., and Kennedy, G.C. (1969). The melting of sodium chloride at pressures to 65 kbar. Phys. Rev., 185: 1135-1140.

Bell, P.M. and England, J.L. (1967). High pressure experimental techniques, in Researches in geochemistry, Vol. 2, edited by P.H. Abelson. New York: John Wiley and Sons.

Bell, P.M. and Mao, H.K. (1971). Subsolidus reactions of jadeite ($NaAlSi_2O_6$) and albite ($NaAlSi_3O_8$). Carnegie Inst. Washington Year Book, 69:168-170.

Boettcher, A.L. and Wylie, P.J. (1968). Jadeite stability measured in the presence of silicate liquids in the system $NaAlSiO_4$-SiO_2-H_2O. Geochim. Cosmochim. Acta, 32:999-1012.

Boyd, F.R., Bell, P.M., England, J.L., and Gilbert, M.C. (1967). Pressure measurement in the single-stage apparatus. Carnegie Inst. Washington Year Book, 65:410-414.

Boyd, F.R. and England, J.L. (1960). Apparatus for phase-equilibrium measurements at pressures up to 50 kilobars and temperatures up to 1750°C. J. Geophys. Res., 65:741-748.

Boyd, F.R. and England, J.L. (1963). Effect of pressure on the melting of diopside, $CaMgSi_2O_6$, and albite, $NaAlSi_3O_8$, in the range up to 50 kilobars. J. Geophys. Res., 68:311-323.

Cohen, L.H. and Klement, W., Jr. (1967). High-low quartz inversion: determination to 35 kilobars. J. Geophys. Res., 72:4245-4251.

Cohen, L.H., Klement, W., Jr., and Kennedy, G.C. (1966-a). Investigation of phase transformations at elevated temperatures and pressures by differential thermal analysis in piston-cylinder apparatus. J. Phys. Chem. Solids, 27:179-186.

Cohen, L.H., Klement, W., Jr., and Kennedy, G.C. (1966-b). Melting of copper, silver, and gold at high pressures. Phys. Rev., 145:519-525.

Getting, I.C. and Kennedy, G.C. (1970). The effect of pressure on the e.m.f. of chromel-alumel and platinum-platinum 10%

rhodium thermocouples. J. Appl. Phys., 41:4552-4561.

Haygarth, J.C., Getting, I.C., and Kennedy, G.C. (1967). Determination of the pressure of the barium I-II transition with single-stage piston-cylinder apparatus. J. Appl. Phys., 38:4557-4564.

Haygarth, J.C., Ludemann, H.D., Getting, I.C., and Kennedy, G.C. (1969). Determination of portions of the bismuth III-IV and IV-V equilibrium boundaries in single-stage piston-cylinder apparatus. J. Phys. Chem., 30:1417-1424.

Heydemann, P.L.M. (1967). The Bi I-II transition pressure measured with a dead-weight piston gauge. J. Appl. Phys., 38: 2640-2644.

Kennedy, G.C. and La Mori, P.N. (1962). The pressure of some solid-solid transitions. J. Geophys. Res., 67:851-856.

Mao, H.K. and Bell, P.M. (1971). Behavior of thermocouples in the single-stage piston-cylinder apparatus. Carnegie Inst. Washington Year Book, 69:207-216.

Newton, M.S. and Kennedy, G.C. (1968). Jadeite, analcite, nepheline, and albite at high temperatures and pressures. Amer. J. Sci., 266:728-735.

Richardson, S.W., Bell, P.M., and Gilbert, M.C. (1968). Hydrothermal study of the kyanite-sillimanite transition. Amer. J. Sci., 266:513-541.

Richardson, S.W., Gilbert, M.C., and Bell, P.M. (1969). Experimental determination of kyanite-andalusite and andalusite-sillimanite equilibrium; the aluminum silicate triple point. Amer. J. Sci., 267:254-272.

Williams, D.W. and Kennedy, G.C. (1969). Melting curve of diopside to 50 kilobars. J. Geophys. Res., 74:4359-4366.

Yoder, H.S., Jr. (1950). High-low quartz inversion up to 10,000 bars. Trans. Amer. Geophys. Union, 31:827-835.

Yoder, H.S., Jr. (1952). Change of melting point of diopside with pressure. J. Geol., 60:364-374.

CHAPTER 8

Internally Heated Pressure Vessels

John R. Holloway

I. Introduction

Internally heated pressure vessels (IHPV) are finding increasing applications in the investigation of systems with a non-condensed phase. Although the concept of an IHPV is a general one, and several dissimilar types are in existence, this chapter will emphasize the description and use of relatively large volume, gas-media types capable of operation to pressures of 10 Kb. Also, while the examples described at the end of this chapter are primarily of studies in experimental geochemistry, the IHPV has obvious use in many types of chemical and physical problems.

In the following pages the individual components of the pressure system will be described with notes on their design and operation. Then some details of the operation of the system will be presented and finally examples of several types of experiments performed in an IHPV will be given. Few accounts dealing with the IHPV are to be found in the literature. Brief mention of them is made by Bell and England (1967) and descriptions of specific systems are given by Yoder (1950) and Goldsmith and Heard (1961). Burnham (1962) first published a description of the vessel type emphasized here and later presented a more detailed description (Burnham *et al.*, 1969).

As is the case with almost all forms of high pressure equipment, the basic concepts used in IHPV systems were developed by P.W. Bridgman. Bridgman's book (1949) and the references therein provide much pertinent information for the more advanced experimentalist. Good introductions to many of the techniques used are given by Bradley and Munro (1965) and Weale

(1967). Recent chemical research at high pressures is reviewed by Weale (1967), and Russian work at high pressures is comprehensively reviewed by Tsiklis (1965).

A. Capabilities and Limitations

Before comparing the IHPV discussed here with other high pressure systems, it must be suitably characterized. By definition an IHPV has a heating element(s) (furnace) contained inside the vessel and, therefore, under the same pressure as the sample. The particular vessel described by Burnham (1962) will be used as a model for the remainder of the chapter. Vessels of this type are available commercially[1] and are in use in several laboratories.

These vessels are capable of holding samples of 30 cm^3 volume at pressures of 10 Kb and temperatures to 1100-1500 C for long periods of time. Temperature gradients in the large samples are slight and pressure measurements accurate. For reasons given below, argon gas is used as a pressure medium; it transmits a perfectly hydrostatic pressure to the sample and does not react with materials exposed to it. The large sample size and the hydrostatic pressure distribution permit the use of complex and delicate apparatus inside the vessel, giving the system considerable experimental flexibility. The large sample volume is especially useful when studying the fluid (gas) phase. Unlike externally heated pressure vessels such as the cold-seal, the maximum working pressure of the IHPV is not dependent on temperature; thus, it performs equally well at 1500 C and 1-10 Kb. Moreover, the sample volume of the IHPV is much greater than that of a cold-seal vessel.

There are, however, some disadvantages to the IHPV and other types of pressure systems are preferable for some

[1] An annotated partial list of suppliers is given in Appendix 8-A.1 at the end of the chapter.

experiments. Specifically, the practical upper pressure limit for the simple IHPV is in the 10-15 Kb range, especially for the large volume apparatus. An IHPV has been developed for use to 30 Kb, but it is difficult to operate routinely, expensive to manufacture and has a small sample size (Birch _et al._, 1957). Thus, the best choice for most studies above 10 Kb is a piston-cylinder or other solid-medium pressure device (see Chapter 7). The IHPV system is moderately expensive ($15,000-$25,000) and requires the presence of an individual with good mechanical ability and technical competence to operate and maintain it. The latter point cannot be over emphasized; in most cases a potential operator should plan to spend at least a few weeks in a functioning IHPV laboratory before attempting to start his own laboratory. Thus, for routine investigations at lower pressures or temperatures in which large volume is not needed, the conventional cold-seal vessel or new molybdenum-alloy (TZM) vessels (Williams, 1966) are simpler to operate and less expensive than the IHPV. (See Chapters 4, 5, and 6).

B. General Description of Gas-Pressurizing Systems

Although very simple in principle, systems for compressing gases to high pressures are mechanically fairly complex, especially when compared to liquid pressurizing systems. This is because gases are very compressible, and, hence, a large volume of low pressure gas yields only a small volume of high pressure gas. Also, gases are hard to contain compared to liquids.

The most common method of compressing gases to high pressure is by means of an _intensifier_, which, as shown schematically in Figure 1 a.), consists of a large piston (usually driven by hydraulic oil at low pressure) acting on a small piston which compresses the gas. The gas pressure is greater than the oil pressure by the ratio of the area of the large piston to the area of the small piston. This ratio is called the

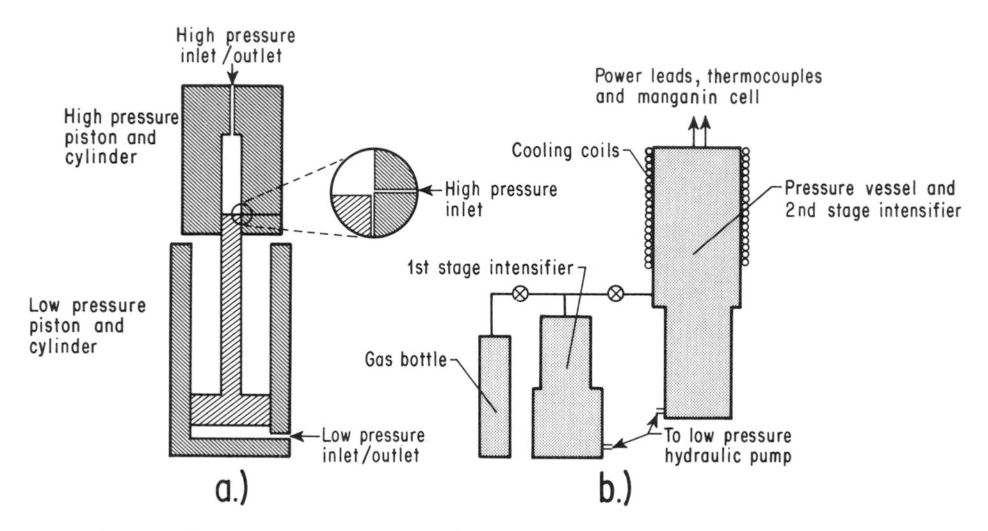

Figure 1. Intensifiers. a.) Schematic cross-section of a piston intensifier. Enlargement shows gas inlet used in integral intensifier systems. b.) Schematic representation of an integral intensifier system.

intensification ratio of the intensifier.

The complete pressurizing system consists of one or more intensifiers connected by a suitable system of valves and gages to one or more pressure vessels. In practice, pressure systems for the IHPV may be divided into two types: those in which the final stage intensifier is connected directly to the vessel, and those in which the final stage intensifier is separated from the vessel by a valve. The first type will be referred to as an integral intensifier system and the second type as a separate intensifier system. Although there is seemingly little difference between the two types there are sufficient differences in their capabilities to warrant the distinction.

1. The Integral Intensifier System

This is the type of system used by Yoder (1950) and in

the Bridgman-Birch apparatus (Birch et al., 1957; Newhall and Abbot, 1968). A schematic drawing of this type of system is shown in Figure 1 b.). The final intensifier and pressure vessel may be part of the same cylinder as in the Bridgman-Birch apparatus, or the final intensifier and vessel may be connected by high pressure tubing. The essential feature is the absence of a valve between the vessel and final stage intensifier. Valves for service at pressures near 10 Kb and above are often unreliable so it is an advantage not to have one at this point in the system. In place of a valve, a small hole in the intensifier wall allows gas to enter the intensifier and vessel when the intensifier piston is at the bottom of its stroke (see insert Figure 1 a.)). As soon as the intensifier piston travels past the hole, the vessel and intensifier are isolated from the low pressure portion of the system. These systems have two major disadvantages compared to the separate intensifier type: (1) The pressure must be achieved with only one stroke of the final stage intensifier and this limits the volume of the pressure vessel used because the volume swept by the intensifier piston must necessarily be small, and (2) one final stage intensifier is required by each pressure vessel which greatly increases the cost of additional vessels. However, the vessel used by Dr. Yoder has performed very reliably for many years and in some instances would be preferable to the system described below. A detailed description of his system has been given by Yoder (1950) and will not be presented here.

2. The Separate Intensifier System

In this system a valve is inserted between the final stage intensifier and the vessel. The addition of a valve introduces one more component which may fail under pressure, but it allows multiple stroking of the intensifier — and hence larger vessel volumes — and also allows additional vessels to be added at a lower cost. The system illustrated

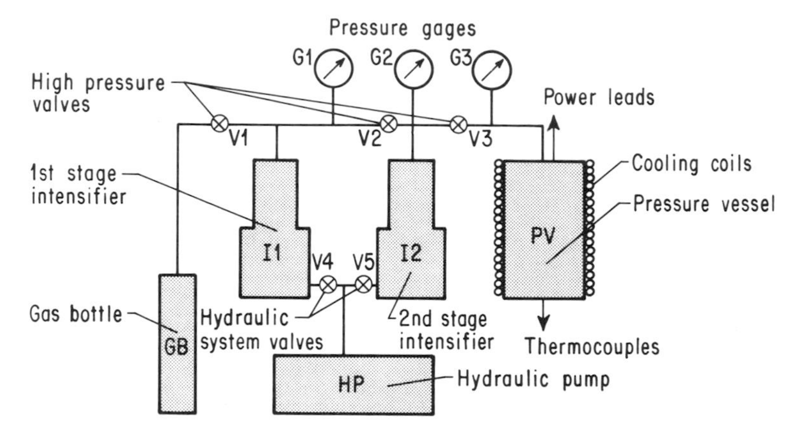

Figure 2. Basic 10 Kb internally heated separate intensifier pressure system.

schematically in Figure 2 will be used as a model for the remainder of the chapter. The steps involved in compressing argon from cylinder pressure to 10 Kb follow:

(1) Starting at GB (which is a commercial argon cylinder) valves V1, V2, and V3 are opened, allowing argon to flow into the first-stage intensifier I1, second-stage intensifier I2, and pressure vessel PV. At this time the P throughout the system is equal to that in the cylinder (about 150 bars).

(2) V1 is closed and V4 opened, transferring hydraulic oil from the hydraulic pump HP to I1, raising the piston in I1 which increases the argon P in I1, I2, and PV. When the I1 piston reaches the top of its stroke V2 and V4 are closed and V1 opened. Now the P is equal in I2 and PV (G2 and G3 indicate the same P) and it is somewhat higher than the argon cylinder P which is now in I1.

(3) Step 2 is repeated, except that V2 remains closed until the P indicated by G1 equals that of G2 and G3, at which time V2 is opened, allowing I1 to increase P in I2 and PV. This cycle is repeated until the P in I2 and PV has

reached the working limit of I1 or the desired P. Typically this P is 2-3 Kb and requires several strokes of I1.[1]

(4) Now with V1 and V2 closed hydraulic oil is directed to I2 by opening V5, raising the piston in I2 and increasing the P in I2 and PV. Argon is relatively less compressible above 2 Kb and P is increased rapidly by I2. When the piston in I2 reaches the limit of its stroke, V3 and V5 are closed, isolating PV from the system.

(5) V1 and V2 are now opened, allowing argon from GB to enter I1 and I2. The P in I2 is now increased to the working limit of I1, following the procedure of step 3 above.

(6) After closing V2 and V4, V5 is opened and the piston raised in I2 until the P indicated by G2 equals that of G3. Now V3 is opened and, as the piston in I2 continues to rise, the P in PV increases.

(7) After closing V3, steps 5 and 6 may be repeated as necessary to bring PV to the maximum working pressure of I2. Once the desired pressure is attained, V3 is closed and PV is isolated from the system and ready for the experiment.

The details of components in the model system will be given in appropriate sections below, along with various alternatives. It should be noted that additional pressure vessels can be added to the system by branching the line between G2 and V3 and adding V_3-G_3-PV units to each branch.

II. The Pressure System Components, Their Design and Operation

A. Pressure Intensifiers

The general aspects of piston intensifiers have been noted

---

[1] Small air-operated, diaphragm pumps with check-valve systems are available which will pump the system automatically to about 1.5 Kb, saving time and wear on I1 when GB pressure is low.

above. Diaphragm compressors are the other type of apparatus commonly used to generate high gas pressure, but they are barely able to deliver high pressure gas in the volume necessary to pressurize a large volume vessel in a reasonable time.

1. The Pressure Medium

The choice of a pressure medium is dictated by two factors: it must remain sufficiently fluid at high pressures to flow through the apparatus and it must remain chemically stable and unreactive under the conditions of pressure and temperature imposed on it. It is the latter factor which eliminates liquids from consideration for use at high temperatures. Silicone oils have been used at higher temperatures than any other liquid, but they decompose at temperatures above 500 C. Virtually all other liquids are either too reactive or unstable at even lower temperatures. In fact the only materials suitable for the pressure medium at temperatures above 500 C are the inert gases and nitrogen. Argon is the nearly universal choice for pressures below 12 Kb; however, at about 12 Kb it freezes at room temperature and so for higher pressures, either nitrogen must be used or all parts of the apparatus kept above room temperature. Unfortunately gas systems are much harder to make leak-free than are systems using a liquid pressure medium, and equipment which holds liquids at pressure perfectly often leaks gases at an unacceptable rate.

2. Piston Intensifiers

Piston intensifiers are manufactured in a variety of sizes and pressure ranges. An excellent review of intensifier design and operation has been given by Newhall (1957). Ideally one wants the largest volume output possible for a given pressure range, but economy may dictate a compromise. Thus, if a large volume intensifier is out of the price range, a smaller model may be used, but the time required to pump the system to the desired pressure will be increased. Even if personnel time is not a factor, working pressure systems almost invariably have slow leaks, so if the pumping rate is too slow the leak rate will equal the pumping rate and no headway will be made. A relatively fast intensifier, for pressures up to 2-3 Kb, would displace a one-half liter volume in about one minute, while an approximate minimum would be an intensifier capable of displacing the same volume in five minutes. For higher pressures, where argon is relatively incompressible, intensifiers can be of much smaller volume, 50 cm^3 being quite satisfactory and 20 cm^3 probably adequate.

In systems using high volume intensifiers, the hydraulic oil is usually delivered by an electric powered rotary pump at a maximum pressure of 200 bars. Thus, the intensification ratio must be rather high, in the order of 10:1 for 2 Kb intensifiers and 70:1 for 10 Kb intensifiers. On the other hand, hydraulic oil for the smaller volume intensifiers can be supplied by an air-operated pump at pressures of 1-3 Kb with correspondingly lower intensification ratios.

3. Diaphragm Compressors

Diaphragm compressors are available for gas service with maximum working pressure of 2 Kb. They have much smaller volume output than the larger piston intensifiers but are comparable in volume output to some of the smaller commercially available piston intensifiers. Diaphragm compressors do have some advantages; however, they are less expensive than piston intensifiers and, more importantly, they do not contaminate the gas they are compressing with lubricant, which is nearly unavoidable with piston intensifiers. Thus, they may be used to compress gases which must be kept chemically pure, and they can pump corrosive gases. A fairly comprehensive description of diaphragm compressors may be found in Wolf and Bowen (1957).

4. Multiple-Stage Intensifiers

As stated implicitly above, it is usually desirable to combine two or more intensifiers into a high pressure system. Use of multiple intensifiers allows combination of features to maximize pumping rate and reliability and minimize cost for a given pressure range. In general the specifications of the intensifier are matched to the properties of the argon gas as closely as possible. Thus, in the low pressure range, a large volume intensifier with a relatively low pressure working limit should be used. In the high pressure region, where argon is less compressible, a small volume intensifier is sufficient.

In the model system (Figure 2) two intensifiers are used. The first stage intensifier displaces about 500 cm^3 per stroke and has a practical working limit of about 2 Kb. The second stage intensifier has a displacement of about 80 cm^3 per stroke and a working limit of about 12 Kb. The model system pressure vessel can be pressurized to 10 Kb in about 20 minutes with the above intensifiers These intensifiers are expensive, however, and smaller components may be substituted by sacrificing pumping rate.

B. Pressure Vessels

A schematic illustration of a large volume IHPV is shown in Figure 3. In principle the vessel is a very simple device. It

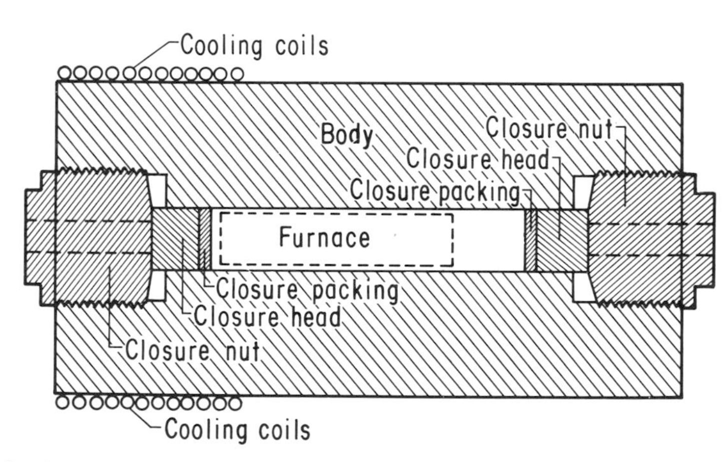

Figure 3. Longitudinal section of a large volume internally heated pressure vessel.

consists of the steel body, closure heads, closure nuts, and coils for cooling water wrapped around the body. A furnace is shown in position inside the vessel. Pressure seals are made by suitable packings between the closure head and vessel body. In the model system vessel, the working space (the pressurized volume inside the vessel which may be used for a furnace and/or other apparatus) is 5 cm in diameter and 45 cm long. The o.d. of the vessel is 30 cm and the length is 90 cm. Details of the vessel components will be presented below.

1. The Vessel Body

The factors involved in pressure vessel design are covered in the general references given at the beginning of this chapter and by Comings (1956) and will not be considered here. For pressures up to 10 Kb vessel cylinders machined from a single steel forging are satisfactory. From theoretical considerations the ratio of the external to internal radii has been shown to be a critical factor in determining the maximum working pressure; the other major factor being the nature of the steel itself. The actual ratios used have been determined mostly by trial and error and are satisfactory for pressures up to 10 Kb. For vessels with a 5 cm i.d., intended for 10 Kb service, a ratio of at least 5:1 is chosen, resulting in a minimum o.d. of 25 cm. However, a 6:1 ratio is usually used for an extra margin of safety. A 5 cm i.d. vessel, therefore, requires a large steel body which is relatively difficult to machine, and hence, expensive. Many experiments do not require so large a volume, and vessels with a 2.5 or 3.5 cm i.d. would have considerably smaller overall dimensions and be less expensive.

Vessels can also be constructed with only one end open. This design is somewhat less expensive, but may be more prone to failure because impurities often concentrated at the center of steel forgings are not removed since the vessel bore does not penetrate the forging completely. Versatility is also impaired in a vessel with a permanently closed end as all lead-ins must be carried through one closure plug.

Alloy steels are available which may be heat-treated to the necessary strength for the vessel body. They should be vacuum-melted in a single billet to reduce impurities. Starting with a steel of sufficient strength, it is most important that the steel be very uniform and contain no defects. The heat treatment must be uniform throughout the body and must be carefully controlled so the steel has just the right amount of hardness and ductility. Very great strength may be achieved in some steels by heat-treatment to great hardness, but pressure vessels of such hardness are prone to brittle fracture and often fail after only a few cycles to high pressure (they fatigue rapidly). Vessels made of relative soft steel can be cycled repeatedly without failure, and usually deform by plastic flow rather than explosive brittle fracture. The steel in the model system pressure vessel is vacuum-melted (CVM) modified SAE 4340,[1] heat treated to Rockwell C-40 to 42 in hardness.

Cooling of the vessel is necessary to maintain the strength of the interior portions when a high temperature furnace is used. It is easily accomplished by wrapping the exterior of the vessel with copper tubing through which cold water is circulated or by constructing a cylindrical jacket with appropriate baffels to force circulation of the coolant.

2. Closure Plugs

The closure plug is perhaps the most important single component of the entire pressure system. It seals the ends of the vessel, contains a pressure port through which argon flows to create the pressure, and contains the electrical leads which allow communication between the vessel interior and the outside world. Each of these functions must be faultlessly performed for an experiment to work.

a. Design of Packing

All currently employed high pressure packings are adaptations of the original Bridgman unsupported-area seal. Because of the importance of understanding the concept of this packing,

---

[1]One composition that has proven satisfactory is: C = .35%, Si = .75, Mn = .95, Ni = 1.85, Cr = .80, Mo = .35, and V = .20.

its operation will be discussed in considerable detail. Figure 4 a.) is a cross-section of a simple Bridgman seal. In this

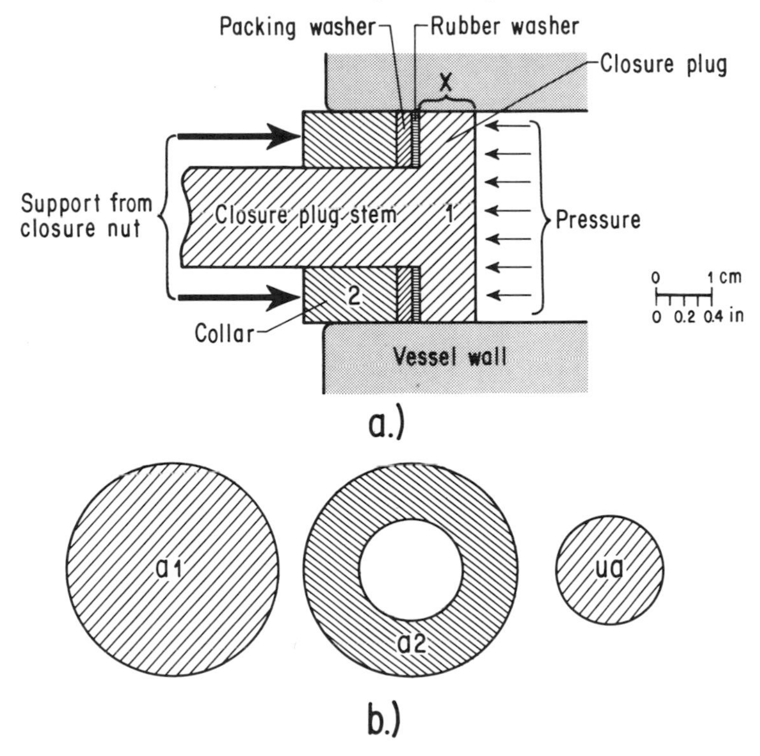

Figure 4. Bridgman unsupported area seals. a.) Longitudinal section. b.) Cross-sectional areas.

arrangement, a collar is supported from the left side by the closure nut (not shown here, see Figure 3). Next there is a packing washer made of a ductile metal such as copper which supports and prevents extrusion of a rubber washer that forms the seal. These washers are sandwiched between the collar and the closure plug. Before pressure is applied to the vessel, an <u>initial seal</u> must be formed by mechanically squeezing the washer outward against the vessel wall and inward against the stem of the closure plug. This is done either by pulling back on the stem from outside the vessel with the aid of a draw tube or by pushing the head against a shoulder in the pressure vessel.

Now when pressure is applied to the vessel, the pressure is exerted over the area of the face of the closure plug, area a1. The force acting on the closure head (trying to push it out of the vessel) is equal to the cross-sectional area multiplied by the pressure. This force is transmitted to the washers and the washers transmit it to the collar. The closure nut which backs up the collar prevents the entire assembly from being blown out of the vessel. Note, however, that the stem of the plug which extends through the collar is <u>not</u> supported by anything, so only the annular area of the collar is available to support the force. If the area of the plug is a1, the area of the collar is a1 - ua = a2, where ua is the area of the stem (the areas are shown schematically in Figure 4 b.)). Because the assembly is static the force (F) acting on a1 is P times a1 and is equal to the force acting on a2. The pressure in the washers may be calculated because they have the same cross-sectional area as the collar, a2. This pressure is $P' = F/a2 = \frac{P \cdot a1}{a2}$. Because a2 is smaller than a1, P', the pressure exerted on the washers will be greater than P, the pressure of the gas in the vessel. This additional pressure ($\Delta P = P' - P$) tends to squeeze the ductile washers out against the vessel wall and in against the closure head stem, thus maintaining a gas tight seal across those surfaces. The reason the seal works so well is because the higher the gas pressure, the greater the additional pressure, ΔP, and the more tightly the washer is pressed against the enclosing surfaces. Remembering that the ΔP exists because the supported area a2 is less than a1 and this arises because the stem of the closure plug is not supported by the closure nut, the derivation of the name <u>unsupported-area</u> seal becomes clear. The ratio of P'/P is directly proportional to a1/a2 or a1/(a1 - ua). The larger the unsupported area ua, the greater the ratio P'/P, and in theory at least, the better the seal. In practice the ratio P'/P must be limited or else P' becomes so large that the ductile washer deforms the pressure vessel wall or pinches off the stem of the closure plug, or both.

The resistance of the closure plug stem to pinching off is a function of the shear strength of the steel and the dimension "X" in Figure 4 a.). The X-dimension is not usually calculated, but is based on experience and intuition.

The simple Bridgman seal is highly reliable and, because of its simple construction, relatively inexpensive. It is not a good seal for large diameter pressure vessels, such as those dealt with in this chapter, because of its tendency to deform the vessel wall. Also, it is extremely difficult to remove a closure plug using a simple Bridgman seal after a high pressure run. This is because the ductile washer remains permanently deformed against the vessel wall.

To overcome the difficulties with the simple Bridgman seal, several modifications of the design have been made. One of the most common is the inverted Bridgman seal, shown in Figure 5 a.). This seal, only slightly more complex than the simple Bridgman, consists of the closure plug supported by a closure nut, which is not shown, and a series of washers and an o-ring. The first washer is made of hardened steel and is beveled on its inner diameter to match a bevel on the closure head. A soft lead washer is placed in front of the steel, followed by an o-ring of rubber or some other elastic material. A retainer ring is screwed onto the closure plug to hold the others in place. An initial seal is made by squeezing the o-ring between the closure plug and the vessel body. This may be done by making the o-ring slightly larger than the vessel i.d. and forcing the o-ring into the vessel along a slight taper at the mouth of the vessel. Once the initial seal is made, gas pressure is exerted over the face of the closure plug and the annular area of the o-ring. As the gas pressure in the vessel increases, the o-ring is forced back against the lead ring deforming the lead and squeezing it against the vessel wall and the closure plug. As pressure increases further, the o-ring shrinks until finally gas is able to leak past the o-ring, but by this time the lead washer has formed a gas-tight seal and is

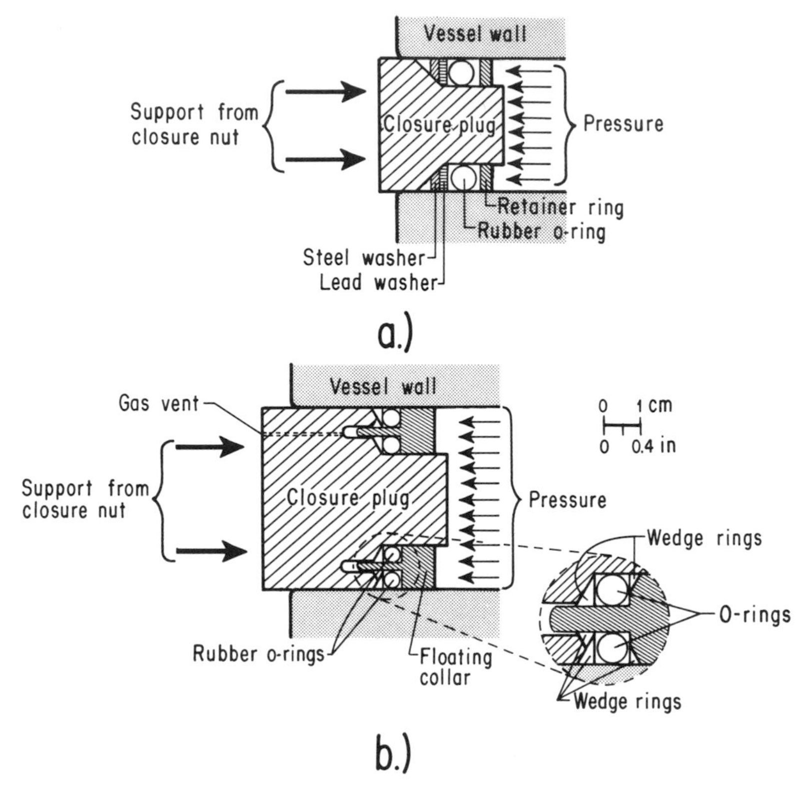

Figure 5. Modified Bridgman seals: a.) Inverted Bridgman. b.) Burnham.

transmitting the force exerted by the gas to the steel washer. This force is applied over the annular area of the front face of the steel washer. Only that part of the steel washer in contact with the sloping part of the closure plug is supported, and an unsupported area exists where the steel washer is not in contact with the plug. As in the simple Bridgman seal, the unsupported area causes pressure in excess of the gas pressure to be exerted on the steel ring, forcing it out against the vessel wall. This seal works well as long as the lead washer forms a gas-tight seal, but because the lead is not unsupported to any degree, failure of the lead washer seal is more common than in the simple Bridgman seal.

A seal that combines the advantages of the unsupported area

principle with ease of operation and reliability has been developed and used very successfully by C.W. Burnham (private communication, 1970). This seal is similar in some respects to one described by Vodar and Saurel (1963). The seal is shown schematically in Figure 5 b.). The closure plug is supported by the closure nut (not shown). Two rubber, or other elastic, o-rings of the same width are contained between the closure plug and a floating collar. Extrusion of the o-rings is prevented by five mild-steel wedge rings (see inset) having 30-60° cross-sections. A gas vent is provided to maintain the space in the groove beneath the floating collar at atmospheric pressure. When pressure is applied to the vessel the floating collar is forced against the o-rings, squeezing them against the vessel wall, the closure plug, and the floating collar. The pressure exerted on the o-rings by the floating collar is greater than the pressure in the vessel by the ratio of the annular area of the floating collar to that area minus the area of the "tongue" of the floating collar (the unsupported area). This overpressure prevents the o-rings from shrinking away from the vessel wall or the closure plug and maintains the seal to pressures of at least 10 Kb. To operate properly, the dimensions of the closure plug and floating collar must be adjusted so that the force on each o-ring is the same. On the release of pressure at the end of a run, the o-rings return to their initial configuration and the closure plug is easily removed from the vessel.

b. Electrical Leads

In practice there are two very different current-carrying requirements placed on electrical lead-throughs. Power leads supply electricity to the heating elements and typically must carry 30-40 amperes of current. On the other hand, leads for thermocouples or other measurement devices need conduct only very small currents.

In order to conduct high current, power leads must be relatively large physically and they should be made of metal with a high conductivity. At the same time they must have high mechanical strength to maintain their shape under high stress. Steel has often been used but beryllium-copper alloys are superior in most cases. A commonly used power lead design is shown schematically in Figure 6. A tapered hole is machined in the closure plug with the largest end facing the vessel interior. A hollow pyrophyllite cone fits into the tapered hole (typically 8°), and the cone-shaped portion of the power lead covered with an Al_2O_3 coating fits into the pyrophyllite cone. The Al_2O_3 coating, applied by plasma-jet, is quite hard and continues to insulate the power lead after, as often happens, the lead has been pushed through the pyrophyllite. However, the pyrophyllite sleeve is porous and the pores must be filled with a silicone-rubber cement before applying pressure to the lead. The Al_2O_3 coating also is porous so a metal boss is left on the power lead which seals against the pyrophyllite sleeve, thereby providing a gas-tight barrier in the Al_2O_3 coating. The initial seal is provided by exerting force on the cone with a teflon sleeve acted upon by a gland nut (not shown). For operation at pressures less than about 6 Kb the Al_2O_3 coating and metal boss may be eliminated.

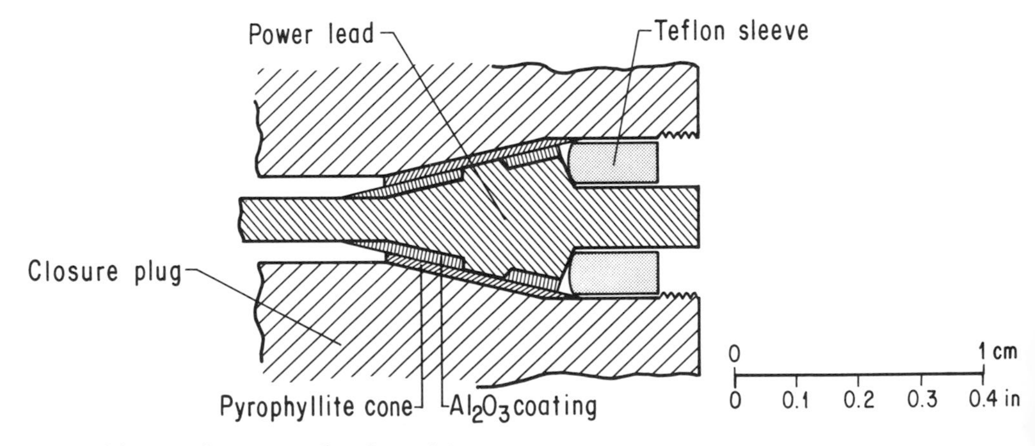

Figure 6. Power-lead packing.

Typically a beryllium-copper power lead designed for 30 ampere, 120 volt service is about five mm diameter at the large end of the cone. These are relatively large and only a few may be placed in the limited space available in the closure head.

Leads for low power purposes, such as electronic sensing devices, may be very small and large numbers of individual conductors are easily placed in a closure head. A good review of several types of construction is given by Downs and Payne (1969). They recommend the sheath-type lead, in which one or more fine wires (either thermocouple, compensated or ordinary copper) are enclosed in an MgO or Al_2O_3 packed metal sheath. These sheathed leads are available from several manufacturers listed in Appendix 8-A.1. In the manufacturing process the metal sheath is tightly swaged onto the MgO filler with the result that the conductors are tightly held in place by friction. The filler is porous, however, and the pores must be filled by some impervious material before the arrangement will seal

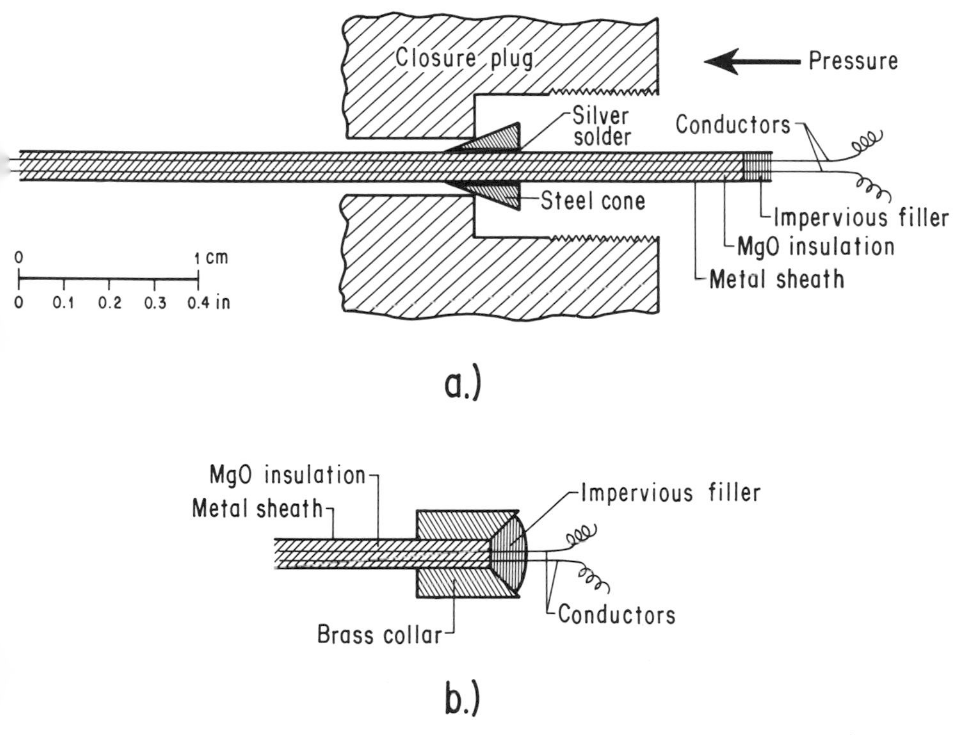

Figure 7. Sheathed electrical leads and seal. a.) Sheath and seal. b.) Alternate sheath termination.

at high pressure. A sketch of a possible arrangement is shown in Figure 7 a.). The conductors are shown passing through a metal sheath, usually 1/8 or 1/16 inch in diameter, which is filled with MgO throughout most of its length. The impervious substance is placed in the end of the sheath facing the interior of the pressure vessel after grit blasting the MgO out to a depth of about 1/8 inch, or alternatively a brass collar can be soldered on the sheath to hold the cement (Figure 7 b.)). Downs and Payne (1969) recommend epoxy cement for the impervious media, but they note that it tends to fracture in service, often breaking the conductor. C.W. Burnham (personal communication, 1969) has used a silicone rubber cement for the impervious media with good success. The seal between the sheath and the closure plug is made by a hardened steel cone attached to the sheath by silver solder. The cone is forced against a shoulder in the closure plug by a nut (not shown). During the initial application of high pressure to the cone, the cone deforms so the nut needs to be retightened before a second application of pressure; thereafter the seal can be expected to work perfectly for long periods of time.

c. Thermocouples

For measurement of temperatures in the range of chromel-alumel thermocouples, sheath-type couples with high purity ceramic insulator-filler (see Chapter 9) are generally much better than the exposed types. The sheath-type of construction and the method of sealing the sheath to the head was discussed in the previous section. The sheathed thermocouple differs from the sheathed lead-through in that the end of the sheath in the thermocouple is welded shut, thus making the sheath impervious to the pressure medium. Many types of thermocouple and sheath material may be obtained in sheathed construction; however, the sheath itself must be usable at the temperatures measured by the thermocouple. For temperatures to at least 1100 C sheaths made of the nickel alloy, "Inconel 600," work well. When used with chromel-alumel thermocouples, the Inconel sheath may be welded integrally with the thermocouple, simultaneously sealing the sheath and forming the thermocouple junction. This construction has proven very successful in many long runs at 10 Kb, although occasionally a bad thermocouple is encountered which leaks immediately or after only a few applications of pressure. In addition to the ease of making a pressure seal through the closure head when using a sheathed thermocouple, the sheath makes a mechanically strong thermocouple which is easy to handle in routine operation.

Unfortunately, when platinum-type thermocouples must be used, either the sheath must be of platinum alloy or the sheath-end and junction must be welded separately. While both of the methods have been used, neither is as reliable as the chromel-alumel junction welded integrally with an inconel sheath. In addition, platinum alloy sheaths are expensive and fragile. Thus, it is probably preferable to use unsheathed platinum thermocouples and the method described in the previous section to bring the leads through the closure head.

d. Pressure Ports

In most modes of operation there is only one pressure port in the IHPV and it allows the gas pressure medium to enter or leave the vessel. The simplest type of pressure port is a small diameter hole bored through the closure plug with a place to attach a cone pressure fitting (discussed in the next section) on the end of the plug facing atmospheric pressure. In some instances this simple type of pressure port cannot be used due to limitations set by the design of the closure plug. In these cases a piece of 3/16 inch (.47 cm) high pressure tubing, threaded at one end, is fitted with a 5/16 inch diameter sleeve that is silver soldered to the threaded tube. This sleeve is then placed in a counter bore in the closure plug and sealed with soft solder.

For some types of experiments it is necessary to have means to introduce liquids or gases other than the pressure medium into the vessel interior. This may be done by replacing the sheath shown in Figure 7 with a piece of thick-walled capillary tubing, or tubing may be connected to cone fittings on either face of the closure head as described above.

In most designs, as in the model system, the furnace power leads and gas-medium pressure port are placed in one closure plug and the thermocouple leads or sheathed thermocouples are placed in the other plug. The model system uses the Burnham closure head design as shown in Figure 5 b.). Two power leads, a pressure port made by soft-soldering a piece of threaded, high pressure tubing, and one sheathed, chromel-alumel thermocouple (used for temperature control) are placed in one head. The other head contains three sheathed, chromel-alumel thermocouples. A longitudinal section of a completely assembled pressure vessel is shown in Figure 1 in Burnham et al. (1969).

3. Valves, Tubing and Fittings

High pressure tubing and fittings are easily obtainable from the suppliers listed in Appendix 8-A.1. Standard fitting types available include crosses, tees, elbows, straight couplings, reducing couplings and adapters. For pressures up to 10 Kb the simple cone fitting works satisfactorily. As shown in the sketch (Figure 8 a.)) a cone is formed on the end of the pressure tubing on which

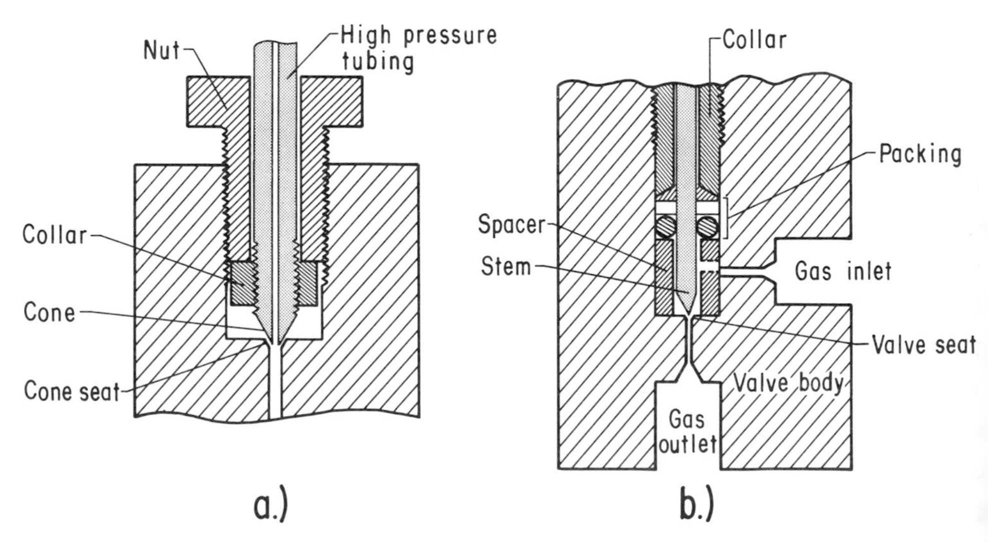

Figure 8. Cone seal on tubing and in a valve. a.) Cone seal on tubing. b.) High pressure valve.

left-hand threads are machined to accept a collar. A right-hand threaded nut transmits force to the cone through the collar, the force being adequate to hold the cone against the seat in the fitting, making the pressure seal. Cone fittings are easily disconnected and reconnected, but repeated tightening of the fitting deforms the cone on the relatively soft stainless steel tubing and the cone must be reformed. Simple hand tools are available for this purpose; an improved coning tool design has been suggested by Davis _et al._ (1969).

The pressure tubing is commonly a series-300 stainless steel; it is thick-walled, the o.d. and i.d. depending on the maximum working pressure desired. Tubing with a working pressure of about 6.9 Kb is available in the common 1/4 inch o.d. size with a 0.0625 inch i.d., and a 3/16 inch o.d. tubing with a 0.025 inch i.d. is available with a working pressure of about 14 Kb. The small i.d. of the 3/16 inch o.d. tubing does not restrict gas flow as long as care is taken not to pinch off the opening at the cone. Of course, small particles of dirt easily clog the tubing so care must be used to keep the system clean.

Valves for use with gases at pressures near 10 Kb have seldom worked as they were intended. Either the valve packing would not hold the high pressure or the valve needle and/or seat would fail after a few openings and closings at high pressures. The reasons for this may become apparent by observing the general schematic of a high pressure valve shown in Figure 8 b.). The essential elements of a needle valve, which is by far the most commonly used type, are the steel valve body, the valve seat in the valve body, the valve stem, the gas inlet, outlet, and the packing. The packing is supported on the pressure side by a spacer and on the exterior side by a collar. Simple o-ring packings are often used at lower pressures, but for pressures above 5 Kb some type of unsupported-area packing must be used. The valve is closed by forcing the cone-shaped stem against the seat. When closed, the gas on the outlet or high pressure side of the valve is confined by the tip of the stem and the valve body.

Once the stem is seated properly there is little chance of leak on the high pressure side (gas outlet) and indeed many commercially available valves will hold 10 Kb on the high pressure side. The trouble comes when high pressures must be held on the low pressure side (gas inlet) of the valve. Gas in this area is confined by the packing around the stem. But the stem must move linearly to open and close the valve and it is difficult to design a packing that is tight enough to contain the gas and yet does not bind the stem so tightly that it is immobilized at high pressures. With manually operated valves it is often so difficult to move the stem through the packing when at high pressure that one cannot tell when the stem is seated. When this is the case, much unnecessary force is placed on the stem and seat, resulting in a broken stem and/or cracked seat. Two innovations have recently been made to improve the situation (C.W. Burnham, personal communication, 1969). Because the rubber o-ring was the packing element which appeared to be binding the stem, teflon sheaths were placed around the o-ring. Teflon has a very low coefficient of friction and makes it much easier to move the stem. To prevent application of excess force to the stem, a large spring which has just the proper force is used to close the valve, a suitable mechanism being used to open the valve against the spring. These modified valves may be opened and closed very gradually at high pressures, permitting small pressure adjustments to be made easily.[1]

4. Heating Devices

The heating devices (furnaces) for use in an IHPV which will be considered below are all of the helically-wound, resistance, element type. A number of designs are available depending on maximum temperature and hot spot volume requirements. A cutaway view of a furnace and a cross-sectional view are both shown

---

[1] These valves are manufactured currently by M. Wilson, Port Matilda, Pennsylvania 16870.

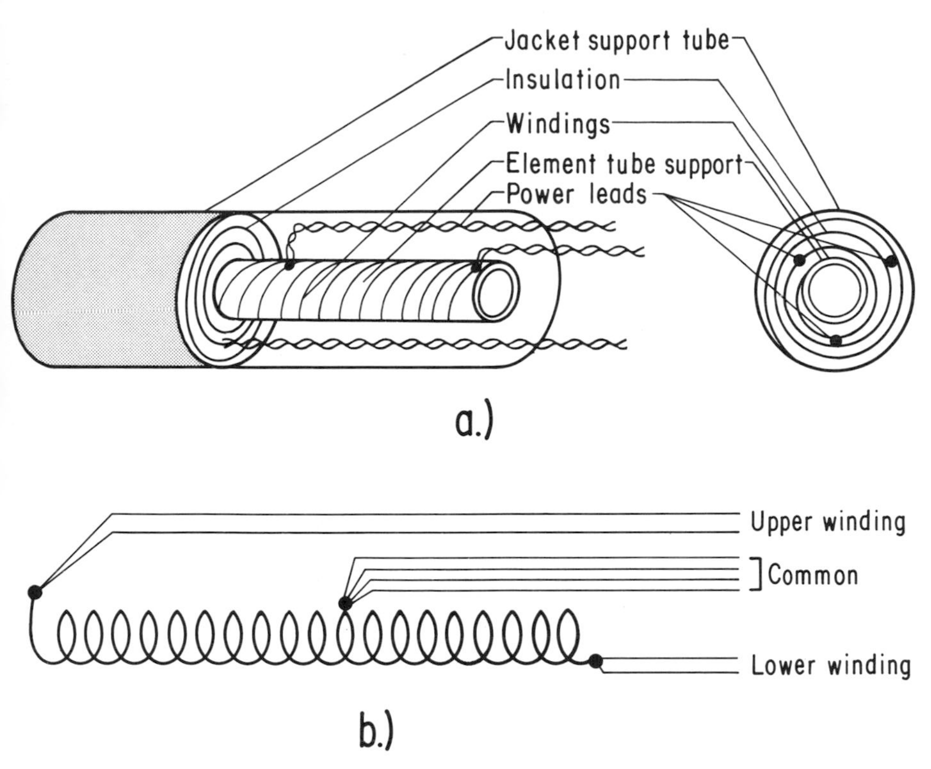

Figure 9. Resistance furnace construction. a.) Cutaway and cross-sectional view. b.) Resistance element connections.

in Figure 9 a.). The diagram shows: the jacket support tube which attaches to the closure head at one end, high temperature fiber insulation, element windings, element support tube and power leads.

a. Design Considerations

Given a limited volume in an IHPV to work with, an obvious general rule is: the higher the maximum working temperature, the smaller the hot spot volume. This rule is dictated by the limited power-handling capacity of the heater elements and the necessity to keep the inner wall of the pressure vessel below about 300 C. To meet these demands requires that the amount of furnace insulation increase as hot spot temperature increases.

To fully utilize the capabilities of the IHPV, the temperature distribution must be closely controlled throughout the largest possible hot spot volume. This means minimizing temperature gradients in equilibrium runs or controlling gradients in steady-state runs. In gas medium apparatus, the temperature distribution is established by the convecting gas. To control the convection and achieve a radially symmetrical gradient, the furnace should be operated so that its long axis is vertical.[1] Also, there should be as few open spaces inside the vessel as possible so the flow of gas will be restricted. (See Chapter 6 also)

Changes in pressure and temperature change the density of the gas medium and this causes the position of the hot spot to change. For example, if the hot spot is at the middle of a furnace at 900 C and 2 Kb, it may be 2-3 cms higher at 900 C and 5 Kb. In order to control the position of the hot spot, and also increase its length, there must be at least two separately powered resistance elements in the furnace. The elements are placed one above the other vertically so the vertical temperature distribution may be changed by changing the relative amounts of power to the upper and lower windings. Three elements in a row would be even better, but would require more power leads and complicate the furnace construction; for most purposes two elements are sufficient. Two resistance elements require only two power leads in the closure head if one end of each element is tied together forming a common ground, as shown in Figure 9 b.). The common is then connected to the vessel so the vessel itself forms the third power lead. This procedure makes the vessel electrically "live" and, even if at ground potential, undesirable ground loops may develop when using sensitive electronic equipment. A preferable alternative is to connect the common to a third power lead, leaving the vessel "floating."

The power requirements of a furnace are determined by insulation efficiency, hot spot size, pressure and temperature. The effects

---

[1]The pressure vessel is mounted on trunions in the support frame which allow rotation from a horizontal assembly position to a vertical operating position.

of pressure and temperature on power consumption in a nitrogen pressure medium are shown graphically by Voder and Saurel (1963) and are similar to the effects in an argon medium. In a vessel with a 2.5 cm i.d. and a furnace with an overall length of 30 cm, a small i.d. furnace (say 1.3 cm) with a 5 cm long hot spot will require from 1000-2000 volt-amps (VA) of power. In a similar vessel, a furnace with a 3.13 cm i.d. (the largest possible) with a 7.5 cm hot spot requires from 3-5 VA. These power requirements and the limited space available for the element windings result in a surface loading of the elements (watts of power/element surface available) considerably greater than the recommended maximum. The high surface loading decreases the lifetime of the elements; nevertheless continuous operation for longer than a year is common. A bifilar element, consisting of two wires twisted together loosely, has about 20% greater surface area per unit resistance than a single wire. Hence, the surface loading for a given power input is about 20% lower and elements constructed in this manner may have a longer lifetime. Selection of the optimum furnace i.d., element material, wire size and number of turns is a complicated process. Once maximum power, voltage and amperage are selected, a functional combination of the above variables is usually determined by considering the resistance per unit length of wire and the total length of wire which can be wound on a furnace of a given size. The elements should be spaced no closer than one wire diameter. Larger diameter wire is preferable when possible because of its greater strength and resistance to oxidation. Specific examples of such calculations will be given below.

b. Construction

With the above requirements and restrictions in mind, step-by-step details of furnace construction are given in this section.

(1.) Sheathed Elements

For operation at temperatures below 900 C the furnace elements may be nichrome, clad in a stainless steel or

Inconel sheath with the same construction as sheathed thermocouples. The insulation between the element and sheath should be Al_2O_3 rather than MgO because of the greater thermal conductivity of Al_2O_3. The sheath gives the windings good mechanical strength and results in trouble-free electrical insulation. The following design was generously provided by Professor G.M. Anderson of the University of Toronto.

A 0.63 cm (1/4 inch) sheathed, B & S gauge nichrome element is wound tightly around a 2.5 cm o.d. x 30 cm long stainless steel winding support tube. Each winding is 15 cm long. The four ends of the winding are brought back to one end of the furnace. To restrict gas convection, that end of the stainless steel tube is welded shut. Gas is prevented from entering the elements by silver soldering a brass collar filled with silicone rubber cement to the element ends (see Figure 7 b.)). One wire from each element is tied together to form the common lead. Now the furnace is wrapped with a high-temperature fiber or cloth insulation until it will just fit into a stainless steel jacket support tube. Each winding has a cold resistance of about 4 ohms and requires about 500 VA at 650 C, 5 Kb. Operating temperatures above 900 C could be achieved by using Kanthal elements and Inconel sheaths, but at the higher temperatures the unsheathed type of construction is probably superior.

(2.) Unsheathed Elements

Unlike the sheathed elements discussed above, the unsheathed element wire must be mechanically supported and electrically insulated. The maximum working temperature is dictated by the physical characteristics of the element and the amount of insulation.

For furnaces of about 1.9 cm i.d. or smaller, commercially available alumina tubes may be used to support the elements and electrically insulate them from the sample holder. These tubes must be coarse-grained alumina with the highest porosity available to prevent entrapment of argon in small pores while at high pressure and temperature, causing subsequent rupturing of the tube on release of pressure. The alumina tubes may be obtained with or without pre-cut grooves for the elements. On ungrooved tubes the windings may be made on a lathe. Another method for ungrooved tubes is to use a piece of soft wire of the same gauge as the element to space the element windings. With any of these methods, after the windings are completed, the element is tied down at the ends, a thin layer of alumina cement is applied and allowed to dry. For the soft wire spacer method, the soft wire is pulled off the tube, and a second thin layer of cement is applied to the windings.

For example, a furnace with a 1.55 cm i.d., 27.5 cm long is constructed in a grooved alumina core with ten turns per inch, the grooves being large enough for 16 gauge wire. Each 13.75 cm winding contains 356 cm of wire which would have a resistance of about 4 ohms if it were 16 gauge Kanthal A-1. (So at a maximum of 120 volts, each winding would carry about 30 amperes producing 3.6 VA power.) The core is wound and the leads in the center tied together for the common. The windings are completely covered with a thin layer of alumina cement and leads long enough to reach the power leads at the closure plug are welded on to the windings. To keep the leads from heating up, a double strand of winding wire should connect the winding ends and four strands should connect the common lead. The core is now wrapped with several layers of high temperature cloth insulation and then the leads brought back to one end of the tube, making sure to provide electrical insulation between the leads and the windings (alumina "spaghetti" may be used to cover the leads for additional protection). Insulation is wrapped over the leads until it will just fit inside the support jacket tube. The support tube, a piece of 5 cm o.d. stainless steel tubing with a 0.15 cm wall, attaches to the power-lead closure plug at one end and the open end of the furnace is flush with the other end of the support tube. The power leads from the furnace are attached to the power leads in the closure plug with mechanical couplings. Large slots cut in the support jacket tube allow access to the couplings when attaching the furnace to the closure plug. This furnace will operate at 1200 C with Kanthal A-1 windings and above 1500 C with platinum windings.

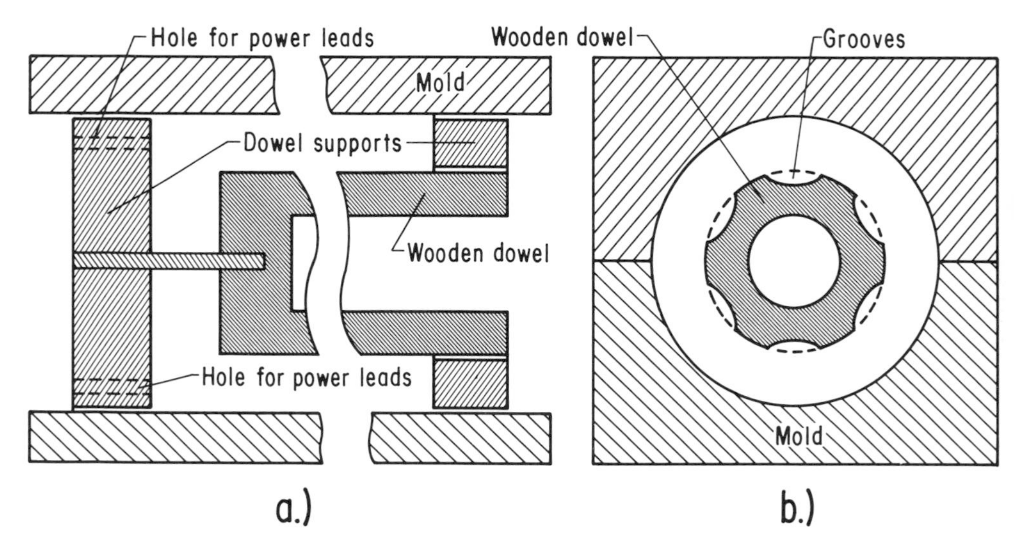

Figure 10. Interior and exterior molds for furnace casting. a.) Longitudinal section. b.) View of sample end.

Special construction is necessary to obtain the maximum possible furnace i.d. (about 3.13 cm) and operating temperatures of 1000 C. Only 0.95 cm is available between the winding and the pressure vessel wall so to make the windings as efficient as possible and leave the maximum space for insulation, no winding support tube is used. Instead the platinum elements are helically wound on a 3.45 cm diameter wooden dowel of straight-grained soft pine. Six 0.23 cm deep, rounded, longitudinal grooves are cut in the dowel as shown in Figure 10. The dowel is soaked overnight in water before winding the elements. After the elements are wound, they are tied down at each end with the same wire as the winding. The windings are covered with alumina cement, making sure to press the cement through the windings into the grooves in the dowel. Two layers of silica-glass cloth (or matte), soaked with a thin slurry of alumina cement, are wrapped around the furnace. Power leads, covered with alumina spaghetti, are welded to the elements and laid against the silica cloth. Two to five additional layers of cement-soaked silica cloth are wrapped over the furnace until it has a diameter slightly greater than two inches. The furnace is pressed between the halves of a split cylindrical mold with the same diameter as the inside of the furnace support tube. The cylindrical mold is clamped in position and dried slowly (takes 24-48 hours). It is helpful to blow a gentle stream of air into the hole in the dowel. Once dry, the furnace is removed from the mold and placed in a cold firing kiln. Often the wooden dowel will have shrunk sufficiently to remove before firing, but if not, the kiln may be heated slowly to 350-400 C and the dowel will burn out if enough oxygen is available. Once the dowel is gone the kiln temperature is slowly raised to 1300 C, held there for 8-16 hours, and then slowly cooled (the firing cycle takes about three days). The alumina cement which filled the grooves in the dowel now forms ribs along the inside of the furnace which prevent contact of the sample holder and windings. In order to save space, a thin brass tube is used instead of stainless steel for the jacket support tube. If the overall length of the furnace is 30 cm and each winding is 15 cm long and there are 48 turns per cm of winding, then each element contains 685 cm of wire. If .0625 cm platinum wire is used to make a bifilar element, each element will have a cold resistance of about 1.4 ohms. This furnace will heat a hot spot 3.13 cm in diameter and 7.5 cm long to 950-1000 C continously.

C. Safety Precautions

1. Personnel

A pressure of 10 Kb is about 3-5 times the chamber pressure in a high powered rifle, and the potential energy stored in a gas medium IHPV at 10 Kb is considerable. Therefore, adequate mechanical shielding and use of common sense by operating personnel is essential. The greatest danger is from failure of small devices, such as power leads and thermocouples, which become high velocity projectiles with high penetrating power. Failure of the vessel itself involves movement of much larger pieces which have little actual penetrating capacity. Shields for the pressure vessel and intensifier should be made from steel plate at least 0.6 cm thick and positioned to form a barrier between the high pressure apparatus and personnel. The shields' mounting should not be rigid, but should allow the shield to yield slightly to dissipate the energy of an impact. To prevent ricocheting, shielding such as thick plywood lining on the steel plate may be positioned in line with the power lead and thermocouple ends when the vessel is in operating position. Long pieces of pressure tubing should be clamped at several locations along their length to prevent whipping of the tubing in event of failure. Valve stems could be extruded violently and should not point in the direction of operating personnel.

2. Apparatus

Safety interlocks should be included in the system to provide automatic protection during unattended operation. The main possibilities for failure are loss of cooling water and malfunction of the temperature controller, resulting in temperature overshoot. Both failures cause overheating of the vessel which might result in permanent damage if not instant failure. An automatic interlock should be installed in the cooling water system which shuts off electricity to the furnace if the water flow drops below a predetermined rate, and the temperature controller should provide automatic shut-off if the temperature overshoots set-point by a fixed amount or if a break occurs in the control thermocouple circuit.

Rupture discs should be installed in the low pressure portions of the system to prevent damage due to overpressure; however, rupture discs do not usually work reliably in the high pressure system and, if used, may give a false sense of security.

It is often desirable, from an experimental standpoint, to mix hydrogen with the argon pressure medium. However, high hydrogen pressures cause serious embrittlement of the steel components in the pressure system, especially at higher temperatures. Hydrogen pressures of a few tens of bars do not appear to harm the vessel or other parts of the system, but the upper limit for safe operation is unknown.

D. Pressure Measurement

Measurement of pressure in the 10 Kb range is relatively easy. Bourdon tube gages of high accuracy and precision are available to about 6.9 Kb (100,000 psi) and of lower accuracy and precision to 10 Kb (150,000 psi). Manganin cells, which utilize the change in resistance of a manganese alloy wire with pressure as a means of measurement have been used at pressures well above 10 Kb. The manganin cells are very sensitive and precise, but for high accuracy must be calibrated occasionally against a secondary standard, such as a calibrated gage. Additional details of pressure measurement are summarized by Bradley and Munro (1965) and by Weale (1967). (See also Chapter 7, particularly Table 7.1)

III. Operating the System

Some comments on system operation have been made in the preceding pages, and many details are dependent on the particular apparatus used and are supplied by the manufacturers. Several items which do not fit elsewhere in the chapter will be discussed below.

A. Preventive Maintenance

A high pressure system requires a good deal of attention if it is to operate at all routinely. Keeping the hydraulic and argon portions of the system dirt-free is perhaps the most important aspect of preventive maintenance. Parts of the system should be carefully cleaned before assembly and filters should be installed in the hydraulic system and in the low pressure argon system.

The bore of the pressure vessel should be kept free of rust and the furnace should be as dry as possible. In damp climates the furnace insulation will absorb enough water to short-circuit the power leads during a run, so the furnace should occasionally be heated under vacuum to completely dry it.

B. Estimating Starting Pressure

The usual method of bringing a run to the desired pressure and temperature is to pressurize the vessel at room temperature and then heat to the desired temperature. However, pressure increases with temperature due to thermal expansion of the argon. Because the argon is not heated uniformly throughout the vessel, pressure increases are difficult to calculate and must be determined empirically for a particular vessel and furnace. The change in pressure with temperature is also a function of pressure. As a rule of thumb one can expect pressure in the model system to approximately double in the range 3-6 Kb when a 3.13 cm i.d. furnace is heated to 900 C. The increase will be greater at lower pressure and less at higher pressure.

C. Pressure Adjustments during Run

Changes in pressure once the run is at high temperature must be made with great care to prevent hot gas from damaging the closure plug or high pressure tubing. Whether or not pressure is to be increased or decreased, the intensifier should always be brought to a slightly greater pressure than that in the vessel before opening the valve to the vessel. Pressure increases may be made fairly rapidly, but the temperature distribution will be upset and a second adjustment of pressure may be necessary once temperature has stabilized. Decreases in pressure must be made very gradually to allow dissipation of heat from the hot gas flowing through the high pressure tubing.

D. Quenching

Rapid quenching is desirable in many experiments and relatively rapid cooling rates can be achieved in an IHPV. The most common method merely involves turning off the electrical power. The cooling rate may be increased by rotating the pressure vessel to horizontal and simultaneously shutting off power. Temperature drops of 100 C in 10-15 seconds may be reached with the latter method. Quenching is usually not isobaric, pressure falling with temperature as the argon cools, but isobaric quenching is possible and it also increases the cooling rate.

E. Leak Detection

There are many places in a gas media apparatus where leaks may occur and isolating a leak may be very difficult, especially if it is slow. Non-corrosive soap solutions may be used conveniently to check fittings. Detecting leaks in electrical lead-throughs and thermocouples in closure plugs with soap solutions may damage the apparatus and is also quite hazardous if the system is under pressure.

A sensitive method of leak detection utilizing the radioactivity of ^{85}Kr has been used by Spetzler *et al.* (1969). About one percent krypton is added to the argon pressure medium and a radiation detector used to locate the leak.

IV. Examples of Experimental Design

Several types of experiments which have been, or could be performed in an IHPV will be discussed briefly in this section; hopefully the examples will illustrate the versatility and potential of the IHPV. Many of the techniques used in the experiments mentioned below are described by Huebner in Chapter 5 of this book.

A. Phase Equilibria Experiments

Determination of stability fields bounded by solid-solid and solid-liquid equilibria do not often require the large volume of an IHPV, but in many cases some of the P,T regions of interest are best achieved with an IHPV. Experiments of this sort are exemplified by the many studies made by Dr. H.S. Yoder, Jr., and his colleagues (for example, Yoder, 1968; Kushiro and Yoder, 1969).

Direct determination of solubility in solid-vapor, liquid vapor, or solid-liquid-vapor, on the other hand, is very dependent on the large volume afforded by an IHPV. Studies of the solubility of solids in aqueous fluids have been made by Anderson and Burnham (1965, 1967) to pressures of 10 Kb and temperatures to 900 C. A good description of the technique, including description of the sample capsules and methods of calculating solubility from the observed data, is given by Anderson and Burnham (1965). Measurements of the solubility of H_2O in silicate melts over a wide P-T region have been made by Burnham and Jahns (1962) who describe the excess water method and compare it with several other methods for determining solubility, and also give a step-by-step procedure for welding sample capsules. Hamilton et al. (1964) determined the solubility of water in basaltic melts and described a weight-loss technique for measuring solubilities. Kennedy et al. (1962) determined both the solubility of silica in H_2O and H_2O in liquid SiO_2 in a study of the upper three-phase region of the SiO_2-H_2O system.

In the study of iron-bearing systems, oxygen fugacity must be controlled to obtain meaningful results. (These techniques are discussed in great detail in Chapters 2, 4, and 5.) In systems containing H_2O the solid buffer phase assemblage technique (Eugster and Skippen, 1967) may be used. At the relative high temperatures common in investigations in an IHPV the solid buffers are rapidly exhausted, however; and control of hydrogen fugacity, and hence oxygen fugacity, may be accomplished with the hydrogen membrane technique (Shaw, 1967) adapted to the IHPV (Piwinskii et al., 1968).

When using a fluid phase containing two volatile components such as CO_2 and H_2O (Holloway et al., 1968) it is often desirable to use a large ratio of fluid to condensed phases to minimize change in fluid composition due to reactions with condensed phases; again the large volume sample size allowed by the IHPV is necessary.

B. Continuous Experiments

The large volume of the IHPV permits introduction of electronic monitoring devices into the high pressure vessel, and these in turn allow measurement of sample properties without quenching from discrete P,T points. Utilizing these continuous methods, one can acquire a large amount of data in a relative short time and, more importantly, measurements can be made which are beyond the scope of quenching experiments.

A recent development of this type is an apparatus for volumetric measurements to 8.5 Kb and 950 C. The apparatus, which uses two IHPV's connected in tandem, is described in connection with measurement of the specific volume of H_2O by Burnham et al. (1969). A precision of measurement of .1-.2% (relative) is possible with the volume apparatus. It is particularly well suited to measurements on liquids and gases, and recently it has been used to measure the partial molar volume of H_2O in an albite liquid (Burnham and Davis, 1969).

Other continuous-type measurements which have been made in an IHPV are: elastic modulii of crystals at elevated T and P (Graham and Barsch, 1969); differential thermal analysis of the high-low quartz inversion (Yoder, 1950); and the change of melting point of diopside with pressure (Yoder, 1952); and measurement of the viscosity of hydrous silicate liquids (Burnham, 1963).

Many other experiments might be performed in an IHPV at elevated P and T; such as electromotive force, thermal gravimetric analysis, and electrical conductivity, but little work of this kind has yet been done. Electrical conductivity measurements have been made by Lebedev and Khitarov (1964) in an IHPV described by Khitarov et al. (1959) but few details of the conductivity apparatus were given.

APPENDIX 8-A.1

Sources of Apparatus

The following is not intended to be a comprehensive listing, but only to indicate availability of rather esoteric equipment from some U.S. manufacturers. Therefore, this list cannot be considered to be an endorsement, nor advertisement by the author or publisher.

Bourdon tube pressure gages; 100,000 psi: 1, 4, 5, 8.
Bourdon tube pressure gages; 150,000 psi: 8.
Diaphragm compressors: 1, 5.
Furnace cores, porous alumina: 7.
Intensifiers for gas media: 1, 2, 3, 5, 8, 9.
Manganin pressure measuring cells: 3.
Pressure vessels: 1, 2, 3, 5, 8, 9.
Sheathed thermocouples and wire: 6, 10.
Tubing, valves and fittings for 100,000 psi: 1, 2, 3, 5, 8, 9.
Tubing and fittings for 200,000 psi: 3.
Valves for 200,000 psi: 3.

The numbers above refer to the following manufacturers.

(1) American Instrument Co., Inc.
(2) Autoclave Engineers, Inc.
(3) Harwood Engineering Co., Inc.
(4) Heise Bourdon Tube Co., Inc.
(5) High Pressure Equipment Co., Inc.
(6) Lewis Engineering Co.
(7) Norton Co.
(8) Pressure Products Industries, Inc.
(9) Tem-Pres Research, Inc.
(10) Thermo-Electric Co., Inc.

BIBLIOGRAPHY:

Anderson, G.M. and Burnham, C.W. (1965). The solubility of quartz in supercritical water. Amer. J. Sci., 263:494-511.

Anderson, G.M. and Burnham, C.W. (1967). Reactions of quartz and corundum with aqueous chloride and hydroxide solutions at high temperatures and pressures. Amer. J. Sci., 265: 12-27.

Bell, P.M. and England, J.L. (1967). High pressure experimental techniques, in Researches in geochemistry, Vol. 2, P.H. Abelson, editor. New York: John Wiley and Sons.

Birch, F., Robertson, E.C., and Clark, S.P. (1957). Apparatus for pressures of 27,000 bars and temperatures of 1400°C. Ind. and Eng. Chem., 49:1965.

Bradley, R.S. and Munro, D.C. (1965). High pressure chemistry. Oxford: Pergamon Press.

Bridgman, P.W. (1949). The physics of high pressure. London: G. Bell and Sons.

Burnham, C.W. (1962). Large volume apparatus for hydrothermal investigations to 10,000 bars and 1500°C (abs). Amer. Ceram. Soc. Program, Seattle, Washington.

Burnham, C.W. (1963). Viscosity of a water-rich pegmatite melt at high pressures (abs). Geol. Soc. Amer. Program 1963 Ann. Mtg.:26A.

Burnham, C.W. and Davis, N.F. (1969). The partial molar volume of water in albite melts (abs). Trans Amer. Geophys. Union, 50:338.

Burnham, C.W., Holloway, J.R., and Davis, N.F. (1969). The specific volume of water in the range 1000 to 8900 bars, 20° to 900°C. Amer. J. Sci., (Schairer Vol.) 267A:70-95.

Burnham, C.W. and Jahns, R.H. (1962). A method for determining solubility of water in silicate melts. Amer. J. Sci., 260:721-745.

Comings, E.W. (1956). High pressure technology. New York: McGraw-Hill.

Downs, H.A., Hastings, J.R., and Waxman, M. (1969). An improved tool for refinishing the conical ends of high pressure tubing. Rev. Sci. Instr., 40:1238.

Downs, J.L. and Payne, R.T. (1969). A review of electrical feed-through techniques for high pressure gas systems. Rev. Sci. Instr., 40:1278-1280.

Eugster, H.P. and Skippen, G.B. (1967). Igneous and metamorphic reactions involving gas equilibria, in Researches in geo-chemistry, Vol. 2, P.H. Abelson, editor. New York: John Wiley and Sons.

Goldsmith, J.R. and Heard, H.C. (1961). Subsolidus relations in the system $CaCO_3$-$MgCO_3$. J. Geol., 69:45-74.

Graham, E.K. and Barsch, G.R. (1969). Elastic constants of single-crystal forsterite as a function of temperature and pressure. J. Geophys. Res., 74:5949-5960.

Hamilton, D.L., Burnham, C.W., and Osborn, E.F. (1964). The solubility of water and effects of oxygen fugacity and water content on crystallization in mafic magmas. J. Petrol., 5:21-39.

Holloway, J.R., Burnham, C.W., and Millhollen, G.L. (1968). Generation of H_2O-CO_2 mixtures for use in hydrothermal experimentation. J. Geophys. Res., 73:6598-6600.

Kennedy, G.C., Wasserburg, G.J., Heard, H.C., and Newton, R.C. (1962). The upper three-phase region in the system SiO_2-H_2O. Amer. J. Sci., 260:501-521.

Khitarov, N.I., Lebedev, E.B., Rengarten, E.B., and Arseneva, R.V. (1959). The solubility of water in basaltic and granitic melts. Geochemistry, 5:479-492.

Kushiro, I. and Yoder, H.S. (1969). Melting of forsterite and enstatite at high pressures under anhydrous conditions. Carnegie Inst. Washington Year Book, 67:153-158.

Lebedev, E.B. and Khitarov, N.I. (1964). Dependence of the beginning of melting of granite and the electrical conductivity of its melt on high water vapor pressure. Geochem. International, 2:193-197.

Newhall, D.H. (1957). Hydraulically driven pumps. Ind. and Eng. Chem., 49:1949-1954.

Newhall, D.H. and Abbot, L.H. (1968). A contemporary version of the Bridgman-Birch 30 Kb apparatus and certain ancillary devices. Proc. Instn. Mech. Engrs., Part 3C, 182:288-294.

Piwinskii, A., Weidner, J.R., and Carman, J.H. (1968). Hydrogen osmosis experiments at elevated temperatures and pressures (abs). Geol. Soc. Amer. Program, 1968 Ann. Mtg:239.

Shaw, H.R. (1967). Hydrogen osmosis in hydrothermal experiments, in Researches in geochemistry, Vol. 2, P.H. Ableson, editor. New York: John Wiley and Sons.

Spetzler, H., Schreiber, E., and Newbigging, D. (1969). Leak detection in high pressure gas system. Rev. Sci. Instr., 40:179.

Tsiklis, D.S. (1965). Handbook of techniques in high-pressure research and engineering, translated by A. Bobrosky (1968). New York: Plenum Press.

Vodar, B. and Saurel, J. (1963). The properties of compressed gases, in High pressure physics and chemistry, Vol. 1, R.S. Bradley, editor, London: Academic Press.

Weale, K.E. (1967). Chemical reactions at high pressures. London: E. & F.N. Spon.

Williams, D.W. (1966). Externally heated cold-seal pressure vessels for use to 1200°C at 1000 bars. Mineral. Mag., 35:1003-1012.

Wolf, R.C. and Bowen, J.C. (1957). Compressing of gases in the pure state to high pressures. Ind. and Eng. Chem., 49:1962-1964.

Yoder, H.S. (1950). High-low quartz inversion up to 10,000 bars. Trans. Amer. Geophys. Union, 31:827-835.

Yoder, H.S. (1952). Change of melting point of diopside with pressure. J. Geol., 60:364-374.

Yoder, H.S. (1968). Akermanite and related mullite-bearing assemblages. Carnegie Inst. Washington Year Book, 66:471-477.

ACKNOWLEDGMENT:

Essentially all of my knowledge of experimental techniques using the IHPV was gained while working with Professor C.W. Burnham at the Pennsylvania State University. Any original ideas presented in the text are his; however, I must take full responsibility for their presentation in this chapter. Professor G.M. Anderson generously supplied a description of his design for a furnace, for which I am grateful. I must also thank N.F. Davis for countless hours of stimulating discussion concerning the design and operation of high pressure equipment. Professors Anderson and Burnham reviewed the manuscript and their many helpful suggestions are appreciated.

CHAPTER 9

Compressibility Measurements of Gases Using Externally Heated Pressure Vessels[1]

D.C. Presnall

I. Introduction

In principle, the determination of the thermodynamic properties of a gas is very simple; one needs only a set of measurements giving the pressure, temperature, and volume occupied by a known mass and composition of gas. However, in practice it is often difficult to obtain highly precise data of this sort. A wide variety of methods has been used for determining the compressibilities of gases and the reader interested in a general review of these methods should consult the discussions by Newitt (1940), Saurel (1957), and Tsiklis (1965).

Experiments are the most difficult and the data are the least precise when it is desired to make measurements where both the temperature and pressure are high. Most of the data collected under these conditions have been determined using a thick-walled bomb of carefully measured and fixed volume which is externally heated by an electric furnace or a thermostatically controlled bath. Auxiliary apparatus consists of a thermocouple for measuring the temperature of the bomb, one or more pressure gages, and apparatus for determining the amount and composition of gas put into the bomb. One modification of this method used by Kennedy (1954) circumvents the necessity for measuring the amount of gas in the bomb or the volume of the bomb. The purpose here is to present a review of this method and its experimental uncertainties and limitations. Recently, a new method that shows great promise uses an internally heated pressure vessel and has

---

[1] Contribution No. 144, Geosciences Division.

been used for measurements in a temperature-pressure range beyond that which can be attained at present using externally heat pressure vessels (Burnham et al., 1969). Holloway describes this method in Chapter 8 of this book.

In the discussion that follows, reference will be made mainly to papers published within the last 20 years. References to earlier workers, many of whom used similar methods, can be found in the bibliographies of the more recent papers or in the review papers by Newitt (1940), Saurel (1957), and Tsiklis (1965).

II. Description of the Method

There are numerous variations on the basic method depending on the pressure-temperature range of interest and the particular gas or gas mixture being studied. Primarily, these variations are concerned with the manner in which the bomb is filled and the method for determining the amount and composition of the gas in the bomb. Consider first those techniques used for one-component gas systems.

A representative experimental arrangement is shown in Figure 1 (Presnall, 1969). It consists of a high pressure portion and a low pressure portion. The high pressure portion includes the bomb contained in a nichrome-wound resistance furnace, bourdon tube pressure gages, a diaphragm-type compressor for pumping hydrogen, and a cold trap for freezing out any grease that may have contaminated the gas from the valve fittings. In order to avoid brittle failure of the bourdon tube gages by contact with hydrogen, the pressure gages are separated from the hydrogen by a mercury U-tube. The low pressure part consists of a 10 liter bottle with an attached manometer for measuring the amount of gas contained in the bomb. Briefly, the experimental procedure is to bring the bomb up to the desired temperature, evacuate it, pump in the desired pressure of hydrogen, wait a few minutes to establish thermal equilibrium, and then expel the gas into the evacuated 10 liter bottle where the amount of gas is determined from the pressure on the manometer, the temperature, and the known

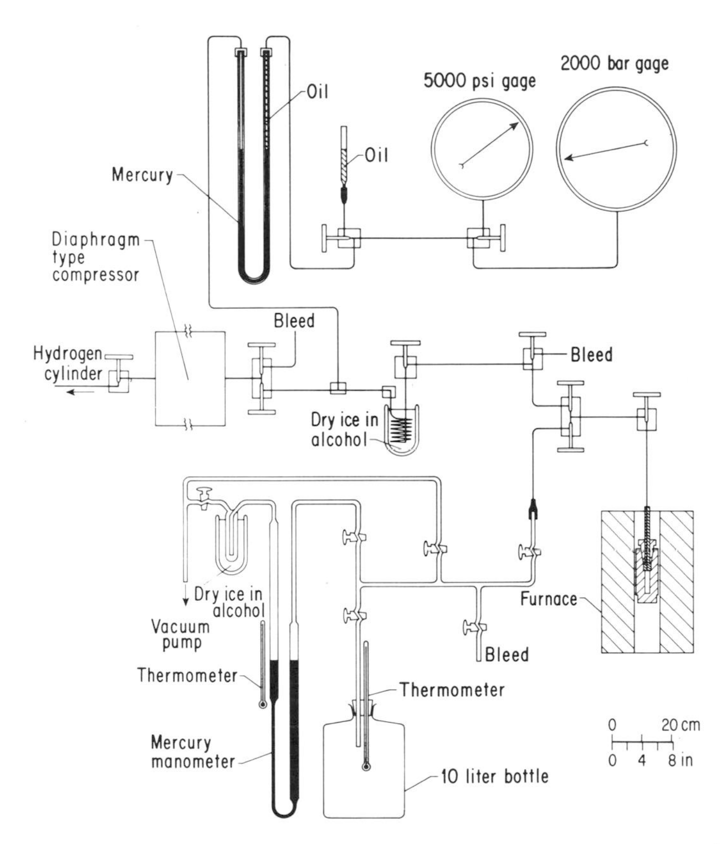

Figure 1. Experimental apparatus for P-V-T measurements on hydrogen. [Reproduced from D.C. Presnall, J. Geophys. Res., 74:6027 (1969) with the permission of the American Geophysical Union.]

volume of the bottle plus connective tubing. By this procedure, each determination of pressure, volume, and temperature for a measured amount of gas must be carried out with a separate filling of the bomb. Sometimes it is possible to eliminate the low pressure gas analysis apparatus entirely. In these cases, a calibrated screw press replaces the diaphragm compressor in Figure 1 and is used to inject or extract measured amounts of gas from the bomb. The screw press consists essentially of a steel cylinder containing a piston that is advanced or retracted by hand along a precision lead screw. This technique is applicable for gases such as CO_2 and H_2O that are liquid at room temperature and moderate pressures (greater than about 70 bars for CO_2). For example, Kennedy (1950) took measurements along isotherms by injecting measured amounts of H_2O in steps into the bomb. Holser and Kennedy (1959) took measurements in the reverse direction by filling the bomb to the maximum desired pressure and then removing with the screw press measured volumes of H_2O in steps. When making measurements on CO_2, Kennedy (1954) found it necessary to keep the screw press in an ice bath so as to avoid changes in the density of CO_2 caused by slight fluctuations in room temperature.

Kennedy (1954, 1957), Holser and Kennedy (1959), and Maier and Franck (1966) took measurements along isochors (constant volume conditions) by first loading the bomb with H_2O (or CO_2 in the case of the 1954 work by Kennedy) while cold and then taking a series of simultaneous measurements of pressure and temperature as the bomb was heated. In making measurements by this procedure, Kennedy (1954) avoided the necessity of measuring either the volume of the pressure vessel or the amount of CO_2 injected into it. For each filling of the bomb he measured the pressure at 150 C and used already existing data on the density of CO_2 along this isotherm. By knowing the density of CO_2 at 150 C for each isochor, he could calculate the density for all other temperatures and pressures along that isochor. Thus, the accuracy of the data depends on the accuracy of the 150 C isotherm. Kennedy (1957) used the same technique for measurements on H_2O. Maier and Franck (1966) did not rely on a previously measured isotherm but instead used a predetermined measurement of the volume of the bomb and weighed the amount of water put into the bomb.

Köster and Franck (1969) followed the method of taking measurements along an isotherm. They increased the pressure by injecting measured amounts of H_2O into the bomb but they worked at pressures so high (up to 10 Kb) that a screw press could not be used. Instead, they used an intensifier to compress either a gasoline-oil mixture or _n_-hexane. The high-pressure line from the intensifier led to a separator containing mercury, and the mercury acted essentially as a fluid piston for driving water out of the separator into the bomb.

Consider now those techniques that have been used for gas mixtures. So far, only binary gas systems have been studied, and these studies include the work of Franck and Tödheide (1959) and Greenwood (1969) on CO_2-H_2O mixtures and Greenwood (1961) on H_2O-Ar mixtures. In these studies, measurements are carried out in basically the same manner as for one-component gases but the methods for determining the amount and composition of the gas are necessarily slightly more complex.

Franck and Tödheide (1959) filled the cold bomb with CO_2 and H_2O from steel bottles and determined the amount of each by weighing the bottles before and after the filling. They then took measurements along an isochor by heating the bomb and measuring the corresponding pressures produced. Greenwood (1961) filled the cold bomb with water, heated it to the desired temperature, and injected argon at high pressures. The gas mixture was analyzed afterwards by expelling the contents of the bomb into an evacuated 10 liter bottle with attached manometer similar to the arrangement in Figure 1. Before entering the 10 liter bottle, the gas passed through a porous plug of glass wool immersed in a dry ice cold trap. This removed the water as ice, which was weighed, and the remaining argon was measured in the 10 liter bottle. In his study of CO_2-H_2O mixtures, Greenwood (1969) took measurements along isotherms. The bomb was first brought to the desired temperature, CO_2 was injected, and its pressure was measured. Then measured amounts of H_2O were injected in steps with a calibrated screw press until the maximum desired pressure of 500 bars was reached. In order to determine the amount of CO_2 injected into the bomb, the gas mixture was analyzed at the conclusion of the run by the same method described above for H_2O-Ar mixtures. This analysis also served as a check on the total quantity of H_2O injected with the screw press.

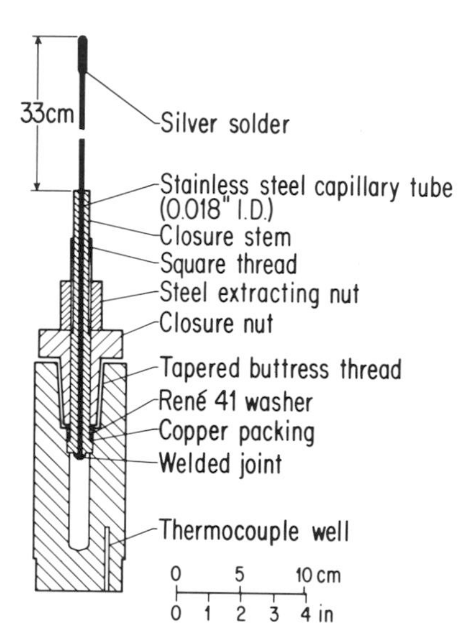

Figure 2. Detail of pressure vessel, made from René 41 alloy. [Reproduced from D.C. Presnall, J. Geophys. Res., 74: 6028 (1969) with the permission of the American Geophysical Union.]

III. Construction and Calibration of Apparatus

A. Pressure Vessel

A typical pressure vessel is shown in Figure 2 (Presnall, 1969). It is cylindrical and is closed by a Bridgman unsupported area seal. The closure nut is made with a tapered buttress thread with a taper of 1 inch per foot, and the extracting nut is made of hardened steel so as to avoid galling of the threads on the closure stem. I have found it convenient to attach the stainless steel capillary tube to the closure stem at its tip by welding. Kennedy (1950) experienced difficulty in welding the tube without closing the capillary hole. Sometimes this occurs but the hole can be drilled out after welding. In the design of Figure 2, the thermocouple fits into a well as shown, but several other arrangements have been used. Franck and Tödheide (1959) placed several thermocouples in diagonal holes drilled in the side of the bomb. Köster and Franck (1969) and Maier and Franck (1966) used a

Bridgman seal on both ends of the bomb and passed a thermocouple through the closure stem of one seal into the center of the gas chamber. Kennedy (1954) achieved the same result by drilling a hole in the bottom of the vessel and welding into it a stainless steel tube, closed at one end, for insertion of a thermocouple into the center of the vessel. This avoided any correction for the temperature difference between the thermocouple well and the interior of the vessel. However, if there is a temperature gradient inside the bomb the temperature at the center is not the average temperature of the gas, so a correction for this difference is still necessary. (See Chapter 6 for more details about argon-filled pressure vessel temperature corrections.)

Metals that have been used for the bomb include the nickel-base alloys René 41 (Presnall, 1969) and Udimet 700 (Maier and Franck, 1966; Köster and Franck, 1969) and the cobalt base alloy S-816 (Greenwood, 1961, 1969). Figure 3 shows yield strengths and tensile strengths for René 41 and S-816 as well as a molybdenum base alloy, TZM. Udimet 700 has not been shown because it is very similar in strength to René 41. These data set limits on the temperatures and pressures at which measurements can be carried out. Since it is necessary to know the volume of the pressure vessel precisely, the volume must be checked regularly when operating near the curve showing the 0.2% offset yield strength. Maier and Franck (1966) and Köster and Franck (1969) noted some permanent deformation in their studies at high pressures.

The relatively new alloy TZM has not been used for determining pressure-volume-temperature (P-V-T) properties of gases and its strength data are shown to illustrate its potential for measurements above 900 C. It should be useful for studying gases such as hydrogen and argon but TZM is easily oxidized and would be useless for gases such as H_2O and CO_2. In order to prevent external oxidation, the bomb would have to be heated in a protective atmosphere or a non-oxidizing sheath. (Williams, 1966, 1968).

Oxidation of the walls of the pressure vessel can be a problem

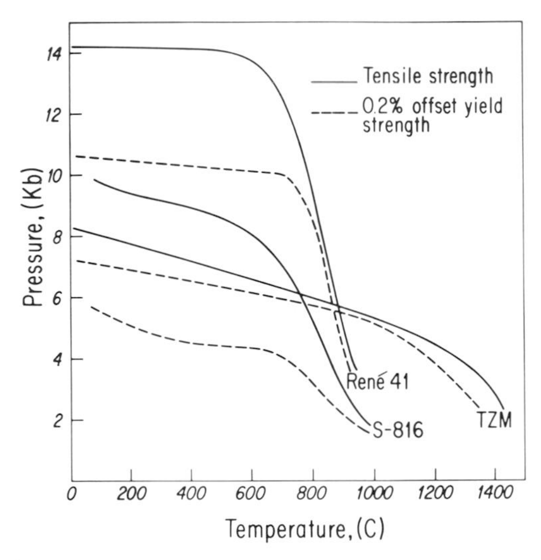

Figure 3. Tensile strengths and 0.2% offset yield strengths of three alloys. The data for René 41 are from the General Electric Co. and are for bars solution heat-treated at 1065 C for 4 hours (air cooled) and then aged at 760 C for 16 hours (air cooled). The data for S-816 are from Simmons and Cross (1954). The data for TZM are for bars in the stress relieved condition and are supplied by the Climax Molybdenum Co. of Michigan.

also with nickel or cobalt base alloys. Kennedy (1957) found that his earlier work on water (Kennedy, 1950) was in error, especially from 800-1000 C, due to reaction of H_2O with Fe, Ni, Cr, and Co in the bomb alloy to produce hydrogen. He was able to suppress the production of hydrogen by placing some CuO inside the bomb. Kennedy's work on CO_2 (Kennedy, 1954) also extended up to 1000 C, and since the equilibrium dissociation of pure CO_2 yields a slightly higher oxygen fugacity than pure H_2O, it might be suspected that his high temperature measurements on CO_2 are also slightly in error. Reaction of CO_2 with the walls of the pressure vessel would produce CO.

B. Volume of the Bomb

As stated earlier, the technique of Kennedy (1954, 1957) does not require knowledge of the volume of the bomb but it is necessary to rely on the accuracy of previous P-V-T measurements at some low temperature. Even if the experimental method chosen requires a determination of the volume of the bomb, it is common practice to use low temperature P-V-T data to determine this volume, and the high temperature-high pressure data are again tied to the low temperature measurements. Ideally, it would be best to use a method for determining the volume that is independent of other P-V-T data, and some investigators do this. However, an independent measurement of volume is not as desirable as it might seem at first. Low temperature data on a number of gases are available that have a precision of about ±0.01% (for example, see the papers by Michels et al., 1951-a, 1951-b), which is about one order of magnitude more precise than can be achieved under the best conditions using the method described in this chapter. Thus, it is not likely that high temperature measurements will be in error because of reliance on low temperature data of this precision. Even if the low temperature data were found to be in error, the high temperature measurements could be revised accordingly.

Two methods have been used for determining the volume of the bomb that are independent of any previous P-V-T measurements. Franck and Tödheide (1959), Maier and Franck (1966), and Köster and Franck (1969) simply weighed the bomb before and after filling it with mercury. They reported very good precision using this method (±0.02-.07%). Greenwood (1969) also found very good precision (±.02%) by weighing the amount of mercury contained in the pressure vessel but he found the volume determined in this way to be about one percent less than the volume determined by several other methods. He believed his determinations using mercury were in error due to trapped air bubbles. It is not being suggested that the volume determinations of Franck and co-workers are in error, but the experience of Greenwood indicates that this method must be approached with caution.

A second method (Greenwood, 1969) that is independent of any previous P-V-T measurements involves the use of a glass cylinder slightly smaller in volume than the bomb. The cylinder is narrowed at both ends to tubes of 4 mm inside diameter, a manometer tube is attached to the lower tube so that it is parallel to the main cylinder, and the whole apparatus is clamped vertically. Using a leveling bulb, mercury is brought to a mark on the lower tube, and the upper tube is attached to the valve of the evacuated bomb. The valve is opened and the leveling bulb is used to equalize the mercury levels in the manometer and the upper tube of the glass cylinder. The volume of the bomb is the volume in the glass cylinder between the upper and lower mercury levels, as determined by weighing of the cylinder with mercury at the two levels. Greenwood found that the precision was $\pm 0.4\%$.

Other methods for determining the volume have relied on previously measured P-V-T data at low temperatures. Kennedy (1950) injected known amounts of water from a calibrated screw press into the bomb at 200 C and 100 bars and calculated the volume from literature data on the density of water at this temperature and pressure. Holser and Kennedy (1959) filled the bomb with water at about 200 C and 200 bars, ejected the water, and weighed it. They claimed a precision of $\pm 0.03\%$. Greenwood (1969) used this same method at 1000 psi and 150 C but claimed a precision of only $\pm 0.5\%$. Greenwood (1969) also determined the volume of the pressure vessel by filling it with argon at 193 bars and 146 C and CO_2 at 147 bars and 148 C. The amount of gas in the bomb was measured by expanding it into a 10 liter bottle with an attached manometer. These methods were both reproducible to $\pm 0.4\%$. Presnall (1969) left the bomb at room temperature and measured the amount of hydrogen it contained at various pressures from 200-1000 atmospheres. Replicate determinations were reproducible within an average deviation of $\pm 0.1\%$ and check measurements using argon and nitrogen were also within this uncertainty. I believe the main reason for this improved precision compared to the identical method used by Greenwood (1969) with Ar and CO_2 is the improved design of the mercury manometer attached to the 10 liter bottle (Figure 1). Greenwood (1969) stated that his manometer was reproducible to ±1.5 mm of mercury, which agrees with my experience using a manometer tube of small diameter. The manometer shown in Figure 1 had an internal diameter of about 1 inch with a small-bore choke between the two arms. Brombacher et al. (1960), give an equation for calculating the size of the choke needed for optimum damping. The large bore in the reading portion of the manometer reduced the correction for surface tension to a negligible amount (see tables by Brombacher et al., 1960), and when the mercury levels were read with a cathetometer against a back-lighted green background, it was found that measurements could be reproduced to ±0.2 mm of mercury. Also, pressure on the manometer was read relative to a vacuum instead of atmospheric pressure, so the correction for variations in barometric pressure could be eliminated. The largest uncertainty in the volume determination of Presnall was in the measurement of pressure with the bourdon tube gage. Use of a dead weight piston would probably allow a volume determination by his method with a precision of ±0.05% or less.

Three corrections are made to the volume of the bomb, one for thermal expansion, one for elastic stretch, and one for the volume of the capillary tube leading from the valve to the bomb chamber. The correction for thermal expansion can be as large as 2-3% and for precise calculations it is best to obtain data on the variation in thermal expansion with temperature. For example, in the case of the alloy René 41, use of low temperature thermal expansion data would produce an error in the volume of 0.2% at 800 C.

The correction for elastic stretch can be calculated from the formula listed where a is internal radius, b is external radius, μ is Poisson's ratio, P is internal pressure, and E is Young's

$$\Delta a = \frac{a^2 P}{(b^2 - a^2)E}\left[(1+\mu)\frac{b^2}{a} + (1-\mu)a\right]$$

modulus (Kennedy, 1954). This formula applies for an infinitely long cylinder and gives too large a correction because of restriction at the ends of the bomb. Fortunately, the elastic stretch correction is small (typically about 0.15% per 1000 bars) and some investigators neglect it entirely. If the determination of the volume of the bomb could be made with a precision of $\pm 0.05\%$, the elastic stretch correction could be determined empirically using one of the very accurate low temperature isotherms of Michels _et al._ (1951-a, 1951-b).

The volume of the capillary tubing leading out of the bomb and the volume in the valve block at the end of this tubing can be determined simply by measuring the internal dimensions of these parts. For a capillary tube of 0.018 inch inside diameter, these two volumes together typically amount to about 0.1 cm^3. Bombs that have been used vary in volume from 10-300 cm^3, so the volume of the capillary stem and valve block can be neglected in some cases but it is a very important correction in other cases. The gas in the capillary stem is in a very strong temperature gradient, so the most accurate method for subtracting off the amount of gas in the capillary stem is to measure this thermal gradient and graphically integrate a plot of gas density versus distance along the stem.

C. Measurement of Pressure

The dead weight piston gage measures pressure directly in units of force per unit area and is the instrument against which all other types of pressure gages must be calibrated. Piston gages are commercially available (for example, from the American Instrument Co.) for use at pressures up to 6 Kb with a precision of $\pm 0.05\%$, and they can be used at much higher pressures with slightly reduced precision (Heydemann, 1967). An alternate and somewhat more convenient method for measuring pressure uses the variation in resistance of a manganin wire with pressure. The accuracy of a manganin coil is very good and is limited largely by the accuracy of the dead weight piston gage against which it must be calibrated. Further details regarding manganin coil gages are given by Michels and Lenssen (1934) and Bridgman (1911, 1935, 1938). Bourdon tube gages are perhaps the easiest to use but are slightly less accurate and are limited at present to pressures of about 6 Kb. The Heise Bourdon Tube Co. makes gages with a precision of 0.1% of the full-scale deflection. As with manganin coil gages, bourdon tube gages always require calibration against a dead weight piston gage. (See Chapter 7 for further pressure calibration data.)

D. Control and Measurement of Temperature

The most common method for generating temperature is to place the pressure vessel in a nichrome or chromel resistance furnace. For temperatures above 1000 C, platinum wire is necessary. In order to help reduce the temperature gradient, Maier and Franck (1966), Franck and Tödheide (1959), and Köster and Franck (1969) used multiple furnace windings, each connected to a separate temperature controller. For temperatures below 600 C, some investigators have immersed the pressure vessel in a stirred thermostatically-controlled bath. For example, Schneider (1949) used a eutectic mixture of sodium, potassium, and lithium nitrates as the bath liquid from 200-600 C and oil below 200 C; whereas Bartlett *et al.* (1928) used a molten eutectic mixture of tin and lead at temperatures up to 400 C. Typical temperature variations inside the pressure vessel are $\pm$0.05-1.3 C (Greenwood, 1961; Presnall, 1969) for vessels contained in resistance furnaces, but the temperature gradient probably could be improved somewhat by using longer furnaces with coils spaced wider in the central portion of the furnace tube than on the ends. Schneider (1949) and Bartlett *et al.* (1928) did not comment about temperature variations using thermostatted baths, but it might be expected that gradients would be lower than when using a resistance furnace.

Temperature measurement is usually made with either a chromel-alumel or platinum versus platinum-rhodium thermocouple. The National Bureau of Standards will calibrate platinum versus platinum-rhodium thermocouples to an accuracy of ± 0.5 C up to 1100 C with the error increasing to ± 1.7 C at 1450 C, and these uncertainties place a limit on the accuracy of the temperature measurement. Whenever possible, bare wire thermocouples protected by 99.7% Al_2O_3 tubing should be used. Sheathed thermocouples made by at least some manufacturers drift excessively, possibly due to the use of impure ceramic between the sheath and the thermocouple wires (Walker et al., 1962).

E. Measurement of the Amount and Composition of Gas in the Bomb

The method for determining the amount and composition of gas in the bomb depends on the particular gas being studied. A method suitable for a substance that is liquid at low temperatures, such as water, could not be used for gases like hydrogen and argon.

One method, suitable for studies on H_2O and CO_2, is simply to weigh the amount of H_2O or CO_2 put into the bomb. Maier and Franck (1966) used a screw press to inject water into the bomb, then refilled the screw press and injected the same volume of water into a weighing bottle. This procedure was repeated as many times as necessary to reach the desired pressure, and the weighed amount of water collected in the bottle was taken equal to the amount of water injected into the bomb. They found that the amount of H_2O could be determined to $\pm .0002$ mole. In their study of CO_2-H_2O mixtures, Franck and Tödheide (1959) first added H_2O and then liquid CO_2 from steel bottles. The amount of H_2O and CO_2 put into the bomb was determined by weighing the steel bottles before and after the additions.

Another method that has been used for H_2O and CO_2 uses a screw press more directly. In this procedure, the screw press is first calibrated at each working pressure by weighing the amount of liquid expelled for each turn of the screw. The screw press is then used directly to either inject or extract measured increments of H_2O or CO_2 from the bomb. This method has been used by Greenwood (1969), Kennedy (1950, 1954), and Holser and Kennedy (1959). Greenwood (1969) found an uncertainty in determining the amount of water of ± 0.0001 mole.

Instead of injecting water by advancing a mechanical piston, as in a screw press, Köster and Franck (1969) used a separator containing mercury that acted like a fluid piston. A pump and intensifier generated pressure on a gasoline-oil mixture or n-hexane, and mercury separated this fluid from water in the bomb. A weighed amount of water was placed in the bomb plus separator and, knowing the volume of the separator, high pressure lines, and bomb, the amount of water in the various parts could be calculated. During a run, measured amounts of water were injected into the bomb by increasing the pressure and thus advancing the level of mercury in the separator. The mercury level was measured by means of a float on the surface of the mercury to which was attached a rod with a ferromagnetic end. The ferromagnetic end extended into the high pressure tubing and its position was determined inductively with a coil around the tubing. Köster and Franck found that the uncertainty in determining the amount of water was about ± 0.003 mole. This device was used at pressures up to 10 Kb whereas the screw press has been used only up to 2.5 Kb (Kennedy, 1950).

For studies on gases such as argon or hydrogen, it is not feasible to use a screw press or the method of weighing, so the amount of gas pumped into the bomb must be measured after the completion of a run. This is done by allowing the gas in the bomb to expand into a large evacuated bottle with an attached manometer. The amount of gas is calculated from the previously measured volume of the bottle plus connective tubing, the temperature, and the pressure recorded by the manometer (Greenwood, 1961, 1969; Presnall, 1969). Presnall found an uncertainty in the determination of the amount of hydrogen of about ± 0.0001 mole. In order to analyze binary CO_2-H_2O and H_2O-Ar mixtures, Greenwood (1961, 1967, 1969) used a cold trap to freeze out the water as it expanded out of the bomb. The water was weighed and the remaining CO_2 or Ar was determined by the method just described.

IV. Summary of Uncertainties

In summarizing the various errors, it seems useful to consider the total error produced in the compressibility factor Z, following the scheme of Greenwood (1969). The proportional error in Z is given by the sum of the proportional errors in P, V, T, and n moles of gas. That is:

$$\frac{\delta Z}{Z} = \frac{\delta P}{P} + \frac{\delta V}{V} + \frac{\delta T}{T} + \frac{\delta n}{n}$$

Consider only the optinum precision that might be achieved. For a measurement taken at 1000 K, we would have:

$$\frac{\delta Z}{Z} = 0.0005 + 0.0005 + 0.0005 + \frac{0.0001}{n}$$

$$= 0.0015 + \frac{0.0001}{n}$$

Thus, as the number of moles of gas decreases to zero, the uncertainty in Z becomes infinite, and the pressure below which measurements would be impractical depends on the volume of the bomb. At higher pressure, the uncertainty in the number of moles becomes negligible, so the minimum possible uncertainty that could be achieved is about ± 0.1-0.2%.

V. Extension of Measurements to Higher Temperatures and Pressures

The limits of temperature and pressure at which measurements can be made are set by the strength of the particular alloy used for the bomb (see Figure 3), and for most gases existing alloys could be used to extend the data considerably. For contamination-free compression of gases or liquids at high pressures, a diaphragm-type compressor is very good because the gas never comes in contact with oil or grease. The American Instrument Co. makes diaphragm-type compressors for use up to 2 Kb and they also sell similar compressors made by the Corblin Co. that can be used up to 4 Kb. The American Instrument Co. also sells an apparatus for contamination-free compression of gases up to 6.8 Kb. It consists of a combination of a 2 Kb diaphragm-type compressor, a 6.8 Kb oil pump, and a cylinder filled with mercury that acts as a separator between the oil and gas. Use of an intensifier would allow this type of arrangement to be used up to the freezing pressure of mercury at 20 C, about 11.5 Kb. For gases such as hydrogen or argon, the amount of gas in the bomb would have to be measured after each run by expanding the gas into a large evacuated bottle of known volume. For gases such as CO_2 and H_2O, use of a bellows-type compressor like that described by Burnham et al. (1969), would allow injection of a large number of increments of gas for a single filling of the bomb.

BIBLIOGRAPHY:

Bartlett, E.P., Cupples, H.L., and Tremearne, T.H. (1928). The compressibility isotherms of hydrogen, nitrogen and a 3:1 mixture of these gases at temperatures between 0 and 400° and at pressures to 1000 atmospheres. J. Amer. Chem. Soc., 50:1275-1288.

Bridgman, P.W. (1911). The measurement of hydrostatic pressures up to 20,000 kilograms per square centimeter. Proc. Amer. Acad. Arts Sci., 47:321-343.

Bridgman, P.W. (1935). Measurements of certain electrical resistances, compressibilities, and thermal expansions to 2000°-kg/cm². Proc. Amer. Acad. Arts Sci., 70:71-101.

Bridgman, P.W. (1938). The resistance of nineteen metals to 30,000 kg/cm². Proc. Amer. Acad. Arts Sci., 72:157-205.

Brombacher, W.G., Johnson, D.P., and Cross, J.L. (1960). Mercury barometers and manometers. U.S. Natl. Bur. Standards Monogr. 8.

Burnham, C.W., Holloway, J.R., and Davis, N.F. (1969). The specific volume of water in the range 1000 to 8900 bars, 20° to 900°C. Amer. J. Sci., 267A:70-95.

Franck, E.U. and Tödheide, K. (1959). Thermische eigenschaften überkritscher mischungen von kohlendioxyd und wasser bis zu 750°C und 2000 atm. Z. phys. Chem. Neue Folge, 22:232-245.

Greenwood, H.J. (1961). The system $NaAlSi_2O_6$-H_2O-argon: total pressure and water pressure in metamorphism. J. Geophys. Res., 66:3923-3946.

Greenwood, H.J. (1967). Mineral equilibria in the system MgO-SiO_2-H_2O-CO_2, in Researches in geochemistry, Vol. 2, P.H. Abelson, editor. New York: John Wiley and Sons.

Greenwood, H.J. (1969). The compressibility of gaseous mixtures of carbon dioxide and water between 0 and 500 bars pressure and 450° and 800°centigrade. Amer. J. Sci., 267A:191-208.

Heydemann, P.L.M. (1967). The Bi I-II transition pressure measured with a dead-weight piston gauge. J. Appl. Phys., 38;2640-2644

Holser, W.T. and Kennedy, G.C. (1959). Properties of water. Part V. Pressure-volume-temperature relations of water in the range 400-1000 C and 100-1400 bars. Amer. J. Sci., 257:71-77.

Kennedy, G.C. (1950). Pressure-volume-temperature relations in water at elevated temperatures and pressures. Amer. J. Sci., 248:540-564.

Kennedy, G.C. (1954). Pressure-volume-temperature relations in CO_2 at elevated temperatures and pressures. Amer. J. Sci., 252:225-241.

Kennedy, G.C. (1957). Properties of water. Part I. Pressure-volume-temperature relations in steam to 1000°C and 100 bars pressure. Amer. J. Sci., 255:724-730.

Köster, H. and Franck E.U. (1969). Das spezifische volumen des wassers bei hohen drucken bis 600°C und 10 kbar. Ber. Bunsengesellschaft phys. Chem., 73:716-722.

Maier, S. and Franck, E.U. (1966). Die dichte des wassers von 200 bis 850°C und von 1000 bis 6000 bar. Ber. Bunsengesellschaft phys. Chem., 70:639-645.

Michels, A. and Lenssen, M. (1934). An electric manometer for pressures up to 3000 atmospheres. J. Sci. Instr., 11:345-347.

Michels, A., Lunbeck, R.J., and Wolkers, G.J. (1951-a). Thermodynamical properties of argon as function of pressure and temperature between 0 and 2000 atmospheres and 0° and 150°C. Appl. Sci. Res., A2:345-350.

Michels, A., Lunbeck, R.J., and Wolkers, G.J. (1951-b). Thermodynamical properties of nitrogen as function of density and temperature between -125° and +150°C and densities up to 760 amagat. Physica, 17, 801-816.

Newitt, D.M. (1940). The design of high pressure plant and the properties of fluids at high pressures. London: Oxford Univ. Press.

Presnall, D.C. (1969). Pressure-volume-temperature measurements on hydrogen from 200°C to 600°C and up to 1800 atmospheres. J. Geophys. Res., 74:6026-6033.

Saurel, M.J.R. (1957). Les equations d'etat des gaz aux hautes pressions. Première Partie. La détermination expérimentale des équations d'état des gaz sous pressions elévées. Méml. Artill. fr., 31:129-184.

Schneider, W.G. (1949). Compressibility of gases at high temperatures. I. Methods of measurement and apparatus. Canad. J. Res., 27-B:339-352.

Simmons, W.F. and Cross, H.C. (1954). Report on the elevated-temperature properties of selected super-strength alloys. Spec. Tech. Publs. Amer. Soc. Test. Mater., 160.

Tsiklis, D.S. (1965). Handbook of techniques in high pressure research and engineering, translated by A. Bobrosky (1968). New York: Plenum Press.

Walker, B.E., Ewing, C.T., and Miller, R.R. (1962). Thermoelectric instability of some noble metal thermocouples at high temperatures. Rev. Sci. Instr., 33:1029-1040.

Williams, D.W. (1966). Externally heated cold-seal pressure vessels for use to 1200°C at 1000 bars. Mineral. Mag., 35:1003-1012.

Williams, D.W. (1968). Improved cold seal pressure vessels to operate to 1100°C at 3 kilobars. Amer. Mineral., 58:1765-1769.

ACKNOWLEDGMENT:

This work was supported by the National Aeronautics and Space Administration (Grant NGL-44-004-001).

CHAPTER 10

The Boiling-Point Technique for the Determination of the Vapor Pressures of Silicate Melts

Louis S. Walter

I. Introduction

Determination of the total vapor pressure of silicate melts was undertaken to provide basic data for the aerodynamic study of atmospheric entry of tektites. These small natural glassy objects (see O'Keefe, 1963, 1966) consist of around 70% SiO_2 plus alumina, lime, magnesia, alkali- and iron-oxides. Although considerable argument exists about their origin, it is generally believed that tektites are formed in either lunar or terrestrial meteorite impacts, and that they entered the earth's atmosphere at high velocities as solid glass. During their passage through the atmosphere, the front surface was heated to high temperatures and experienced significant ablation (Chapman and Larson, 1963; Walter and Adams, 1967).

The extent of heating depends upon the distance between the front surface and the "stagnation point" which, if the vapor pressure of the material is high, is displaced forward from the leading surface of the body. Information about the vapor pressure of tektite material is therefore crucial for such aerodynamic calculations. However, it is not necessary to obtain data on the partial pressures of the individual species in the gas, and a high degree of precision is not required of the data.

Precise vapor-pressure data for complex systems involving several components are now usually obtained by mass spectrometry, in which the vapor evolved at high temperatures is analyzed. The total vapor pressure is then determined by summing the partial vapor pressures of all the evolved gas species. This procedure is the basis of the Langmuir and Knudsen techniques of vapor-pressure measurement. However, these techniques require a knowledge of the gaseous species in the vapor. Furthermore, the Langmuir method, in which gas evolves freely from the sample, requires a knowledge of the vaporization coefficient. These techniques and others are described more fully by Kubaschewski and Evans (1958). Alcock and his coworkers (particularly Peleg and Alcock, 1966) have devised several more modern methods of total vapor-pressure determinations at high temperatures.

The principle underlying the present technique (Walter and Carron, 1964) is that boiling occurs when, to a first approximation, the vapor pressure of the melt equals the ambient pressure. This, basically, was the technique employed by Ruff and Schmidt (1921) in their pioneering efforts to determine the vapor pressure of SiO_2. Ideally, the vapor-pressure curve (P_v vs. T) can thus be determined by varying the ambient pressure (in a vacuum chamber) and observing the temperature at which boiling first occurs at a given pressure. However, several experimental problems exist:

(1) Formation of bubbles during boiling depends partially on the presence of nucleation sites such as sharp minute protuberances in the crucible. The lack of such sites can result in superheating in the melt before bubbles form.

(2) In multi-component systems, the composition of the vapor is generally not the same as that of the liquid. Thus, as boiling continues, both the composition of the melt and its vapor pressure will change.

(3) Atmospheric volatiles (*e.g.*, H_2O, O_2, SO_2, N_2) can be adsorbed onto the surfaces of the grains of sample material before melting, causing premature boiling.

(4) Boiling may be inhibited by the viscosity of the melt, particularly in the case of silicates which are quite viscous.

(5) Bubbles forming at the bottom of the crucible will experience a slightly higher apparent ambient pressure due to the added weight of overlying material.

(6) The surface tension of the melt will also inhibit boiling,

producing the effect of an additional ambient "pressure" on the liquid.

As explained in the following sections, effects produced by the first four factors can be largely removed by suitable adjustment of the experimental conditions. The effects of the last two conditions can be calculated and the experimental data modified to take them into account.

II. Experimental

The crucible material selected for these experiments must, of course, be capable of withstanding the extreme temperatures at which silicate melts boil, and cannot react with the melts at these temperatures. It must also have a vapor pressure low in comparison with silicate melts, and it must not produce changes in the composition of the gas surrounding the melt, particularly in the partial pressure of oxygen. These considerations immediately eliminate the most common crucible materials (e.g., Pt, ZrO_2 and W). It was found that, for the temperature range of interest (up to 2200 C), iridium met these requirements. The crucibles were therefore made of 1/2 inch discs cut from four-mil thick iridium sheets. These discs were first annealed in an oxidizing gas flame; a depression was then made in them by squeezing metal spheres of increasingly smaller diameters into their centers (annealing being performed in between each pressing step). The resulting iridium dish, containing the powdered sample, was placed on a ZrO_2 substrate which was in a quartz-glass pedestal in the vacuum system (see Figure 1). Heating was accomplished by means of an induction heater — the coils surrounding the vacuum system which contained the pedestal and crucible.

Temperature measurement and observation of the sample during heating were both performed with a telescopic optical pyrometer. An image of the sample was viewed in a 45° prism which was placed above the optically smooth flat top of the vacuum system. The system was calibrated by observing the apparent pyrometer temperature determined on the iridium crucible at the time of melting

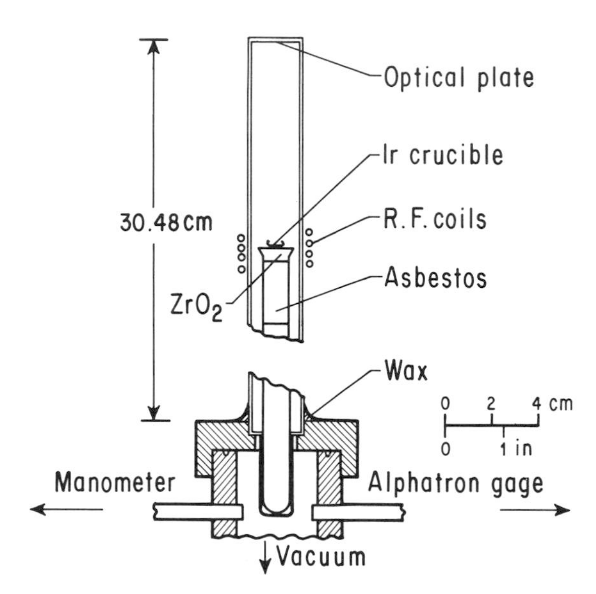

Figure 1. Experimental apparatus for the determination of boiling-point curves.

of various standard oxides (e.g., $CaSiO_3$; $MgAl_2O_4$; etc.). This calibration made it unnecessary to calculate the effects of iridium emissivity and spectral changes in the optical system. A scale was included in the optical system so that the size of objects in the crucible could be estimated.

An alphatron gage was used for pressure measurements below 10^{-3} atm. However, in the high-pressure range of greatest interest and significance for melts of tektite composition, a simple U-tube manometer was used.

III. Results and Calculations

Determinations were carried out on powdered samples of a natural philippinite tektite which had approximately 70 wt. percent SiO_2. Philippinites with this silica content generally have the following weight percentages of other oxides: Al_2O_3 - 15; FeO - 5; CaO - 3; MgO - 2.5; K_2O - 2.5 and Na_2O - 1.5. Tektites are unusual in that their iron is almost entirely in the ferrous state and they contain virtually no water.

For each determination approximately 100 mg of powdered tektite material was placed in an iridium crucible and the temperature was raised until the powder fused. With further increasing temperature, a point was reached at which a few bubbles formed. A slight decrease in the temperature (10-30°) at this point would cause the bubbles to decrease in size and, finally, to disappear. It was often possible to cause a bubble to "pulse" in size by slightly increasing and decreasing either the temperature or ambient pressure. Runs were usually made at constant pressure and many runs were made in the pressure range from 10^{-3} atm to 1 atm.

The results of these determinations — the "boiling-point curve" — are shown as the lower curve in Figure 2. The brackets on the

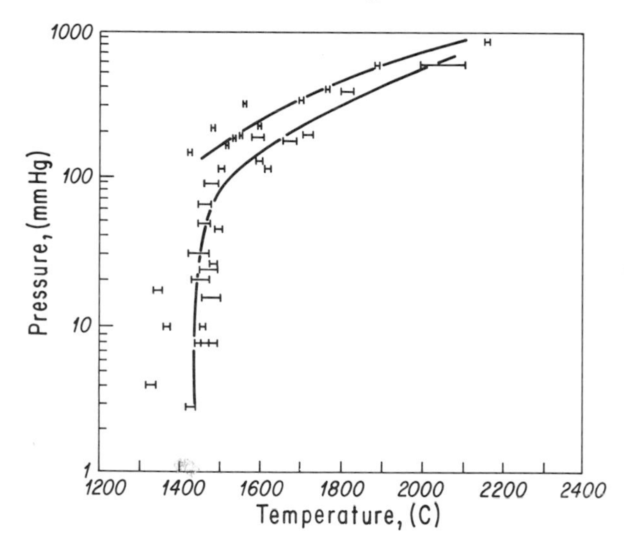

Figure 2. Lower curve: experimentally determined boiling-point curve for philippinite glass. Upper curve: vapor pressure curve calculated from the lower curve by adding calculated effect of surface tension.

data do not indicate limits of error, but instead represent the range of temperatures through which bubbles were observed to persist in the melt, (neither disappearing nor boiling effusively). If a melt was held for too long at high temperatures (i.e., more than 10-15 minutes), it was observed that the boiling temperature increased. This effect is undoubtedly due to vapor fractionation and a resultant decrease in the vapor pressure of the melt. In addition to the ambient, pressure is exerted on the bubbles from two sources: (1) the weight of overlying liquid and (2) the surface tension of the melt. The column of liquid above the bubbles was never greater than 2 mm in height. At 2.5 g/cc, the weight of liquid would be .5 g/cc or .0005 atm. In the pressure range of interest, this factor is negligible.

The pressure due to surface tension can be calculated from the expression:

$$P_s = \frac{2\sigma}{r}$$

in which σ is the surface tension and r is the radius of the bubble. Bubble radii were determined from size calibrations of the optical system. The value used for σ, (380 dynes), was determined by Chapman (1963) who derived this value for tektite glass at high temperatures from the contact angle of the melt with a substrate. The value of σ was assumed to be independent of temperature, because the variation of the surface tension in the temperature range studied is small enough to fall well within the limits of error of the present technique. The value of P_s calculated in this way is added to the boiling-point curve (Figure 2) in order to obtain the upper vapor pressure curve.

IV. Discussion

Having described the experimental method and results, the precautions listed in the introduction can now be discussed:

(1) The effect of the availability of nucleation sites for bubble formation is a problem. However, it was demonstrated that bubbles, once formed, could be made to shrink and disappear by either lowering the temperature or increasing the ambient pressure. In these experiments, therefore, equilibrium conditions obviously exist, and nucleation is not a significant problem. Nucleation sites for the condensing vapor are readily available at the walls of collapsing bubbles. Confidence in vapor-pressure results rests on one demonstration of reversible equilibrium between bubbles

and melt.

(2) It was obvious that the boiling temperature of the melt increased as the duration of the experiment increased. As previously stated, this was due to vapor fractionation and enrichment of the melt in the less volatile species. The increase in boiling temperature with time is obvious and readily determinable, and all the results presented here were obtained within the first 15 minutes of the runs, before significant vapor fractionation had taken place. Vapor fractionation is, however, a strong function of temperature, and runs at higher temperature must be considerably shorter than runs at lower temperature. Errors arising from such fractionation were kept to a minimum by increasing the temperature of the crucible very quickly — a procedure which is possible using a powerful induction heater, thin crucibles, small charges and a system with low thermal inertia. It is believed that errors produced by this method are less than other uncertainties inherent in the technique itself.

(3) If powder is used as a starting material, gases adsorbed onto the grain surfaces might appear as bubbles at the beginning of melting. If the amount of gas greatly exceeds the solubility limits for the melt, such an effect would be recognized because it would be impossible to make the bubbles shrink and disappear by lowering the temperature or increasing the pressure. On the other hand, small amounts of adsorbed gas may be soluble in the melt, and, to check for such an effect, some runs were performed by using one or two large fragments of the starting material in place of the powdered sample. (Such large fragments were not generally used because the reduced area of contact between the sample and crucible makes it difficult to attain desired temperatures quickly.) The results using single fragments were quite similar to results in which powdered material was used. An alternate method of eliminating the effect of adsorbed gases is to warm the powdered starting material — preferably in vacuum — before each run in order to drive off adsorbed gases.

The related problem of original volatile components in the sample material was minor in the present case because tektites are usually low in volatile constituents. However, it is possible that 100 ppm or less of water in the tektite could be the major constituent of the bubbles formed during heating. Even if this is the case, the <u>total</u> vapor pressure of the defined starting material is independent of such fractionation affects. Stated another way, the total pressure is specified regardless of the fact that pH_2O is very close to P_{total}. The only precaution is that, if H_2O is an important vapor component, vapor fractionation affects will be quite severe and the total vapor pressure will vary significantly with minor changes in volatile composition.

(4) The effect of melt viscosity is to induce sluggishness in the formation or collapse of bubbles, thereby reducing the accuracy of the observations. There appears to be no easy solution to this problem, although the effect of viscosity can be reduced by careful temperature control and sharp experimental observations. In some cases, notably pure SiO_2, viscosity is so high that bubbles cannot form and the material "steams" from the upper surface, rendering the boiling-point method useless.

The other precautions were treated as calculations in the preceding section.

The boiling-point technique for vapor-pressure determinations is classic. It is relatively simple in concept and design. Results obtained have considerable inherent uncertainty and the compositional range of materials for which it is applicable may be limited. However, the difficulties involved in other methods of vapor-pressure determination also result in large uncertainties and their complexity makes it attractive to attempt the boiling-point method first when making total vapor-pressure measurements of high-temperature melts.

BIBLIOGRAPHY:

Chapman, D.R. and Larson, H.K. (1963). The lunar origin of

tektites, NASA Tech. Note TN D-1556.

Kubaschewski, O. and Evans, E.L. (1958). Metallurgical thermochemistry. London: Pergamon Press.

O'Keefe, J.A. (1963). Tektites. Chicago: U. of Chicago Press.

O'Keefe, J.A. (1966). The origin of tektites. Space Sci. Rev., 6:174-221.

Peleg, M. and Alcock, C.B. (1966). Application of the torsion technique for the measurement of vapor pressures in the temperature range 1500-2500°C. J. Sci. Instr., 43:558-563.

Ruff, O. and Schmidt, P. (1921). Die damfdrücke der oxyde des siliciems, aluminiums, calciums und magnesiums. Z. Anorg. u. Allegem. Chem., 117:172-190.

Walter, L.S. and Adams, E. (1967). Vapor pressure of natural tektite melts at high temperatures and its application to aerodynamic analysis. J. Geophys. Res., 72:3717-3728.

Walter, L.S. and Carron, M.K. (1964). Vapor pressure and vapor fractionation of silicate melts of tektite composition. Geochim. Cosmochim. Acta, 28:937-951.

CHAPTER 11

Experimental Techniques in Dry Sulfide Research

G. Kullerud

I. Introduction

The chemistry of the common sulfide minerals is relatively simple inasmuch as most of them are composed of only two major elements, while some contain three, and only a few contain four or more.

Knowledge of the compositions and relationships of the mineral phases found in ores is important to geologists and metallurgists alike. The former need sulfide data in order to locate and exploit new deposits, while the latter require sulfide data to extract the metals from the ore.

Systematic investigations of synthetic sulfide systems during the last decade have yielded information about the thermodynamic stabilities of some of the common minerals and mineral assemblages as well as of the solid solutions they form with each other. Applications of the experimental findings to ores have already yielded considerable evidence concerning the temperatures, pressures, and chemical environments during ore formation. Thus potentialities of systematic investigations on geologically significant systems have been clearly demonstrated.

The experimental methods and the basic equipment that has been designed for investigations of the phase relations among the common sulfide minerals have not been described in detail in the literature. At this time, when a number of laboratories are in the planning stages of sulfide research programs, an account of some of the most used equipment and of the approaches employed is particularly useful. For this reason we shall discuss in this chapter the various types of common apparatus and methods used in sulfide research.

II. Rigid Reaction Vessels

Because of the reactive nature of sulfur there is a very limited choice of materials suitable to serve as containers for sulfide-type reactions. Pure silica does not react with sulfur, at least not below 1100 C, and is, therefore, used extensively for reaction vessels in sulfide research. Silica glass is a rigid material, and since the containers cannot be completely filled with the sample, vapor is an inherent phase in such experiments.

The vapor pressures inside the tubes are not known accurately. The maximum pressure attainable over a metal-sulfur system is known to be less than that over pure sulfur at the same temperature (see, for example, Allen and Lombard, 1917). The vapor pressure in runs where an excess of sulfur is present, to produce both liquid and vapor, can be estimated from the pure sulfur vapor pressure curve (Kullerud and Yoder, 1959). Unfortunately, even the vapor pressure curve for pure sulfur is not well known over the entire range of pressures attained in silica glass tube experiments.

The commonly used silica tube designs can be classified as simple tubes, tube-in-tubes, and thermocouple well or differential thermal analysis (D.T.A.) tubes.

A. Simple Tubes

The majority of the sulfide experiments to date were carried out in simple tubes prepared from high-purity silica glass tubing. The glass is heated over an acetylene-oxygen or natural gas-oxygen flame until it softens sufficiently to close the tubing. The closed end is then fire polished. Short tubing may by this method be closed at one end, and thus a single simple tube results from this operation. Long tubing may be heated in the middle and will, when slightly drawn during the closing process, produce two short tubes, each with one fire-polished, closed, and rounded end (see Figure 1, A. and B.).

The weighed starting materials are next placed in the tube.

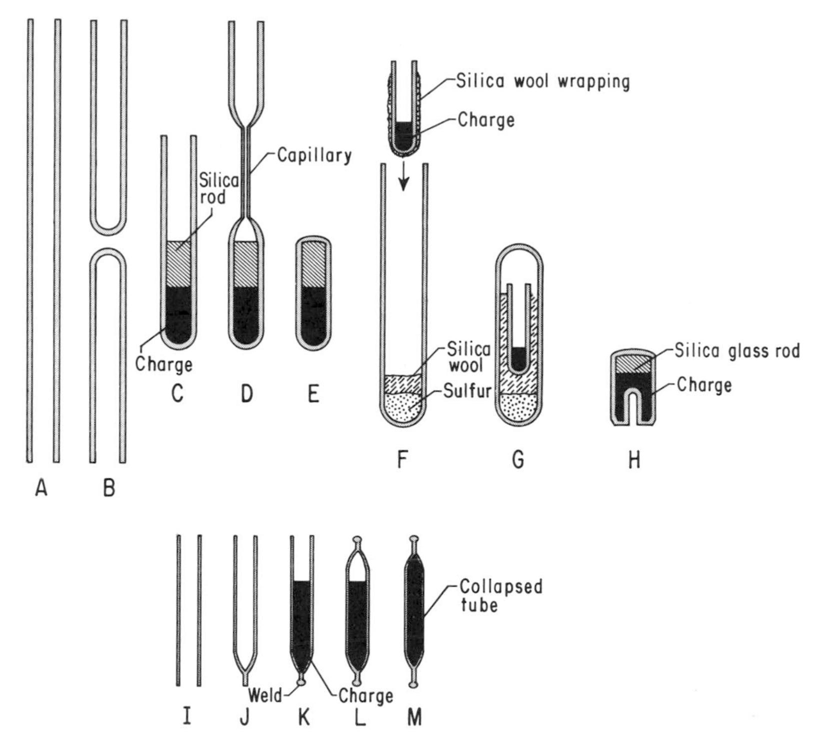

Figure 1. Reaction vessel designs: A.-E., simple silica tube arrangement with glass rod; F.-G., tube-in-tube reaction vessels; H., differential thermal analysis tube design; I.-M., collapsible tube arrangement.

Whenever loss of material to the vapor phase during experimentation is critical, the free space in the tube may be effectively reduced by placing closely fitting, short silica glass rods over the charge (Figure 1, C.).

The lower portion of the silica tube is then wrapped in a wet cloth which extends to about 1/8 inch from the top of the silica rod, that is to about 1/8 inch below the point where the tube will be sealed off. The wet cloth serves to prevent heating and oxidation of the starting materials during subsequent procedures. For many years the next step in the preparation of the simple tube experiments was to heat the tube at the upper end of the rod sufficiently to draw a capillary (Figure 1, D.). We have more recently found that it is not necessary nor always desirable to draw such a capillary. The tube is next connected with a vacuum pump and the system is evacuated. This has to be done carefully so as not to lose any of the starting materials in the process. The rate of evacuation is controlled by a stopcock on the vacuum line of the pump. The evacuation is always facilitated by the closely fitting silica rod which restricts the rate of air flow out of the tube. When the starting materials are very fine grained, a fine capillary provides additional support in the effort to avoid loss during evacuation. After obtaining the desired vacuum (0.02 mm Hg or better), the tube is heated rapidly with the vacuum pump maintaining the vacuum. The wet cloth remains in its place and the tube is rotated over the flame to distribute the heat evenly to a narrow zone on the tube circumference. The hot silica in this process collapses onto the upper end of the silica rod (Figure 1, E.). The silica rod, therefore, not only serves to reduce the vapor space drastically but also facilitates evacuation and shields the starting materials from being heated and volatilized. In the instances when the capillary still is a must, this silica rod in addition prevents oxidation, since the capillary is drawn when the tube is open to air.

Tubes of internal diameter from 3-7 mm and with wall thickness of 0.5-2.0 mm are used in most experiments. It was found that tubes of 3 mm inside diameter and 1.5 mm wall thickness, when carefully prepared and sealed, withstand the pressure of saturated sulfur vapor even at 1000 C.

A number of sulfide systems contain phases that are stable only at elevated temperatures and that cannot be quenched even by rapid cooling. The properties of such phases can only be studied in the temperature range at which they are stable. X ray powder diffraction patterns of such phases can be obtained at elevated temperatures by heating the material in closed silica tubes having a wall thickness sufficiently small to permit entrance and escape of x rays. Commercially available simple silica tubes with an outside diameter of 0.3 mm and with a wall thickness of 0.01 mm are employed for this purpose. These tubes withstand the pressure of sulfur saturated vapor up to about 700 C.

B. Tube-in-Tubes

This arrangement is commonly employed for the determination of composition or cation-to-anion ratios of certain compounds. Material of accurately known composition and weight is placed in a short open-end tube. This tube is next inserted into a larger tube (see Figure 1, F.) containing sufficient sulfur to produce saturated vapor during the run. The sulfur is covered by a layer of silica glass wool, which also is used as a liner of the large tube to keep the small tube in a central position. The silica glass wool prevents contact of the inner tube with liquid or solid sulfur and with the outer tube, but permits sulfur vapor to react with the material in the inner tube. The outer tube is heated and drawn into a capillary, and the system is evacuated and sealed in a way similar to that described for the simple tubes (see Figure 1, G.). Weighing of the inner tube after each run determines the amount of sulfur consumed in the reaction.

C. Thermocouple Well Tubes (D.T.A. Tubes)

This tube design, shown in Figure 1, H., is used for all differential thermal analysis experiments on sulfide-type materials. The silica-glass sample container has a thermocouple well parallel to the tube wall and extending about 6-8 mm into the tube. One of two identical platinum vs. platinum-10% rhodium thermocouples, connected by a gold-palladium wire, (65 Au - 35 Pd) is inserted in the well to be as close as possible to the charge. The other thermocouple is placed at a distance of about 5 mm from the sample container. In the many cases where loss of material to the vapor phase is critical, the free space can be reduced effectively by placing a closely fitting silica glass rod over the charge. However, the vapor space is usually reduced by introduction of a small amount of powdered quartz which also serves as an internal standard by virtue of its α - β inversion of 573.1 C. Evacuation and sealing of these tubes are performed in the same manner as for the simple tubes. Thermocouple "well tubes" of 5.0 mm inside diameter and 2.0 mm wall thickness containing saturated sulfur vapor are run as high as 1040 C without difficulty and thus withstand an internal pressure of about 120 bars even at this high temperature.

Determinations of phase relations in the sulfur-rich portions of certain systems, for instance that involving iron and sulfur, necessitated D.T.A. experimentation to temperatures above 1100 C. This was accomplished by placing the D.T.A. tubes and thermocouples inside a Stellite pressure vessel. By introduction of an inert gas, such as argon, into the Stellite vessel, the internal pressure in the D.T.A. tube is readily counteracted by inert gas pressure.

III. Collapsible Reaction Vessels

In order to investigate the sulfide systems in the absence of vapor it is necessary to employ reaction vessels that will not react with sulfur and will collapse by application of external pressure. Considerable success has been attained in the synthesis of hydrous minerals of high water pressures using sealed collapsible metal tubes. This technique was developed many years ago at the Geophysical Laboratory and was found directly applicable to the sulfide studies. Gold does not react with pure sulfur above 240 C (Kullerud and Yoder, 1959). Gold tubing can be welded to form reaction vessels that collapse readily under pressure, and, therefore, has been used extensively for experimentation in the pressure-temperature regions where vapor is absent. Unfortunately, during the investigations at the Geophysical Laboratory, gold has been found to react appreciably with certain sulfides and arsenides at elevated temperatures, and its usefulness as a reaction vessel is for this reason considerably limited. Efforts made to date have not been successful in locating another material that has a high melting point and still can be welded, and that in addition will collapse under pressure and will not react with the components of the sulfide-type systems. The gold reaction vessels, which usually are about 20 mm long, are made from high purity commercial gold tubing with an inside diameter of 2.6 mm and a wall thickness of 0.2 mm (Figure 1, I.). One end of the tube is next closed by pinching it tightly shut (Figure 1, J.) and welding it with a DC carbon arc (Figure 1, K.). The materials to be investigated are then weighed into the tube and the other end of the tube is pinched shut and welded[1] (Figure 1, L.). The gold tubes are carefully weighed on completion of the sealing operation and weighed again at the termination of the experiment. Thus leakage during experimentation is duly recorded. These tubes collapse readily in the pressure apparatus, transmitting any applied pressure to the material inside the tube (Figure 1, M.). Thus the external and internal pressures are closely the same; the only difference being the small pressure required to cause the tube to collapse.

IV. Characteristics and P-T Limitations of Reaction Vessels

The various kinds of reaction vessels were designed for different purposes. There are definite limitations to the usefulness of each kind of tube. Some experiments can only be performed

---

[1] Evacuation of these collapsible tubes is not required. The tube volume after welding, but before application of external pressure, is about $(1.3^2 \times \pi \times 10) \simeq 50\ mm^3$. Less than 50% ($25\ mm^3$) of this volume actually consists of air; we shall assume, however, that air occupies 50% of the volume and that it behaves as an inert gas, although the oxygen in the air certainly will react to form insignificant amounts of oxide. During experimentation at, for instance, 2000 bars and 1000 K (727 C) the $25\ mm^3$ of air which existed at 1 atm and at room temperature has diminished to a volume of about $0.04\ mm^3$, whereas the solid materials have undergone a negligible volume change. Under these conditions, therefore, the air amounts to only $(\frac{0.04}{25} \times 100) = 0.16$ volume percent and is negligible for most investigations.

successfully when taking place in tubes of special designs. It is therefore necessity to realize the limitations inherent in each design before the results of experimental efforts can be evaluated adequately and correctly. (See Chapter 12 for additional discussion.)

A. Presence of Vapor

All rigid tubes contain a space that gives vapor opportunity to form. This space must exist because the tubes cannot be filled completely with material. Thus vapor occurs as a phase in all rigid tube experiments. The presence of this phase naturally limits the use of rigid tubes to the P-T region containing vapor, as is shown for example in the region below the (Cu_9S_5) CuS + L + V and (CuS) Cu_9S_5 + L + V univariant curves of Figure 2.[1] It is noted that the univariant curves for the three reactions CuS $\rightleftarrows$ Cu_9S_5 + V, CuS + V $\rightleftarrows$ L, and Cu_9S_5 + V $\rightleftarrows$ L can all be determined by rigid tube experiments, provided vapor pressures can be measured. Silica glass tubes have been used as high as 1200 C without reacting measurably with sulfur or sulfides. Since temperatures beyond 1100 C only occasionally are required during explorations of sulfide-type systems, temperature alone does not put severe restrictions on the use of silica tubes. The pressures developed over sulfide systems are sometimes appreciable and will in tubes containing saturated sulfur vapor reach 1 atm at 444.6 C, 25 atm at about 750 C, and 100 atm at about 1000 C (Kullerud and Yoder, 1959). The ability of the tubes to withstand internal pressure increases significantly by an increase in the ratio between wall thickness and inside diameter of the tubes. However, rapid chilling in cold water to quench the phases stable above 1000 C sometimes results in cracking of even the very heavy walled tubes used to contain

[1]Each of the four univariant curves, which originate in the invariant point C, is specified by the phase (enclosed in parentheses) that is absent.

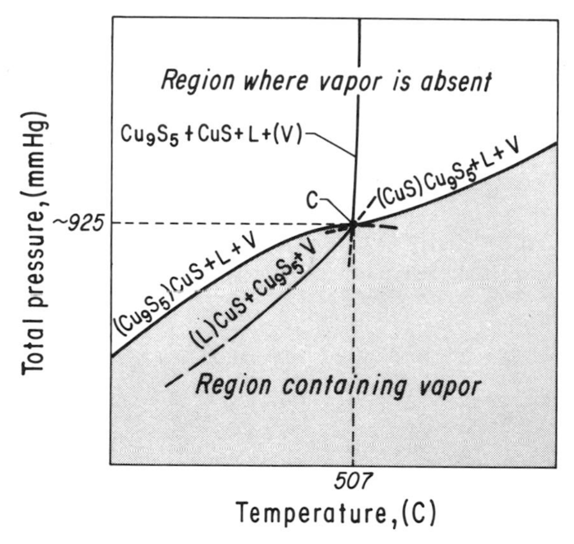

Figure 2. Invariant point in the Cu-S system at 507 C, at which the four phases covellite + digenite + liquid + vapor coexist. The univariant curves (Cu_9S_5) CuS + L + V and (CuS) Cu_9S_5 + L + V separate the vapor-containing region (below the curves) from that where vapor is absent (above the curves) (after Kullerud, 1957).

saturated sulfur vapor. The pressure range is extended very significantly when the tubes are supported by external pressure. This can be done by inserting the tube in a cold-seal type bomb and using H_2O or a gas such as argon as a supporting pressure medium. It is a serious handicap with this procedure that the tubes cannot be rapidly cooled to quench their contents at the termination of the experiments.

B. Absence of Vapor

The collapsible tubes do not contain vapor when exposed to an external pressure sufficiently large to squeeze the reaction vessel walls against the charge and thus eliminate the free space in the tubes. Pressure-temperature regions that do not contain vapor, therefore, may be explored by this method. In Figure 2 is shown the P-T region that can be explored by use of collapsible tubes. It is noted that the univariant curve for the reaction $CuS \rightleftarrows Cu_9S_5 + L$ is the only curve of the four surrounding the invariant point C that can be determined by such experiments.

V. Preparation of Experiments

The individual experiments, whether they take place in rigid silica tubes of various designs or in collapsible tubes, normally contain weighed amounts of chemicals (usually referred to as starting materials). Various types of weighing techniques are employed, depending on the accuracy that is required. The method by which the components are weighed accumulatively on the balance pan is sufficiently accurate for many experiments in addition to being fast. After weighing, the mixture of chemicals is swept by means of a camel's-hair brush through a funnel and into the tube. Owing to static electricity, a portion of the chemicals adheres to the tube and funnel walls. This is in part freed by tapping, and the remainder is cleaned out with a pipe cleaner. The usefulness of good quality, lint-free pipe cleaners cannot be overemphasized. It was also found that cleaning of the tube with acetone (both inside and out) just before weighing and loading the charge greatly reduces the loss of material (essentially sulfur) by eliminating static electricity. By this type of weighing procedure the loss is not known, and although it usually is small it adds uncertainty to the composition of the run. A test that was performed by weighing the tube before and after inserting the mixture of chemicals revealed that as much as 0.8 mg of material of a 100 mg total was lost when fine-grained materials were used. When better than one percent compositional accuracy is required, the clean tube is first weighed empty and then weighed again after addition of each desired reagent and after the walls have been meticulously cleaned with pipe cleaner. With this technique and employing a semi-micro "Gram-atic" type balance, which, with care, will weigh within 0.1 mg, compositions within $\pm 0.1\%$ of that desired (when using about 100 mg runs) are obtained comparatively quickly. By using about 400 mg of material in each experiment the uncertainty in composition is readily decreased. Much higher compositional accuracy is obtained by the same procedure when a good analytical balance is used instead of the "Gram-atic" balance. The compositional uncertainty by use of, for instance, a Beckering balance is easily less than 0.01% in 400 mg runs. Accurate weighings in this manner are time consuming, but are necessary whenever exact compositions are required. Preparation of an experiment may take as little as 10 minutes when the chemicals can be weighed rapidly in an accumulative manner on the balance pan of a "Gram-atic" balance. When the chemicals are weighed in the tube on the same kind of balance, preparation of an experiment may require about 20 minutes, and when the analytical balance is used for the same kind of weighing one hour may easily be required for preparation of one experiment. Because of the time element it is advisable to evaluate the necessary compositional accuracy and let that decide which weighing method should be used.

VI. Starting Materials

Theoretically, pure chemicals are non-existent and the materials used for synthetic studies, therefore, do not possess thermodynamic properties identical with those of truly pure materials. The validity of application of experimental results obtained on impure chemicals to the theoretically pure systems is, therefore, largely dependent on the kind and magnitude of the impurities. For this reason all chemicals should be quantitatively analyzed to ascertain their degree of purity as well as the identities and amounts of the unavoidable impurities. Hundreds of examples can be found in the literature of usage of terms such as: high purity, sufficiently pure, reagent grade, 99.9% pure, etc. Such chemicals are, without further analyses, undesirable for phase-equilibrium studies. This is demonstrated by the following example: Material containing 99.9% iron is readily available. It can be shown that, if the 0.1% impurity consists of materials such as nickel and cobalt, which are chemically similar to iron, the results of the phase studies are similar to those obtainable with theoretically pure iron. However, if the 0.1% of foreign material consists of carbon or phosphorus, the determined phase relations will probably display no more than incidental similarity to those of the pure system.

VII. Furnaces

The loaded reaction vessels are heated in various types of furnaces, depending on the nature of the individual experiment. Horizontal furnaces are used to heat most of the simple silica tubes, the thermocouple well tubes, and the tube-in-tubes. Relatively few of the simple silica tubes, those of 0.3 mm outside diameter and 0.01 mm wall thickness, are heated in a vertical position, in a special furnace operating in vacuum inside an x ray diffraction high-temperature powder camera. The collapsible tubes are heated inside Tuttle or cold-seal pressure vessels (Tuttle, 1948, 1949), which again are placed inside vertically mounted split furnaces. The vertical and split furnaces are commercially available and will not be described further. (See Chapters 4, 5, and 6) However, the horizontal, cylindrical, electrical furnaces were designed and constructed for sulfide investigations at the Geophysical Laboratory.

Figure 3 shows this furnace arrangement in cross section. The furnace has a metal shell 18 inches long and 10 inches in diameter. The ends are made of Transite, and magnesia powder serves as insulation.

The furnace core consists of two concentric alundum tubes. The inner tube, which has an inside diameter of 1 inch, is centered inside the larger outer tube and is in the central part of the furnace covered by a cylindrically shaped nickel jacket 1/16 inch thick and 6 inches long. This jacket serves to distribute the heat uniformly over its length inside the furnace. The outer tube is wound with No. 18 Nichrome wire; the central 10 inches have six turns per inch, and a 3 inch section on either side has ten turns per inch. The total resistance at room

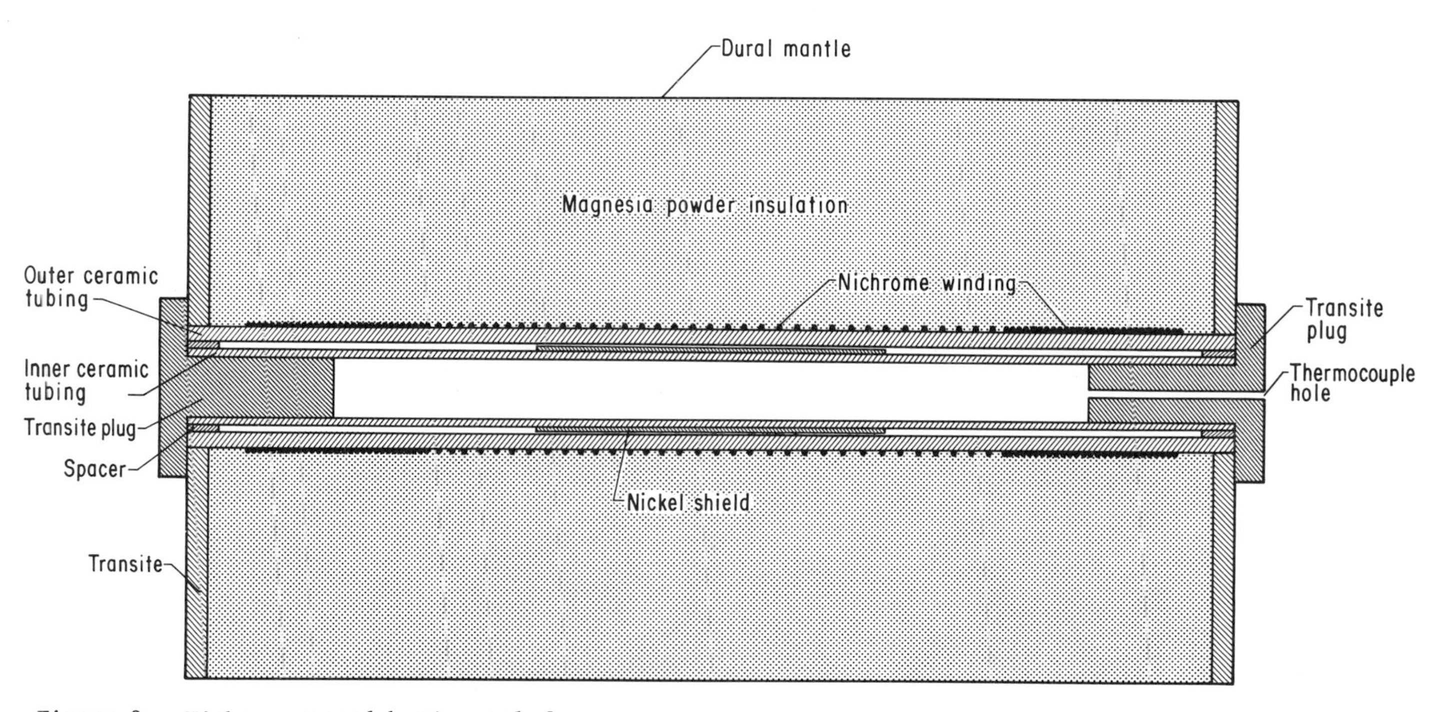

Figure 3. Nichrome wound horizontal furnace.

temperature is about 24 ohms. The ends of the alundum tubes are closed by Transite or firebrick plugs (see Figure 3). The one or more thermocouples used for temperature reading and recording are inserted into the furnaces through the rear plug.

If the furnace is electrically heated with a source of constant voltage, and if its surroundings remain unchanged, equilibrium will be obtained, after some time, between the electric energy supplied to the furnace and the heat energy lost to its surroundings. The temperature now remains constant at any one point in the furnace, but varies slightly from one point to another. Because of the horizontal furnace arrangement, the centro-symmetrical winding, and the centrally located nickel jacket, the maximum temperature (hot spot) is always recorded close to the center of the furnace (i.e., about 9 inches from each end). When the temperatures are measured at various distances on both sides of the hot spot a typical temperature distribution is recorded such as that shown by the curve in Figure 4. Relatively sharp drops in the temperature occur at and beyond the ends of the nickel jacket, which ideally extends 3 inches on either side

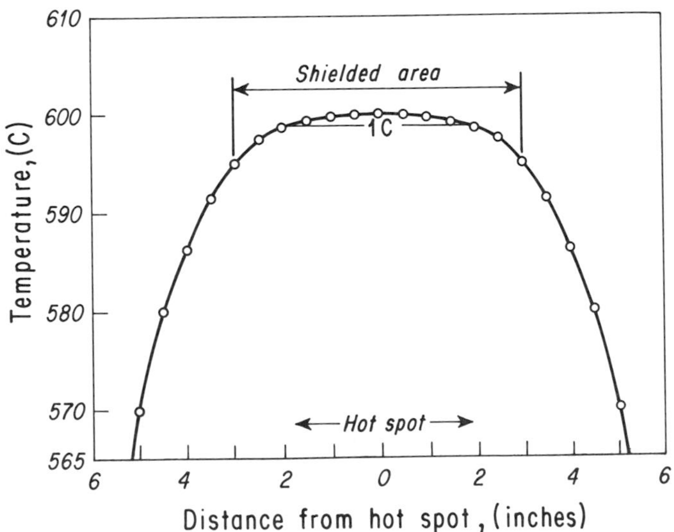

Figure 4. Temperature distribution at hot spot temperature of 600 C in a horizontal furnace with Ni-shield.

of the hot spot. It is noted that the temperature stays within 1 C of that of the hot spot over a total distance of 3.5 inches at 600 C. At higher temperatures this distance decreases and at lower temperatures it increases. In Figure 5 are given the results

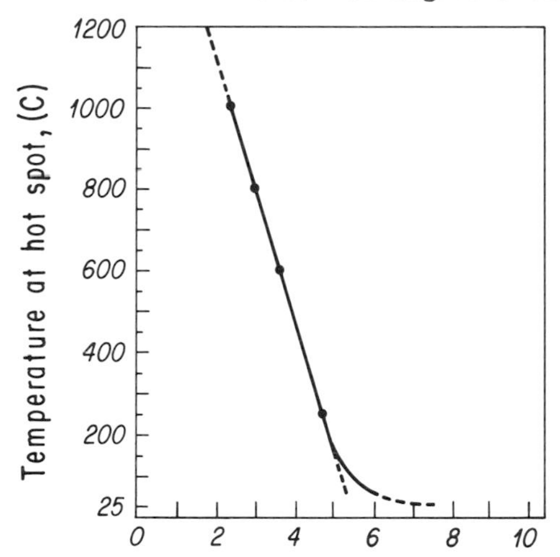

Figure 5. Curve showing the length in inches of zone of constant temperature (± 1 C) between 25 and 1200 C.

of measurements from 75-1000 C of the distances within which the temperatures stay within 1 C of that of the hot spot. Here distances (D) in inches are plotted along the abscissa, and temperatures (T) at the hot spot are plotted along the ordinate. It appears that the relationship between D and T is, for all practical purposes, linear between about 150 and 1000 C. Below 150 C the curve becomes hyperbolic and approaches 25 C (room temperature) asymptotically. The curve has not been determined above 1000 C.

The silica reaction vessels are commonly about 1 inch long and hardly ever exceed 2 inches in length. Thus, by placing the center of such a tube at the hot spot, the temperature variation over the length of a 2 inch tube is less than 1° even at 1000 C. The sharp temperature decrease occurring at each end of the nickel jacket is utilized in some experiments in which a temperature gradient is desirable or even required. Thus, by using long simple silica tubes without inserted rods, temperature gradients of 50 C or even 100 C may readily be obtained. Figure 4 shows that by using a 3 inch tube and placing the inner end 2 inches, and the outer end 5 inches from the hot spot a temperature gradient of 25 C is obtained over the length of the tube. Steep gradients of this kind promote growth of crystals in the cooler end of the tube and are for this reason very useful.

A. Temperature Control and Measurement

The voltage of the commercially available AC electric current normally varies considerably over a 24 hour period. Such current will, when used to heat a furnace through a variable transformer (Variac), produce temperature variations that during a day may easily in some geographic areas exceed ±50 C. This may be a sufficiently close temperature for some purposes, such as the synthesis of compounds stable over a temperature range exceeding that occurring in the furnace, but is wholly inadequate for most purposes. Much improved temperature control can be obtained when the raw current is regulated, by one of the numerous types of high-quality commercial voltage regulators, before being distributed to the furnaces. Variac controlled regulated current will keep the temperature in the furnace constant within ±3 C. This is satisfactory for many purposes, such as determinations of solvus curves. Instead of central voltage control, however, each furnace may be equipped with commercial temperature regulators of various designs. These will normally, when operating on raw current, maintain the temperature inside the furnace constant within ±3 C, and within ±1 C when preregulated current is used.

Control of the furnace temperature within ±1 C even when using raw current is obtained with temperature regulators built and extensively used at the Geophysical Laboratory. In an early design, a Variac, built into the regulator chassis, is driven to a balanced position by means of a two-phase electric motor. This motor is actuated by an out-of-phase voltage amplified from an unbalanced Wheatstone bridge. The 6 volt bridge circuit consists of a manually variable resistance and a platinum sensing element (resistance). The sensing element is wound in with the Nichrome winding in the furnace in order to detect without delay any temperature change at the heat source. Even a very slight variation in the temperature, therefore, will change the resistance on the sensing element side of the Wheatstone bridge. This will be compensated for by the motor driving the Variac to a higher or lower position, depending on whether the temperature variation is due to loss or gain of heat.

Recent advances in electronics have led to designs of solid state temperature controllers which employ silicon-controlled rectifiers (Hadidiacos, 1969). The new regulator permits the temperature for an experiment to be set even from a cold start without overshoot. Temperatures up to 1000 C can be readily maintained within ± 1 C. The parts for such regulators cost as little as $125, which compares very favorably with costs of about $600 for parts for the earlier design discussed above.

Furnace temperatures are usually measured with chromel-alumel[1] (28 gauge) or platinum-platinum 10% rhodium (Pt/Pt10%Rh) thermocouples. Chromel-alumel thermocouples are very desirable at temperatures below about 700 C. The absolute electromotive force produced at the chromel-alumel junction is approximately four times as large as that produced at the Pt/Pt10%Rh junction. For this reason it is easier to read accurate temperatures, below 700 C, and to detect small temperature variations with chromel-alumel than with Pt/Pt10%Rh thermocouples. Chromel-alumel thermocouples will oxidize measurably in air in a few weeks at 700 C, at 800 C in a few hours, and at 900 C in a matter of minutes.

[1]Chromel consists of a nickel-chromium alloy, and alumel consists of a nickel-aluminum alloy.

Such oxidation results in uncertainties in the temperature readings. The measured electromotive force at the chromel-alumel junction at constant temperature increases with time as a function of oxidation. Consequently, if an oxidized and an unoxidized thermocouple are situated side by side in a furnace, the apparent temperature obtained from measuring the electromotive force of the oxidized couple will be higher than that obtained from the unoxidized couple. This difference may at 800 C in rare cases exceed 15 C. On the other hand, it was found that a chromel-alumel thermocouple heated at 600 C in air in a furnace for three years did not oxidize measurably, as indicated by the electromotive force, which at the termination of this period was identical with that of a new thermocouple. Chromel-alumel couples when used in the split furnaces or in the horizontal furnaces appear to be increasingly influenced by the AC current in the furnace windings when the temperature exceeds 750 C. It appears that at this temperature the current through the furnace windings (about 3 amperes) is sufficiently strong to produce a significant AC induction current in the thermocouple. This secondary current, while not affecting the absolute electromotive force of the couple, is carried through the thermocouple leads to the galvanometer or Nul Point indicator circuits, and reduces the sensitivities of these instruments drastically. Thus, for instance, while a temperature change of $1/4^{\circ}$ below 700 C brings the indicator needle completely off scale, a change of 10° at 900 C does not deflect the needle more than about 1/5 of the scale.

In experiments employing cold-seal or Tuttle pressure vessels the thermocouples are inserted into a hole drilled in the Stellite and are, therefore, in direct contact with the metal during the run. (See Chapters 4, 5, and especially 6.) Pt/Pt10%Rh thermocouples readily form solid solution with Stellite even at moderate temperatures. This reaction influences the electromotive force and renders the thermocouples useless. Chromel-alumel thermocouples, however, do not react measurably with the Stellite and are used exclusively for this kind of experimentation. Because of rapid oxidation above 750 C the thermocouples should be discarded after each run exceeding this temperature.

The temperature in each furnace is printed at about 5 minute intervals on a continually moving chart of a 12 channel temperature recorder, providing complete records of the thermal history of all experiments. These temperature readings are usually reliable within ± 3 C. For more accurate work, however, the temperatures are measured by means of accurate potentiometers and a cold junction at 0 C (ice-water bath) as temperature reference.

VIII. Quenching Techniques

The phases formed at elevated temperatures in silica and gold tubes are commonly assumed to be preserved at room temperature when the reaction vessels are chilled rapidly from the temperatures of experimentation. The phase relations existing under equilibrium conditions at elevated temperatures are then deduced by investigation of the chilled sample. Many published phase diagrams are based on the results obtained in this manner. Cooling at such a rate that the phases, stable at the temperature of the experiments, are preserved and remain unchanged at room temperature for a period ample for their identification is referred to as quenching. Phases that can be preserved unchanged are, in accordance with this terminology, called quenchable, and those that change are labeled non-quenchable. It is not always possible to decide by room-temperature identification whether the rapid chilling resulted in quenching of the high-temperature phases. For this reason, investigations at temperature by D.T.A. and by x ray diffraction are often necessary preliminaries to serious efforts of phase-relation determinations by "quenching type" experimentation. It has now been discovered[1]

[1] It may be mentioned that ore minerals such as chalcopyrite, bornite, digenite, chalcocite, heazlewoodite, and many others at elevated temperatures occur in non-quenchable crystalline modifications.

by one or both of these methods that non-quenchable transitions are not at all uncommon in sulfide type systems.

In the majority of cases where quenching is possible, rapid cooling of the material in the rigid reaction vessels is usually accomplished by quickly removing the silica tubes from the horizontal furnaces, using long tongs, and plunging them directly into a cold water bath. In this way the temperature is lowered very rapidly to that of the room; for instance, cooling from 800 C takes place in 3-5 seconds.

The vapor pressure in the rigid silica tubes falls during the cooling at a rate dependent on the temperature and degree of filling of the tube.

Cooling of the material in the collapsible reaction vessels is accomplished more slowly since these tubes are heated inside Stellite cold-seal or Tuttle bombs, which because of their mass do not permit rapid chilling. It is possible to cool the material in a gold tube inside a Tuttle bomb from 800 C to below 100 C in 35 seconds by applying a stream of water directly on the bomb. When cold-seal type bombs are employed, cooling from 800 C to below 100 C requires 2-3 minutes because of the larger mass of steel to be cooled.

The pressure in the cold-seal and Tuttle bomb is maintained constant during the cooling because the system is open to the pressure reservoir. The rates of cooling of the materials contained in collapsible tubes inside the two types of pressure vessels are much slower than the rate of cooling of the silica tubes. Because of these slower cooling rates, reactions due to cooling are more commonly observed in the sulfides investigated using pressure equipment, in particular cold-seal bombs, than in the rapidly chilled sulfides synthesized in silica tubes.

A. Preparation of Polished Sections

One or more polished sections are made on representative samples of the quenched materials. It is desirable to be able to investigate the results of the experiments within minutes of opening the reaction vessels.

Conventional methods for obtaining powder x ray diffraction patterns in less than 1 hour from smear mounts on glass slides have been perfected years ago and are in common use (Adams and Rowe, 1954, review methods). The conventional procedures for preparation of polished sections, however, require considerable amounts of material and could not be adopted for the sulfides for this and the following two reasons: the time involved and the heating required for mounting of the sample.

Many sulfides oxidize very rapidly in air, others undergo transitions after a few hours at room tempeature, and others attempt to adjust to the new conditions by delayed equilibrations involving development of exsolution textures. The time, usually several hours, needed for completion of conventional polished sections poses a serious handicap on the progress of any synthetic investigation, because the results of metallographic studies on one series of experiments form the basis for the next. Thus, by using conventional polished sections, many hours must pass between the completion of one set of experiments and the start of a new set.

The preparation of conventional polished sections in Bakelite requires both heat and pressure, which are commonly applied in the presence of air. Short (1940) points out that one method calls for heating at 135 C under a pressure of 35,000 psi for several hours, whereas another (the Harvard method) specifies 130 C and 15,000 psi for 14 minutes. Under such conditions many sulfides will oxidize very rapidly and will react in an effort to satisfy the new equilibrium demands imposed by addition of air and a mounting medium and by exposing this complicated new system to considerable temperatures and pressures. Mounting of polished sections by this method affects the compositions and textural relations of common sulfides such as pyrrhotite and pyrite, and produces new phases such as covellite when certain copper-iron sulfides are involved.

Investigations of various plastic cold-setting media revealed that "Caulk Kadon" resin for restorative dentistry (Kadon liquid and Kadon powder) is a well suited medium for mounting of polished sections. The sample is inserted in a hardened drill rod[1] steel cylinder, shown in Figure 6 a.), on top of the closely fitting

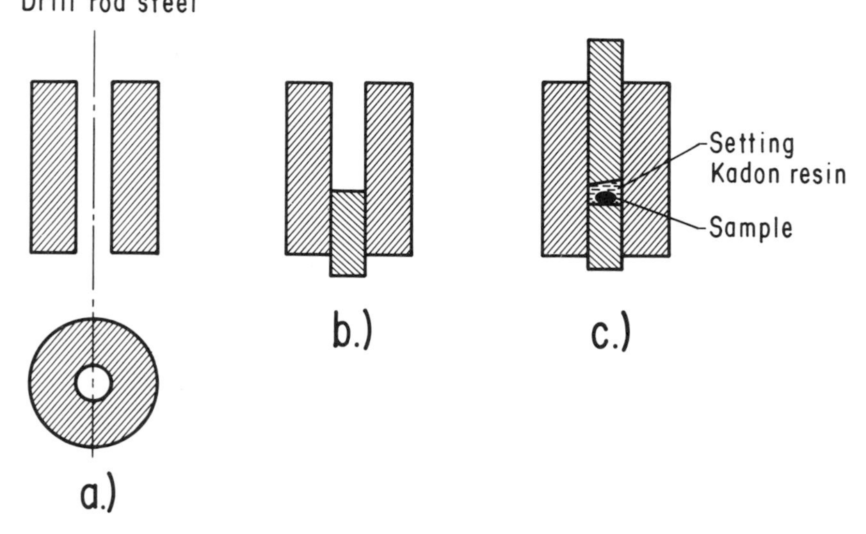

Figure 6. Press for mounting of polished sections: a.) cylinder of drill rod steel; b.) cylinder with lower piston rod inserted; c.) upper and lower piston rods in position after insertion of sample and Kadon resin.

lower piston rod also made of drill rod steel (Fibure 6 b.)). Next Kadon plastic powder is placed on top of the sample and is dropped onto the powder. The upper piston rod, which is also made from drill rod steel and which is beveled as shown in Figure 6 c.), is next inserted into the cylinder. The piston rods are forced against the sample in a small vise for about 5 minutes, during which the plastic sets to a solid cylindrical pellet. The bottom of the pellet is smooth, flat and normal to its axis, and the top is also smooth and flat but (because of the beveling) not quite

[1]Drill rod steel is a cold drawn carbon tool steel which contains about 1.2% C, 0.2% Mn, 0.025% P, 0.025% S, and 0.25% Si.

normal to its axis. This pellet is attached near one end of a standard 1 inch by 1 3/4 inches microslide by means of Vinylite plastic, as shown in Figure 7 a.). When the Vinylite is still soft the slide is overturned and gently pressed toward a glass

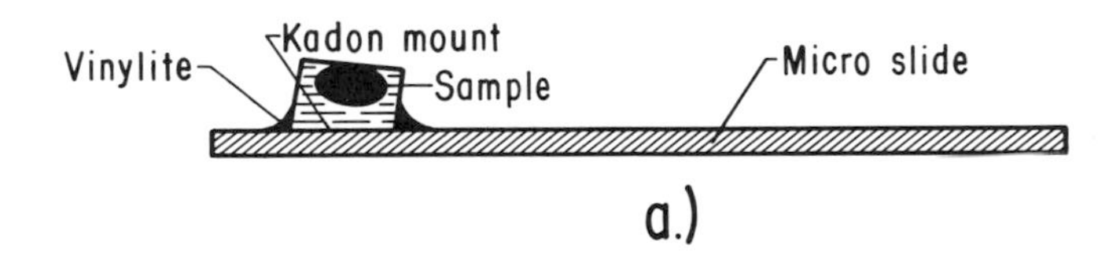

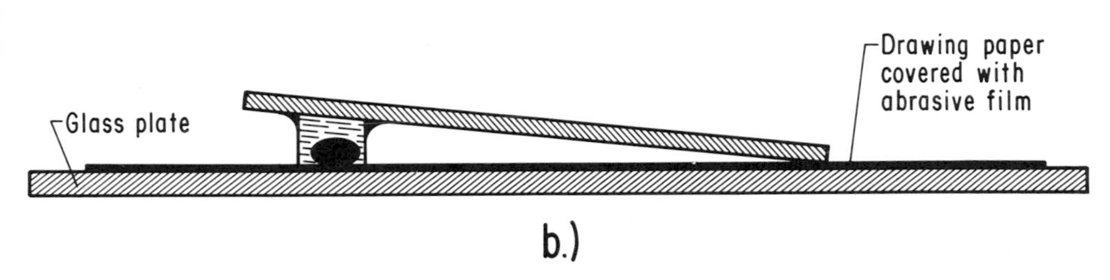

Figure 7. a.) Pellet, consisting of sample mounted in Kadon resin, glued to a standard size microslide; b.) polishing on drawing paper covered with film of abrasive and backed by a glass plate. For preliminary grinding the drawing paper is replaced by emery paper.

plate. Because of the beveling, the lower pellet face will be nearly parallel to a polishing surface when the glass slide is turned over as shown in Figure 7 b.). In this position grinding of the pellet, to obtain sufficient exposed material on an even surface, now may proceed on fine sandpaper or emery paper. Then, polishing is accomplished by using first American Optical Company "AO Emery" M305 and then Linde Fine Abrasive A5175 on drawing paper (or even 5 x 7 inch filing cards) held firmly against a glass plate to assure flatness. In some mounts containing mixtures or intergrowths of relatively hard and soft minerals (as, for instance, pyrite and pyrrhotite) scratches will form almost invariably during the preparation of the polished sections. These scratches are commonly caused by plucking of small particles of the harder mineral. These fine hard grains are, of course, especially damaging to the softer parts of the polished sections. Although most sections can be studied adequately even when deep scratches occur, it is commonly preferable to prevent the occurrence of scratches in photomicrographs because of the danger of misinterpretation. These scratches will often disappear if the final polishing is done by using ultra-fine Fe_2O_3 rouge and by applying a very slight pressure on the slide to prevent further plucking.

B. Identification of Phases

The products of quenching type experiments are identified at room temperature by their macroscopic physical properties, optical properties in reflected light, and x ray powder diffraction patterns. Phases that are stable at elevated temperature but have been found to be non-quenchable are identified by means of x ray powder diffraction patterns taken at the temperature of experimentation. The thermal and compositional stability limits of the phases identified by this method are determined by differential thermal analyses in all cases where the heat effects of the phase changes are sufficiently strong to be recorded.

1. Physical Properties

The quenched phases are frequently recognized by their color and, when crystal faces or even single crystals occur, by their crystal morphology. The magnetic behavior of many compounds is often a very useful means of identification. A small hand magnet often serves as a valuable tool in distinguishing magnetic from essentially non-magnetic phases.[1] The hardness of a mineral may sometimes be of help in its

[1] As an example may be mentioned pyrrhotite (magnetic) from pyrite (non-magnetic).

identification. The relative hardness of the various compounds (or minerals) present in polished section is usually demonstrated by the character of the scratches or by the relief between adjacent phases produced in the section during polishing.

2. Optical Properties

The products of all quenching experiments are routinely investigated in polished sections by means of reflecting microscope techniques. The phases in such sections are identified by their optical properties. Less than 1/100 of one percent of a phase with distinctive optical properties can readily be detected by this method. It is commonly observed that color differences between phases are considerably enhanced when polished surfaces are viewed in oil immersion. For this reason, we employ oil immersion not only at high magnification in the customary way, but also at low and medium magnifications. When it is desirable, the relative proportions of the phases occurring in these sections are determined by a "point-counting" technique similar to that described by Chayes (1949). Many minerals that appear to be almost identical optically may be distinguished by various staining techniques, which have been reviewed by Short (1940).

3. X ray Patterns

The quench products also are routinely identified by their x ray powder diffraction patterns made at room temperature. Most minerals have very distinct x ray patterns, allowing fast and sure identification. These patterns are obtained by using either a diffractometer with a lacquer-acetone smear mount on a glass slide or a powder camera with the material mounted in cylindrical rods. Reflections appearing on diffractometer charts or powder films are, when accurate interplanar spacings are required, measured in relation to the reflections of an internal standard which is mixed with the sample. The standard and unknown reflections are measured to 1/100 inch on the diffractometer charts with a metal scale and vernier constructed at the Geophysical Laboratory. The reflections on the films are measured to the nearest 1/20 mm.

Comparison of the interplanar spacings, of well known materials, determined by using diffractometer charts (goniometer speed $1/2^{o}$ per minute and chart speed 1 inch per degree) and by using films from 114.56 mm diameter powder cameras (where 1 mm = 1°) shows that the first method gives the more accurate results. In addition, the x ray charts and determinations of reflections are by this method completed in 1 or, at the most, 2 hours, whereas commonly the film exposure alone requires 10-15 hours when the powder camera method is used.

The x ray powder patterns for most known sulfide-type minerals are given in the literature; at least in the older literature, however, the data are usually incomplete and inaccurate. For this reason a file containing indexed x ray powder charts or films of synthetic compounds, as well as of natural minerals, is invaluable.

BIBLIOGRAPHY:

Adams, L.H. and Rowe, F.A. (1954). The preparation of specimens for the focusing-type x ray spectrometer. Amer. Mineral., 39:215-221.

Allen, E.T. and Lombard, R.H. (1917). A method for the determination of dissociation pressures of sulphides, and its application to covellite (CuS) and pyrite (FeS_2). Amer. J. Sci., 43:175-195.

Chayes, F. (1949). A simple point counter for thin-section analysis. Amer. Mineral., 34:1-11.

Hadidiacos, C.G. (1969). Solid-state temperature controller. J. Geol., 77:365-367.

Kullerud, G. (1957). Annual Report of the Director of the Geophysical Laboratory. Carnegie Inst. Washington Year Book, 56:195-197.

Kullerud, G. and Yoder, H.S. (1959). Pyrite stability relations in the Fe-S system. Econ. Geol., 54:533-572.

Short, M.N. (1940). Microscopic determination of the ore minerals. U.S. Geol. Survey Bull., 914.

Tuttle, O.F. (1948). A new hydrothermal quenching apparatus. Amer. J. Sci., 246:628-635.

Tuttle, O.F. (1949). Two pressure vessels for silicate-water studies. Bull. Geol. Soc. Amer., 60:1727-1729.

CHAPTER 12

Investigations in Hydrothermal Sulfide Systems[1]

H.L. Barnes

I. Introduction

In experimental studies of sulfides, methods involving hydrothermal fluids are now common and becoming increasingly important. Aqueous fluids must be present, of course, in experiments designed to investigate hydrothermal solutions coexisting with sulfide assemblages but this only partially accounts for the popularity of hydrothermal methods. In addition, reactions among minerals can often be catalyzed by aqueous solutions and, consequently, such fluids can be used to achieve equilibrium at lower temperatures than possible in anhydrous experiments. Furthermore, solutions of properly chosen compositions may be used to control and buffer sulfur fugacity at any desired value (below sulfur-saturation) instead of just those where mineral pairs, such as pyrite + pyrrhotite, are stable.

Experiments on hydrothermal sulfide systems are capable of providing <u>precise</u> data on either the aqueous fluid composition or on mineral compositions and stabilities, but not on both simultaneously. If the aqueous fluid is of primary interest, then for accurate analysis, samples must be withdrawn from a reaction vessel through a valve system. In such an experiment, quenching is not efficient and there is a probability that the minerals present would react during cooling. Alternatively, if the compositions and phase relations among the minerals are of primary interest, then a small sealed container encapsulating both the minerals and the solution can be used for the experiment.

---

[1] Contribution No. 70-38, Ore Deposits Research Section, College of Earth and Mineral Sciences, The Pennsylvania State University.

In this case, very rapid quenching is possible so that virtually no change in mineral compositions takes place during cooling. However, in both types of experiments, the principal problem is the inertness of the container, both to sulfidation and to corrosion. We shall examine these problems in detail in this chapter.

To the hazards intrinsic in experiments at high pressures and temperatures on any materials, sulfide systems add the possibility of poisoning from leaking hydrogen sulfide, as well as the possibility of an explosion due either to leaking of the hydrogen often present in such reduced environments, or from failure of the pressure vessel from hydrogen embrittlement or excessive pressure. For these reasons, the experimenter completely inexperienced in these pressure systems would be well advised to learn the requisite techniques in a laboratory where experience with sulfides could be acquired comparatively painlessly. It is not possible to include here all the precautions and detailed procedures mandatory for success with hydrothermal sulfide experiments.

II. Control of Chemical Environment

In an aqueous solution at any specific temperature, the fugacities of oxygen, hydrogen, and sulfur and the pH are intimately interrelated and are not generally independent variables. However, the relations between fO_2, fS_2, and pH, which are shown on Figure 1, are schematically correct for a large range of P-T conditions (Barnes and Czamanske, 1967). The stabilities of the oxides and sulfides common in ore assemblages depends on fO_2 and fS_2 of the hydrothermal environment, for example:

$$3FeS_2(\text{pyrite}) + 2O_2(g) \leftrightarrow Fe_3O_4(\text{magnetite}) + 3S_2(g), \qquad (1)$$

and their solubilities depend on these fugacities plus pH and concentration of complexing ligands. For these reasons, fS_2 is doubly important. Many details of these basic chemical relations are discussed in "Geochemistry of Hydrothermal Ore Deposits," edited by Barnes (1967) and only those of direct experimental consequence are considered here.

A. Sulfur Fugacity

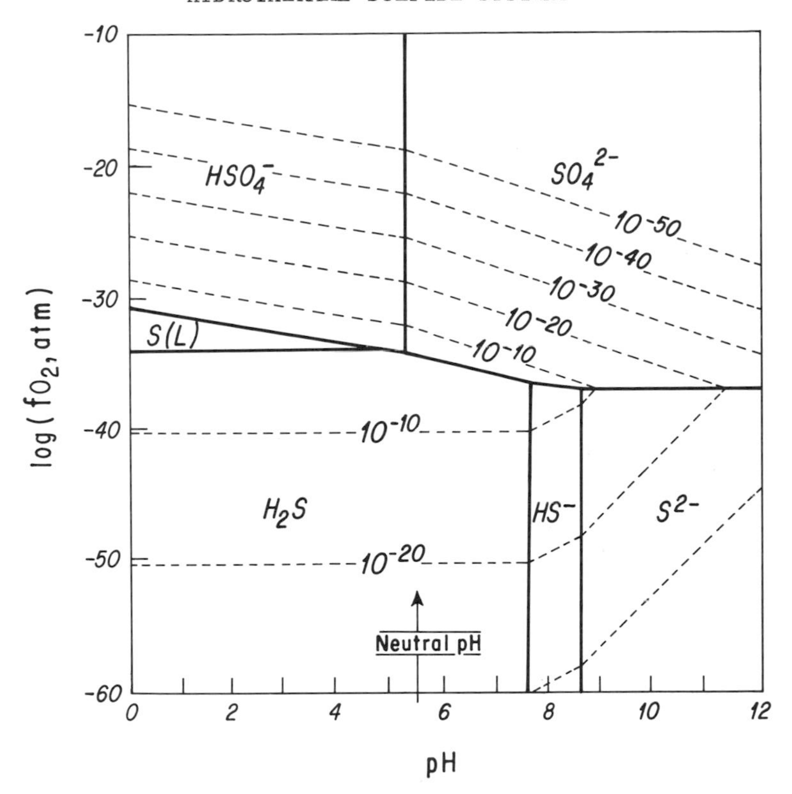

Figure 1. Equilibrium relations between oxygen and sulfur fugacities and pH at 250 C where the total activity of all sulfur-containing aqueous species is 0.1. Fields where aqueous species are predominant are delineated with solid lines and contours of S_2 fugacity are dashed.

A specific fS_2 can be realized at any one temperature through proper choice of fO_2, pH, and ΣS, where ΣS is the total of the activities of all sulfur-containing aqueous species. When H_2S is the predominant aqueous species pH has little effect on fS_2 but in the alkaline region where HS^- or S^{2-} are predominant, fS_2 varies over about 20 orders of magnitude, decreasing to the higher pH's. At constant pH and ΣS, fS_2 increases linearly with fO_2 where the sulfide species, H_2S, HS^-, or S^{2-}, are predominant, but decreases rapidly with fO_2 where the bisulfate or sulfate ion are predominant. Note that the same fS_2 may occur, and some mineral buffer pairs (such as Hg + HgS, Barnes et al., 1967) may be stable, at the same pH and two quite different oxidation states, one where aqueous sulfide ions are predominant and one where sulfates predominate. At constant pH and fO_2, fS_2 also increases as a function of $(\Sigma S)^2$.

Usually it is impossible to calculate the exact combination of pH, fO_2, and ΣS required to give a specific fS_2; either the thermodynamic data are unavailable or the uncertainties in the data are too large. However, the fS_2 can be achieved by successive approximations with a series of experiments. Consider the following two examples.

Assume that the assemblage bornite (Cu_5FeS_4), pyrite (FeS_2), and chalcopyrite ($CuFeS_2$), had been synthesized in the dry state but that the product was too finely crystallized for further experiments. From extrapolation of vapor pressure measurements at high temperatures on the anhydrous system, we estimate that fS_2 is about 10^{-9} for this assemblage at 250 C. In order to recrystallize it at 250 C, we first select a pH where the solubilities will be large for better catalysis of the crystallization, perhaps pH 8. In our first attempt at recrystallization using a small volume (relative to the volume of the minerals) of 0.1 molal NaHS solution, we find that the product is entirely pyrite + bornite. Clearly, the fS_2 of the solution was too high and the minerals were unable to buffer it at the necessary lower value. In the next experiment, the ΣS (or aH^+) should be lowered until all three sulfide minerals become stable.

A second example is a situation where the stability limit of one or more sulfide minerals is being investigated at one temperature. In such experiments, buffering of fS_2 could be advantageously controlled by the solution composition using solutes considered later in this chapter. In this case, the volume of solution must be large in comparison to the volume of minerals present. A broad range of fS_2, can be explored most easily in sulfide solutions simply by a series of experiments at one ΣS and at pHs from neutral to highly alkaline.

B. Oxygen Fugacity

Control of fO_2 in sulfide solutions is automatic unless either the metal lining of the reaction chamber acts as a buffer by producing hydrogen:

$$Me + H_2S(aq) = MeS + H_2(g) \quad (2)$$

and

$$H_2O = H_2(g) + 1/2\ O_2(g) \quad (3)$$

or some oxidizing agent, such as air, is included in the experiment. Normally, the fO_2 of H_2S solutions is the result of the reaction

$$H_2S(aq) + 4H_2O(\ell) = HSO_4^- + H^+ + 4H_2(g) \quad (4)$$

generating some hydrogen which then equilibrates with water as in reaction (3). At higher ΣS, liquid or solid S may form rather than HSO_4^-, but the net effect is still to buffer fO_2 above about 250 C where S is not inert in oxidation-reduction reactions. Similar reactions can be written for water plus either HS^- or S^{2-} but they equilibrate with H_2 plus SO_4^{2-} instead of HSO_4^-. The resulting values of fO_2 lie in a narrow band two log units below those along the sulfide-sulfate boundaries shown on Figure 1.

C. Useful Solutes

The aqueous complexes formed by many sulfides may be sufficiently soluble for catalysis of solid state reactions in solutions where H_2S, HS^- or S^{2-} species are predominant, but maximum solubility of these complexes often occurs where NaHS or other alkali bisulfide concentrations are high and H_2S concentration is significant. These bisulfide solutions are simply prepared by bubbling H_2S through a CO_2-free sodium hydroxide solution until the pH drops to the desired value:

$$H_2S(g) + NaOH(aq) = H_2O + NaHS(aq). \quad (5)$$

With an excess of H_2S this reaction quantitatively converts NaOH to NaHS so that the amount of NaHS formed is equivalent to the quantity (not concentration) of NaOH present in the original solution. The resulting pH is fixed by the ratio of H_2S to HS^- through the equilibrium:

$$H_2S(aq) = H^+ + HS^-. \tag{6}$$

It is possible to measure this pH directly in solutions near room temperature, using standard pH electrodes, but these electrodes will become poisoned if left in sulfide solutions for a few hours. At higher temperatures, the pH can be calculated from the measured H_2S pressure, experimentally fixed HS^- concentration, equilibrium constants for H_2S and NaHS, and activity coefficients.

Besides the pH buffer, H_2S - HS^-, which is useful in near neutral solutions, there are several other buffers which are compatible with sulfide solutions. The sulfate buffer, HSO_4^- - SO_4^{2-}, is effective in moderately acid solutions, the NH_3 - NH_4^+ buffer in neutral to acid solutions depending on temperature, and the H_2CO_3 - HCO_3^- buffer in weakly acid to weakly alkaline solutions. The optimum buffering capacity occurs where the two buffering species are at equal activity which is found at the pH equal to the negative log of the appropriate ionization constant. These constants have recently been compiled and discussed by Marshall (1969).

Extremely low sulfur fugacities can be controlled readily by using high alkalinities in solutions without any added sulfide ions except those contributed by slight dissolving of the sulfide minerals (see Figure 1). For example, NaOH can be used as the solute and the resulting OH^- ions are often effective complexing agents to generate the necessary solubility. The fS_2 can be readily adjusted by varying the NaOH concentration and thereby fixing the pH.

Control of fS_2 using alkali halide solutions as solvents is not as effective as the above solutions because there is only the weak buffering action resulting from changing ΣS in response to changes in alkali halide concentrations but without the major pH effect present at high alkalinities. However, because chloride and other complexes are often appreciably soluble, it is sometimes useful to combine two solutes such as sodium chloride and sodium hydroxide, one to form soluble complexes and the other to control

fS_2.

At temperatures below about 100 C, metastable ammonium polysulfide solutions, $(NH_4)_2S_X$, are useful to fix high values of fS_2 and also to complex some metals (Clark and Barnes, 1965). The polysulfides dissociate at higher temperatures, however.

Amine solutions, such as NH_4I, are often very effective solvents, but with increasing temperature, they become acidic through reactions such as

$$NH_4I = NH_3 + H^+ + I^- \tag{7}$$

which leads to severe corrosion of reaction vessels. In addition, control of fS_2 by the solution is minor and only over a narrow range so that effective buffering must depend on the mineral assemblage.

D. Gas Pressures

1. H_2S

The maximum concentration of H_2S attainable in subcritical aqueous solutions is limited by the formation of either the solid hydrate, $H_2S \cdot 6H_2O$, or of an immiscible, H_2S-rich, sulfide liquid, as shown on Figure 2. The quadruple point, 29.5 C, 324.7 psi (22.1 atm)[1] is the upper stability limit for the sulfide solid which means that if the H_2S pressure is above 325 psi in the reaction vessel, then tubing connected to the vessel must be maintained above 30 C to prevent filling with the solid. This problem is important particularly for small diameter lines leading to pressure gages where clogging with the solid can lead to low readings, but the solid can be easily melted with the aid of a heating tape.

If sufficient H_2S is added to water in a reaction vessel to form the solid and a vapor phase, then on heating, the pressure rises rapidly along the univariant curve, $S_S - L_A - G$, while an

---

[1]Many pressures in this chapter are given in psi units because of the low overall range of pressure involved. A units conversion table is given in Chapter 4 as Table 4.1.

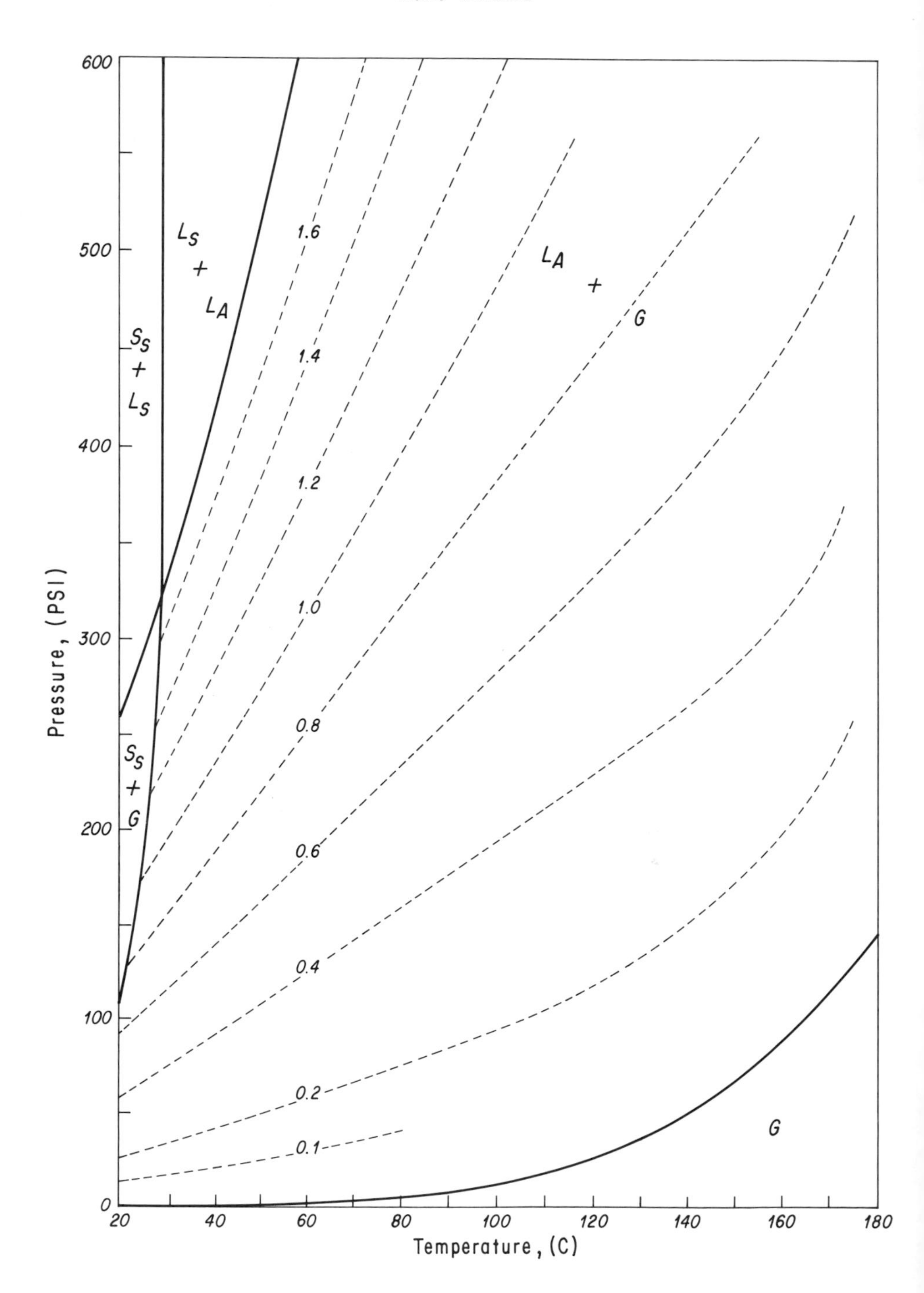

600
500
400
300
200
100
0
Pressure, (PSI)
20
40
60
80
100
120
140
160
180
Temperature, (C)
L_S + L_A
S_S + L_S
S_S + G
L_A + G
G
1.6
1.4
1.2
1.0
0.8
0.6
0.4
0.2
0.1

increasing amount of S_s dissolves in L_A. At the quadruple point, all remaining S_s melts to L_s and on further heating with a vapor present, the system follows the L_s - L_A - G curve with rapidly increasing pressure and minor dissolving of L_s in L_A. This curve terminates at the critical point, 100.2 C, 1306 psi where L_s and L_A become miscible.

The effect of solutes such as NaHS is to increase the size of the field L_A + G at the expense of surrounding fields but the change is apparently less than 2 C along the S_s - L_A - G and L_s - L_A - G curves. Ionic solutes are not readily accepted into solid solution in the sulfide solid nor in solution in the aprotic (non-conducting) sulfide liquid.

In an isochoric (constant volume) experiment, the P-T path on heating follows virtually along a constant aqueous composition contour (Figure 2) if starting with L_A + G and the volume of L_A exceeds that of G. If S_s is present, the system follows S_s + L_A + G until all S_s dissolves and then moves along a constant composition of L_A. If sufficient S_s is present to saturate L_A to 30 C, then S_s melts to L_s and the experiment follows first S_s + L_A + G, then L_s + L_A + G, and finally a composition contour in L_A + G.

Heating an aqueous solution causes expansion of a magnitude that may be surprising to neophytes (Figure 3). For example, water expands in volume about three times between 25 C and its critical point at 374 C. Because aqueous solutions have very low compressibilities, it is foolhardy to risk an explosion by heating a rigid vessel until it becomes full of liquid, unless provision is made to limit the pressure by controlled bleeding or expansion. The maximum expansion likely in an H_2S-saturated aqueous solution

---

Figure 2. Phase relations of the system H_2S-H_2O. Solid lines bound the phase regions and dashed lines are contours of H_2S concentration in the aqueous solution in moles per liter. S_s = sulfide solid hydrate, L_s = sulfide liquid, L_A = aqueous liquid, and G = gas phase.

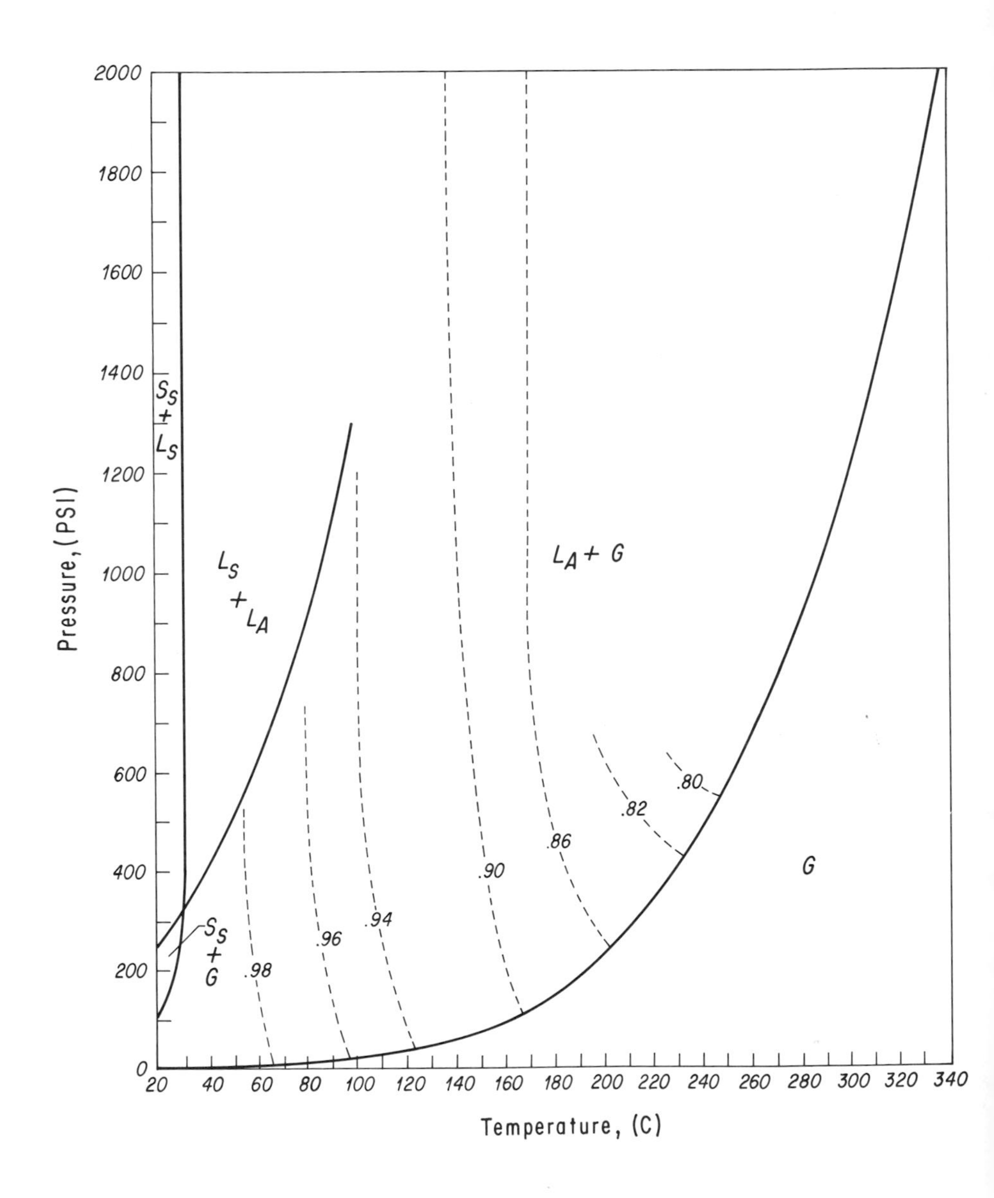

Figure 3. Densities of H_2S-saturated water in the system H_2S-H_2O. S_s = sulfide solid hydrate, L_s = sulfide liquid, L_A = aqueous liquid, and G = gas phase.

can be estimated from Figure 3, because dissolving of ionic solutes causes a decrease in volume below that expected for the purely binary solution.

2. SO_2

The pressure of sulfur dioxide may also be important in hydrothermal sulfide systems. SO_2 is the predominant gas species in equilibrium with sulfate ions to at least 300 C, and because it is relatively soluble (as sulfates) compared to H_2S, the pressures are low at a given concentration. However, at higher temperatures of about 500-600 C, depending on oxidation state, the reaction

$$H_2S(aq) + 2H_2O \rightarrow SO_2(g) + 3H_2(g) \quad (8)$$

may generate high SO_2 pressures. For example, rising temperature increases pressure over $FeS_2 + Fe_{1-X}S + Fe_3O_4 + H_2O$ more rapidly above 575 C where SO_2 is predominant than below this temperature where H_2S is predominant (Barnes and Czamanske, 1967). The equilibrium change in gas species from H_2S to $SO_2 + H_2$ in an isolated reaction vessel implies no inherent change in the oxidation state; such a change can only result from reaction with added oxidizing or reducing agents.

3. H_2

If a hydrothermal system contains an effective reducing agent, then hydrogen is produced by reaction with water. For example, reaction (8) can release large quantities of hydrogen at temperatures of a few hundred degrees. Other sources are the metals or some common mineral assemblages (those including pyrrhotite, for example) may generate pressures large enough to burst some hydrothermal systems even at room temperature. Obviously, calculations are advisable to evaluate this possibility, and runs under virgin conditions should be carefully monitored.

Hydrogen may also escape without any apparent leakage. This small molecule is difficult to contain, particularly at high temperatures, due to diffusion through the metals of any reaction vessel. For typical hydrothermal vessels, H_2-loss becomes important generally above about 500 C. The result is that any run not buffered in oxidation state may gradually oxidize if held for long periods at high temperatures. Diffusion can be minimized by using lower temperatures, shorter runs, buffered systems, and vessels with thicker walls and made of less permeable alloys.

Another insidious effect of hydrogen under pressure is to react with alloys, especially the high strength types usually used in reaction vessels. These reactions may be of several types but all lead to a loss in ductility and frequently of strength and are called hydrogen embrittlement. Several causes of this phenomenon may be found in a discussion by Rogers (1968) who shows that both structural state and composition of metals influences the extent of embrittlement. Because hydrogen embrittlement cumulatively affects pressure vessels, periodic testing is useful to avoid catastrophic rupture in vessels often used near their stress limits with reduced systems.

III. Equipment Design

A. Vessel Materials

The principal obstacle to successful experimentation with hydrothermal sulfide systems is the container. Ideally, it should remain completely inert and unreactive to hydrothermal fluids internally and air externally but with sufficient strength for use at high pressures and temperatures. Of the stainless steels, those with highest Cr-content are less reactive with sulfide solutions. Both 310 and 316 stainless steels have proven to be satisfactory with the former being more inert and the latter having better mechanical properties for use at higher temperatures. The easily machined stainless steels, 303 and 304, become tarnished by sulfide solutions more readily than 310 or 316.

A vessel can be protected by using a liner or by plating or coating the interior. Dickson and his associates (1963) and Weissberg (1970) have developed a design using flexible liners of teflon or gold and have used it for several excellent studies. However, such liners require support by a fluid between the liner and vessel wall in contrast to a plating; therefore, a much more elaborate plumbing system is needed to nearly exactly balance the internal and support pressures particularly as temperature changes or during sampling. Because liners flex under any differential pressure, they also tend to develop pinhole leaks that are extremely difficult to locate and repair. Consequently, plating is preferable for most applications. In addition, recent improvements in readily available commercial solutions for coating with teflon or for plating with gold, platinum, rhodium, etc. allow these to be applied in the laboratory when needed,without the delays of shipping vessels out for treatment.

Chromium plating may effectively reduce reaction of bomb walls with aqueous sulfide solutions. This plating should have minimum crystallinity (to reduce permeability), a thickness in excess of .003 inches, and no undercoating of copper. The quality of the surface finish appears not to be of major importance. Reaction vessels of 316 or 310 stainless, chrome-plated, have been used with concentrated sulfide solutions to 400 C and only near this temperature does tarnishing become important. However, weakly acid or acid chloride solutions can readily dissolve chrome plating even near 25 C.

For extremely corrosive solutions, including acid halide solutions, teflon-coating is useful up to about 290 C for sulfide systems. The coating must be applied to a roughened surface and involves spraying on, then baking first a primer, then a series of enamel coats aggregating about .005 inches thickness. There is a minor release of an acidic compound, perhaps from the "accelerator" solution used in the primer, beginning at about 200 C but the quantity is unimportant except for solutions with completely unbuffered pH. Above about 290 C, teflon degenerates to produce a monomer which hydrolyzes giving HF and other toxic compounds.

At moderate temperatures, Au and Pt plating or liners will sulfidize by reactions such as

$$Au + HS^- \rightarrow AuS^- + 1/2\ H_2(g) \qquad K_{250\ C} \simeq 10^{-3.3} \qquad (9)$$

and

$$Pt + H_2S\ (g) \rightarrow PtS + H_2(g) \qquad K_{250\ C} \simeq 10^{+4.3} \tag{10}$$

unless conditions are kept sufficiently reduced with appropriate H_2 pressures. Weissberg's results (1970) show that gold liners require strongly reducing conditions to prevent corrosion, about 10 atm H_2 at 25 C and about 100 atm at 250 C, to keep Au solubility below 50 ppm with concentrated (5 m) bisulfide solutions. Both high H_2 pressures and very low sulfide concentrations are necessary to prevent forming PtS from Pt liners in low temperature runs. In chloride solutions, even mildly reducing conditions prevent corrosion of a gold liner. When corrosion takes place in chloride solutions, the reaction is principally one of forming soluble complexes such as

$$Au + 4Cl^- + 3H^+ \rightarrow AuCl_4^- + 1.5\ H_2(g). \tag{11}$$

For experiments above 500 C, only gold or platinum have so far been found to be sufficiently inert to sulfide solutions. Tests of other relatively stable metals, for example, Ta and Ti, show that although they may neither dissolve nor sulfidize, they can become useless due to hydrogen-embrittlement; most metals simply form a sulfide. At temperatures above 500 C and oxidation states fixed by the quartz-fayalite-magnetite buffer, Au is insoluble in chloride solutions but slightly soluble in dilute sulfide solutions; with the higher oxidizing conditions of the hematite-magnetite buffer, Au becomes somewhat soluble also in chloride solutions.

B. Simple Closures

The closure of a reaction vessel to be used with sulfide solutions must be reliable over repeated cycles to required P-T conditions and cannot include a seal which would be threatened by corrosion. To minimize costs, it should not be mechanically complex nor require frequent remachining and replacement of gaskets. Two simple designs which have satisfied these criteria in several years use are shown in Figures 4, 7 b.) and 8. Both seals tolerate high temperatures so that the entire reaction

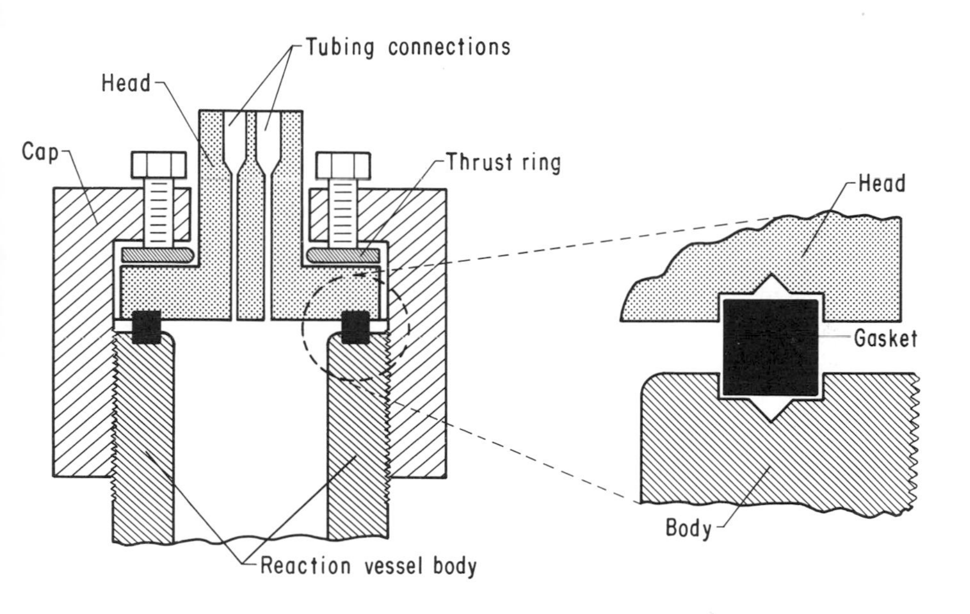

Figure 4. A simple flat-gasket seal for hydrothermal reaction vessels for low pressure and temperature experiments.

vessel may be isothermal.

The design shown on Figures 4 and 7 b.) is reliable within the P-T limits of Figure 6 if a gasket of 304 stainless steel is annealed prior to use. The gasket, if not marred when removed from its grooves in the head and body, can be annealed again for reuse repeatedly until it becomes too thin to keep the body and head separated. The thrust ring and bolts should be of high strength alloys but thermal expansion must be considered so that loading on the gasket is not reduced by higher temperatures. The gasket should be loaded by gradually and symmetrically tightening the bolts in the cap using a torque wrench until the gasket becomes elastically compressed. Independent seals are made on each edge of the V-shaped grooves in the gasket seats, doubling reliability.

The closure design of Figure 8 is similar but includes a delta-gasket and isolation of the tubing connection to lower temperatures so that the vessel can be used to higher pressures and temperatures, to 600 C and 15,000 psi. Above this temperature, the bolts slowly spread where they contact the thrust ring. For this reason, the bolts should be ground prior to use so that they are slightly coned and also so that the end contacting the thrust ring is flat. All threads must be coated with a good high temperature lubricant such as "Never-Seez" or "Fel-Pro" to prevent seizing and to prolong thread life.[1]

The gasket provides an unsupported area seal which tightens as internal pressure increases. This gasket, of 304 stainless steel, must be large enough so that it contacts the beveled portion of the seat, near the base, to make an initial seal. If the gasket can be rocked slightly in its seat, it is safe to use it again. Normal life has been 8-12 runs. Only after several years use does remachining of the seats become necessary. Consequently, the equipment cost per experiment is minimal.

C. Agitation

Reaction rates are increased significantly, particularly between gases and solutions if run materials are actively mixed during an experiment. Several methods have been used. A simple stirrer is effective but requires a gas-tight stuffing box for sealing around the drive shaft, thereby limiting the P-T range of application. To remedy this problem, magnetically driven impellers were designed which were entirely enclosed in the reaction vessel and produced either a rotary motion or mixing with a plunger action. The magnetic drive is limited by the thickness of the vessel which reduces the field intensity at the impeller;

---

[1] These are available from Fel-Pro, Inc., Skokie, Illinois 60076 or Never-Seez Compound Corp., 2910 S. Eighteenth Ave., Broadview, Illinois 60155.

furthermore, effective impetus to the impeller requires the use of ferromagnetic materials. Such materials require plating or coating for protection from hydrothermal fluids and the impeller is particularly subject to both mechanical and chemical corrosion. In addition, the upper limit of ferromagnetic behavior is limited to temperatures below the Curie Point, 350 C for Ni, 770 C for Fe, and 1131 C for Co.

Mechanical agitation of the entire reaction vessel avoids the above problems and normally may use motions of rocking (shaking) or dashing. The dashing motion is most efficient for gas entrapment by a solution (Snyder et al., 1957) but requires more energy and is intrinsically less durable and more complex mechanically.

For maximum agitation in rocking autoclaves, the angle through which the vessel is moved and the rate must be properly matched, typically at 30° of arc at 36 cycles per minute. Less important factors are the length/diameter ratio of the vessel and the fraction of the vessel filled; the viscosity and surface tension of the fluid also have some effect but are fixed by the research. For evaluation of these design factors, the reader is referred to Hoffman et al. (1948).

In non-isothermal experiments, convection can stir the contents of a reaction vessel if the temperature gradient decreases upwards. If the top of the vessel is warmer so that there is no convective mixing, significant concentration gradients will eventually result, particularly in supercritical fluids, due to the Soret effect (see Harned and Owen, 1958).

D. Control of Temperature and Pressure

Isothermal conditions are very difficult to achieve in reaction vessels used over a range of temperatures if only a single furnace winding is used, because heat loss varies with temperature and is rarely the same from both ends of a vessel. For this reason, dual furnace windings are useful, particularly for vessels over 4-6 inches long, to adjust the temperatures of each end either for eliminating gradients or for adjusting them precisely. Only a single temperature regulator is needed if the control thermocouple is placed at the vessel wall and between the two windings and if power input is proportioned between the two furnace windings. Convenient proportioning is provided either by variable transformers or by a pair of silicon-controlled rectifiers (SCR slaves).

Besides the great flexibility of this type of temperature control system, there is a bonus safety feature. Temperature controllers occasionally fail to operate properly so that overheating is possible. If the output of the proportioning units is minimized so that the controller is fully on over 95% of the time to maintain the set temperature, then failure of the controller can only supply a few percent increase in power and a nominal rise in temperature. This precaution is particularly valuable where run materials include H_2S, the vessel is large, and the controllers are senile.

Where corrosion becomes a problem thermocouples or thermistors can be placed in wells projecting into the interior of a vessel in wells drilled into the body or cap of a reaction vessel, or, least satisfactorily, in external contact with the vessel. Any temperature gradients between the vessel contents and the measuring thermocouples, or inaccuracy of the thermocouples, can be detected simply by heating a pure liquid for which the P-T properties are well known. Precise measurement of the pressure should agree with the vapor pressure of the fluid at the measured temperature. Water is useful for this purpose particularly from 200-374 C where the measured pressures, about 225-3200 psi, are large enough to be accurately measured with ease.

The accuracy and durability of pressure gages are impaired if sulfide solutions are allowed to fill their bourdon tubes. Two types of pressure transducers can be used to isolate a gage from the fluids of a hydrothermal vessel, using either a floating piston or a diaphragm plus bubble-free water between the isolator and the gage.

The floating piston method (as in Figure 8) uses a piston which moves freely within a polished cylinder in response to a pressure differential. The piston is fitted with one or two o-ring gaskets set in circumferential grooves which seal between the piston and the cylinder walls so that fluid does not pass the piston but also so that friction between the piston and the walls is minimized.

For an accurate pressure measurement, a correction must be made for the hysteresis caused by this friction. If a small-displacement, manual pressure generator (or alternatively, one or more valves having stems with at least 5 mm travel) is connected to the system between the isolator and gage as in Figure 8, then water can be pumped in, or removed, to force the piston to move. A maximum pressure reading, taken while water is being pumped in, should be averaged with the minimum value read while water is extracted. Hysteresis tends to decrease as pressure increases due to compression of the o-rings which also reduces the reliability of the seal. To minimize diffusion of H_2S, butyl o-rings are effective but these, and other materials commonly available cannot be used at high temperatures, consequently the isolator must not be too close to the hot reaction vessel. The piston transducer may be used to pressures over a kilobar but the useful range is limited by the particular size of o-rings used, due to friction at low pressures or leakage at high pressures. In normal use, the piston is positioned near the vessel end of the cylinder to prevent corrosion of the cylinder.

The diaphragm transducer is similar to an aneroid barometer in design (Figure 7 a.)). No hysteresis could be detected within ± 1 psi when using a diaphragm of .002 inch-thick 304 stainless steel with an exposed 2 inch diameter area. Such a diaphragm is simply sealed into a stainless steel housing with two opposed o-rings. Although this type leads to more accurate pressure measurements than the piston transducer, comparatively there are three disadvantages. Because larger areas are involved, the housing must be robust even for pressures of a few thousand psi and the upper pressure limit is lower. Also, there is a larger volume of liquid from the reaction vessel that is trapped in the transducer thereby affecting the composition in the vessel. In addition, the thin diaphragm may fail under long term exposure to corrosive fluids and remain undetected because there is no test for its integrity while under pressure.

E. Volume Measurements

The volume of a reaction vessel can be determined within 0.1-0.3% if isothermally filled with clean water. One procedure is to weigh a flask of deaerated water and then to carefully and completely fill the vessel with this water so that no bubbles are entrapped. From the difference between the beginning and final weights of the flask + water and the density of water at the correct temperature, the volume of the vessel can be calculated. The buoyancy correction (about 0.1%) for weighing in air is necessary and evaporation during the procedure must be averted. A second method depends on emptying the water from a filled vessel into a preweighed flask for weighing. Because it is difficult to remove all water without leaving droplets on the surface or trapped around the gaskets, the second method is more troublesome. (See Chapter 9 for other techniques of volume measurements in such vessels.)

It is possible to measure the volume of liquid in many subcritical systems while at elevated P and T and with reasonable accuracy. The principle depends on measuring the angle of inclination of the vessel at the precise position where an electrode, sealed into the vessel and gradually tilted until it touches the solution, is effectively shorted to the vessel walls (see Figure 5 insert). By adding successive aliquots of a salt solution from a standardized pipette as in Figure 5, a calibration curve can be developed to relate this angle of inclination to solution volume. In this example, accuracy is limited to ±10 ml

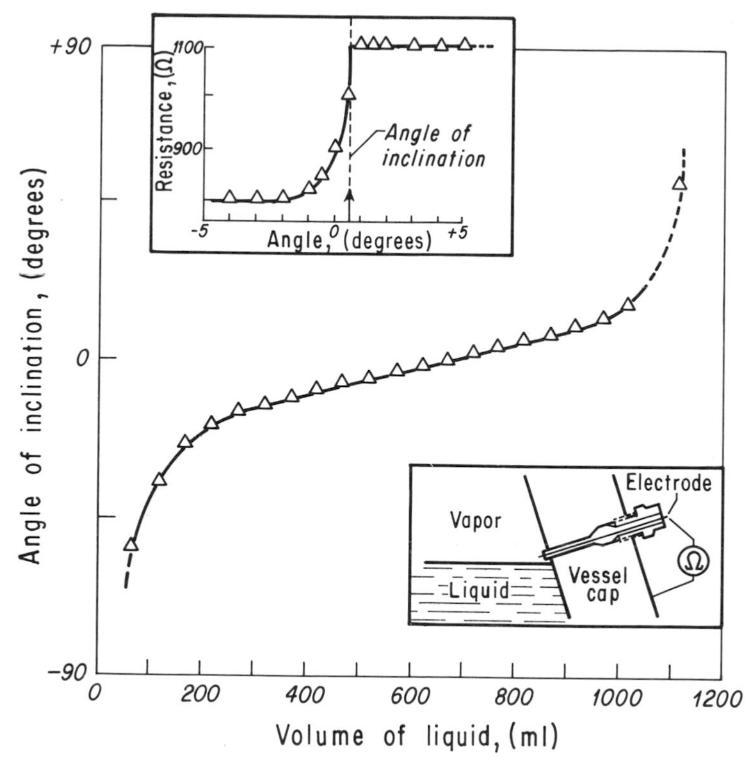

Figure 5. A calibration curve for the volume of liquid corresponding to each angle of inclination of a reaction vessel where an electrode contacts the surface of the liquid. The vessel contained 1150 ml and internally is approximately 3.3 inches in diameter and 8 inches long. (From Haas, 1966)

for 100-1000 ml of liquid due to uncertainty in the measurement of the angle of inclination of the furnace enclosing the reaction vessel. An electrode of tungsten wire has been little affected by use in halide, hydroxide, sulfide, and weakly acid solutions to over 400 C; however, the electrical insulator is more of a problem. Teflon, glass, ceramic, and soapstone insulators are limited either to use below 290 C (teflon) or to solutions of low alkali content and better insulators are needed. In some applications, Lucalox (G.E. Co.) which is sintered alumina and may be custom-shaped, is an excellent insulator, but its low thermal expansion causes problems in sealing it in pressure vessels.

Complete P-V-T studies can be made of hydrothermal systems using these methods because the volume of the gas phase can be calculated from the total volume of the vessel less the volume of liquid phase.

F. Hazards of Hydrothermal Research

Two dangers have already been discussed — those of hydrogen embrittlement and also of overfilling a vessel so that thermal expansion fills it with a nearly incompressible liquid — and there are others which merit attention, such as possible mechanical defects in a vessel. Before use with toxic fluids, all vessels should be cold tested and, because deterioration is gradual and often subtle, they should be periodically retested. Operating conditions obviously should be conservatively within rated P-T limits when H_2S, radioactive materials, or other poisons are involved. For example, Figure 6 shows the safety limits in use in my laboratory for a common size of vessel (4 3/8 inches outside diameter by 3 5/16 inches inside diameter and 8 inches inside depth, about 1100 ml capacity, and gasket as in Figure 4 and 7 b.)) of 310 or 316 stainless steel when used with sulfide solutions. Operating conditions are well within test conditions and are conservative for these vessels if free of structural or mechanical defects.

Because failure of a vessel is always a possibility, vessels should either be oriented so that the weakest segment (normally the closure) is aimed in a direction away from areas frequented by personnel, or shielding should be arranged. This precaution is particularly important if quenching, or even rapid cooling, of a hot, pressurized vessel is planned because the added stress is significant. Quenching in cold water of a typical large volume reaction vessel containing sulfide solutions at high temperatures and pressures is simply foolhardy and should not be attempted.

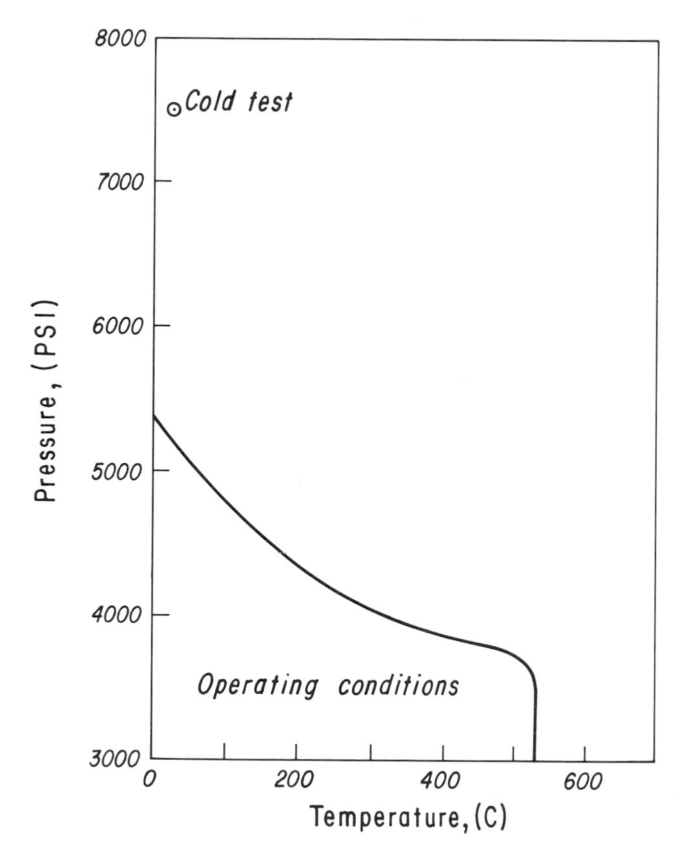

Figure 6. Operating limits for a 1.1 liter reaction vessel (as described in text) when used with toxic fluids.

1. Characteristics of H_2S

The maximum tolerable level for extended exposure to H_2S is currently 10 ppm by volume but its odor is detectable at levels as low as about 5 ppb. This initial sensitivity to the smell of H_2S deteriorates in minutes, however, and olfactory fatigue can easily lead to overexposure without awareness. H_2S when inhaled is quantitatively absorbed and rapidly accumulates in the bloodstream from which it is only slowly removed by oxidation. Concentrations above 50 ppm are dangerous to the eyes due to rapid absorption and above 500 ppm lead to acute physiological reactions resulting in collapse within seconds and without premonition. The victim may recover if promptly moved to fresh air and respiration restored but the rescuer should either not inhale in the H_2S-rich area or use a gas mask designed for use with this gas.

Exposure to H_2S causes some distinctive responses. For low levels over extended periods, one finds an increased sensitivity so that the vapor released on opening a freshly boiled egg is distasteful, and walking by a sewer opening may cause nausea. On exposure to very high concentrations there is a warning provided by contraction of the nasal membranes. The conditioned response should be thoroughly indoctrinated in anyone working with sulfide systems to instantaneously stop breathing and leave the area whenever this contraction begins.

Because H_2S is 19% heavier than air, it settles readily to the floor. Only powerful drafts are capable of flushing a laboratory of H_2S and these are more effective if intake vents are at floor level. H_2S carried to the roof of a building by a hood system may cascade down over the exterior of a building, particularly in calm, humid weather, thereby alienating other occupants.

Large quantities of H_2S can be discarded most easily and quickly by burning to SO_2. It is advisable to have the necessary equipment always ready during runs with H_2S so that the run can be dumped quickly if a large leak should develop. Tubing from the vessel should lead first to a trap, to collect any aqueous solution, and then to a stainless steel tube held in the flame of a bunsen burner. The burner is necessary for complete combustion because H_2S is only flammable at 4.3-45% by volume in air and, by itself, is easily extinguished by a variable flow rate. As a result of H_2S readily diffusing through plastics and some rubbers, there will be an odor from loss through the tubing unless it is of metal. The expansion of H_2S on leaving the vessel may be sufficient to intermittently seal the exhaust tube with frozen $H_2S \cdot 6H_2O$. Consequently, one should be doubly certain that the pressure in the vessel is actually down to 1 atm and not simply rely on the cessation of flow, before opening the vessel.

Spent sulfide solutions, such as that left in the vessel, or in the trap after burning off H_2S, should not simply be poured into the sink. If the waste solution is neutral or alkaline, it will be acidified by most other sewage and the resulting release of H_2S may cause nearby basement sinks to bubble H_2S from their drains. These solutions can be discarded in sealed containers, by evaporation in the back of a hood, or by voluminous dilution <u>before</u> discharging down the drain.

H_2S burns in air with a luminous light blue flame. The product, SO_2, is neither as toxic, nor is the nose as sensitive to it as to H_2S so that dilution with air in a hood is generally a satisfactory method of disposal.

H_2S is supplied commercially as a condensed liquid in gas cylinders. Its vapor pressure, about 250 psi at room temperature, is not high enough so that the gas phase alone can provide the necessary volumetric concentration for many experiments. For these, liquid H_2S can be condensed through pressure tubing from a gas cylinder into a reaction vessel. This is a hazardous procedure if done without forethought. The cylinder may be heated gently with heating lamps (or less conveniently with a water bath) but only after all valves between it and the reaction vessel are open, because otherwise expansion of the liquid might rupture the cylinder. For the same reason, the lamps should be turned off and the cylinder partially cooled before closing it off from the cooler reaction vessel. Another hazard arises from the fusible plugs sealed into many gas cylinders. Because these are designed to fail at about 74 C (165 F), <u>cylinders should never be heated above 52 C</u> (125 F).

2. Testing for Leaks

Before H_2S or any toxic gas is charged into a vessel, risks of initial leaks can be eliminated by monitoring the decay of a vacuum in the pressure system. First, a vacuum of less than 100 microns must be attained and then a valve closed isolating the vacuum pump from the pressure system. If, after about 15 minutes, there is no measurable decay of the vacuum, then the system is at least gas-tight to 1 atm.

If leaks are present, they can readily be located at pressures greater than 1 atm by painting connections with a syrupy soap solution so that gas bubbles will reveal the defect. This procedure requires an internal gas pressure and is effective only up to temperatures where the soap solution boils away.

IV. Investigations of Fluids

Several methods have been used to determine the composition of aqueous fluids while at high P-T conditions. Ideally, the composition should be measurable without removing a sample for analysis. Attempts to do this include using radioactive tracers, conductivities, and absorption spectra. However, a combination of insensitivity, nonspecificity, thermal limitations, and cost severely limit the utility of these methods and most research on hydrothermal fluids depends on separating samples from the fluid in a reaction vessel.

Samples can be withdrawn during a high P-T run either by dynamic (continuous) sampling or by static withdrawal of an aliquot of fluid. The dynamic method involves pumping a fluid into a reaction vessel slowly enough so that the desired temperature and reaction period can be maintained while simultaneously leaking an equivalent volume through a restricted vent. An advantage is that large samples can be collected so that analysis is relatively easy. Unfortunately, the quality of the sample is always in doubt. Equilibrium is not assured because the reaction cannot be unequivocally reversed. In addition, simply slowing the flow rate, until concentration in the effluent is unaffected by the rate, only demonstrates that a steady state has been achieved, perhaps at a metastable pseudoequilibrium between mineral surfaces and the fluid. Furthermore, there is a very steep P-T gradient just inside the vent and it would be remarkable if there were no deposition here or if no particles were carried through as the opening is adjusted to control flow rate when it changes in response either to corrosion or deposition in the throat. Solubilities measured using the continuous flow method are often higher than those measured by other methods.

A. Static Sampling

These methods depend upon withdrawing a sample after the solution has been in residence in the reaction vessel for some period, rather than simply flowing through it. If the sample is small compared to the volume of the phases in the vessel, then the extraction can be virtually isobaric. If the sample is removed quickly into preheated or thermally non-conducting tubing, then the process may also be nearly isothermal.

Utilizing relatively small samples allows a series of samples to be withdrawn without seriously perturbing the P-T-X conditions in the vessel. These samples can be analyzed for several purposes, including the following: (1) to check for precision of sampling among replicates taken over a short interval, (2) to determine rates of equilibration over a longer interval, or (3) to check for evidence for equilibration by approaching the P-T conditions of sampling first from lower and then from higher temperature.

Figure 7 shows the plumbing of a system designed for static sampling of hydrothermal runs.[1] Detailed procedures for operation are given in Barnes (1963). In brief, uncontaminated weighed quantities of solids, H_2S or other gases, and H_2O can be put into the vessel as follows. The sealed vessel containing solids (sulfides, NaOH, NaCl, etc.) is first evacuated and then a gas is condensed into the vessel through the sampling valve. Finally, water is pumped from a weighed quantity in the reservoir, through the resin column (unless a solution is pumped instead of water) and into the vessel so that the residual water in the reservoir can be weighed to find the amount pumped by difference.

[1]This is similar to units manufactured by Tem-Pres Research Division, Carborundum Company, 1401 S. Atherton St., State College, Pa. 16801.

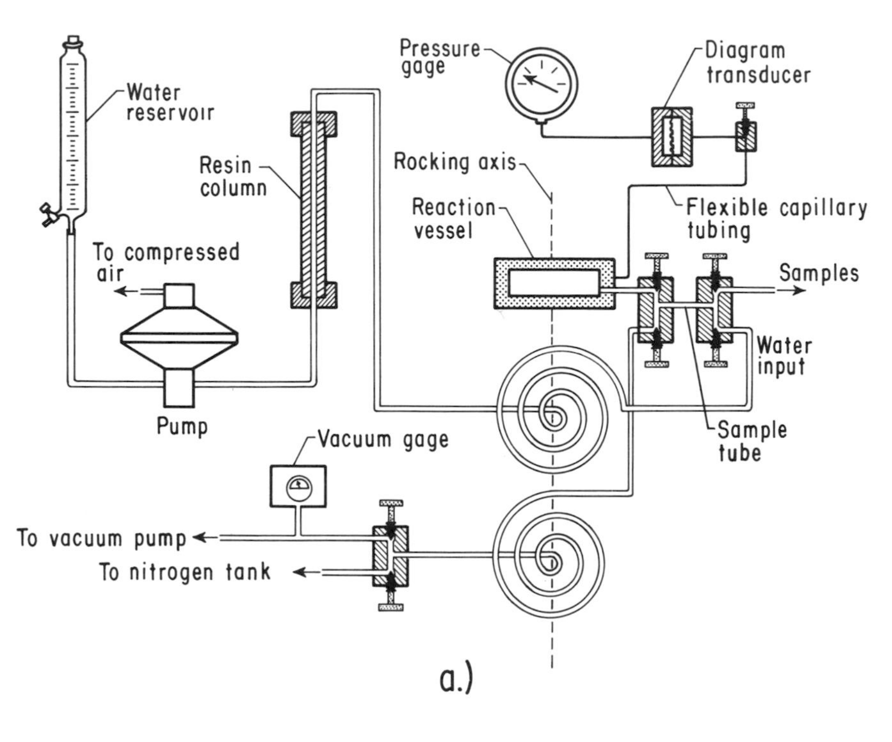
Water reservoir
Pressure gage
Diagram transducer
Resin column
Rocking axis
Reaction vessel
Flexible capillary tubing
To compressed air
Samples
Water input
Pump
Sample tube
Vacuum gage
To vacuum pump
To nitrogen tank

a.)

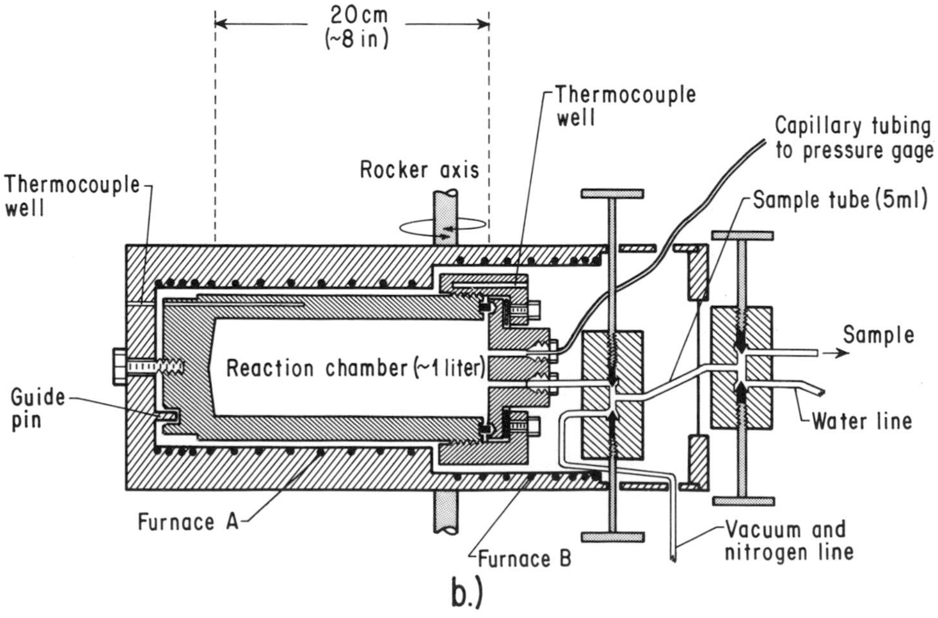
20 cm (~8 in)
Thermocouple well
Capillary tubing to pressure gage
Rocker axis
Thermocouple well
Sample tube (5ml)
Sample
Reaction chamber (~1 liter)
Guide pin
Water line
Furnace A
Vacuum and nitrogen line
Furnace B

b.)

Samples of liquid can be taken by tilting the vessel down, submerging the exit port, or tilting the vessel up to collect the gas phase. To sample, the sample tube is first evacuated, then filled by opening it to the vessel and finally the contents are flushed into a container using a flow of nitrogen. The amount of the sample removed can be determined volumetrically by calibration of the sample tube volume. Inasmuch as a gas phase might separate under some conditions during sampling of a liquid, it is preferable to collect liquid samples, including any gases, for weighing. Such two phase samples, or single phase gas samples, can be condensed without spattering into a flask cooled by liquid nitrogen, or an acetone-dry ice bath, if done slowly. More easily for sulfide fluids, a concentrated solution of NaOH + H_2O_2 will nearly instantly absorb H_2S vapor and can be used to prevent any loss. Strongly alkaline solutions will absorb CO_2 and similar reagents can be found to absorb most gases.

Gases above 300-400 C, or supercritical fluids, need to be cooled rapidly during sampling to prevent spattering and loss as a vapor. An effective method is to pass the fluid through a few feet of stainless steel capillary tubing immersed in ice water (about 0.020 inch i.d. tubing does not clog). This procedure allows direct metering of fluid from the reaction vessel into a container for weighing and any precipitate is forced to flow along with the fluid by the high pressure in the vessel.

B. Filtering

The principal difficulty in sampling is complete separation of particulate matter in the fluid in the reaction vessel from the sample plus any included precipitate which might form during

---

Figure 7. A hydrothermal system for determining fluid compositions at moderate temperatures and pressures modified after ← Barnes (1963) with permission from Economic Geology. a.) Schematic diagram of components and pressure lines. b.) Detail of the pressure vessel design.

the extraction. Imperfect separation causes intolerable scatter in analyses of replicate samples and, as a result, is easily recognized. Depending on whether the solids in the vessel are colloidal or in suspension, the problem can at least be minimized.

Colloids are charged particles sufficiently tiny that they remain in metastable equilibrium and will not settle out of solution. For example, metallic sulfides are generally peptized by H_2S, but addition of an electrolyte neutralizes these and other colloids allowing them to flocculate or precipitate. Only 0.01 m KI was sufficient to reduce extreme scatter in measurements of Ag_2S solubility in H_2S-rich aqueous solutions to a range equivalent to the analytical precision. If an electrolyte is to be added to flocculate a colloid, only reagents should be considered which will not decompose at high temperatures, which will not form complexes or otherwise interact with the primary components of the experiment, and which are largely ionized so that only small concentrations are needed.

Suspensions occur more frequently in hydrothermal experiments and may result from using too finely divided a powder in a run or from disaggregation of run solids due to reactions or to the stresses of high temperatures and pressures. After experiments on the aqueous solubility of sphalerite in which grains only greater than 1 mm diameter were used, dust-like particles commonly were found in the vessel even though pressures were only a few hundred psi and temperatures were 100-200 C. A partial cure is to halt agitation in order to allow the suspension to settle. This can be done effectively in subcritical fluids by first tilting the reaction vessel up so that the exit port is above the solution, then waiting 5-15 minutes for settling, and finally slowly and smoothly submerging the port and waiting another 5-15 minutes before sampling. Another 5-15 minute wait is useful between successive samples. In spite of these precautions, an occasional particle will be lodged in or near the exit port and may be flushed into the sample tube resulting in an aberrantly high analysis. Because particles are most likely to be entrapped during rocking, the first of a series of samples is most likely to be affected.

With some run compositions, suspensions are apparently inevitable and may even be required for equilibration. A few solids, such as pyrite or hematite, are so slow to react that they develop a surface rind armoring particles from further reaction unless the particles are very small. In such systems, or in others where the settling rate is too slow, meaningful samples can be taken only by filtration.

Filtering of samples is easy if the rate of precipitation in the sample is slow enough that the sample can be extracted and then externally passed through filter paper without loss, except for particles from the reaction vessel. Some precipitation may be permitted on the filter paper if this precipitate is much more soluble than the particles so that a reagent can be used to wash it through the filter.

Internal filtering within the reaction vessel or in the sample tube is a last resort because the filters, being small, tend to clog with solids and also to recrystallize so that flow into the sample tube is throttled, boiling results and more precipitation occurs. If internal filters cannot be avoided, fritted metallic discs are most convenient; of the many types available, those with 15 or 20 micron pore size have been useful when made of 316 stainless steel for sulfide systems and of nickel where any contamination by iron could not be tolerated.

Fritted filters, of a diameter only slightly larger than the internal diameter of pressure tubing, can be placed in the end of a tube after reaming. When the tubing is then connected, the filter is held tightly in place by the usual cone seal. Filters located in the upstream end of the sample tube (Figure 7 b.)) can be removed for examination and cleaning during a run — a distinct advantage. Filters can also be inserted in the tube connecting the first valve to the reaction vessel but not in the vessel-end where the filter would be subject to high temperature corrosion.

V. Investigations of Solids

Here we are concerned not primarily with composition of fluids phases, but rather with equilibration between solids or with crystalline morphologies. Recrystallization of solids, when dependent for a mechanism only on solid state diffusion, is often extremely slow; somewhat faster is *in situ* recrystallization involving very local fluid transport generated by chemical potential differences due to non-equilibrium solid compositions or high surface energies, such as in powders less than about 1 micron in grain size. A kinetically much more rapid and easily controlled process is the dissolution, transport and deposition of solids through a fluid phase across an environmentally-controlled chemical potential gradient. Accordingly, the fluid phase is a potent catalyst for recrystallization if the solids are not insoluble and, equally important, if the solubilities change markedly with temperature. Variables other than temperature can be used to regulate the potential gradient, *i.e.*, differences in pressure, pH, oxidation state, etc., but these are not as easily controlled as in a chemically closed system with a temperature gradient.

In large reaction vessels, such as that in Figure 7 b.), steep temperature gradients are not possible even when heating only at one end. Both convection and the thermal conductivity of the necessarily thick vessel walls at least in part iron out gradients. For example, a very large gradient for the vessel of Figure 7 b.) is only 20-25 C at 250 C. A "cold-finger" extending into the vessel provides more flexible adjustment of gradients and also localizes deposition of solids. The "cold-finger" is constructed of two concentric tubes, the outer of which is closed at the tip and sealed in its midsection into a port of the reaction vessel using a standard cone seal. The inner tube carries a flow of water or air into the vessel to the closed tip of the finger with return flow between inner and outer tubes. Temperature at the tip can be monitored using a thermocouple inserted along the inner tube. An application of the "cold-finger" technique to crystal growth is illustrated in Scott and Barnes (1969). Nevertheless, if many hydrothermal temperature-gradient runs are anticipated in a research project, particularly with small volumes, then it is more practical to use a vessel designed for this purpose.

A. Temperature-Gradient Vessels

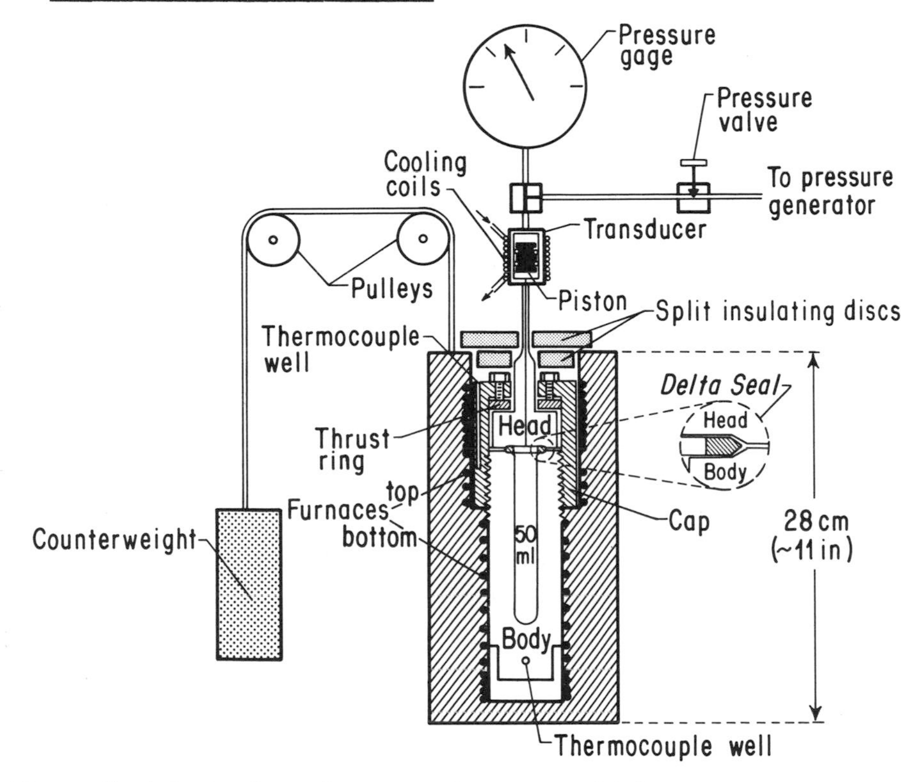

Figure 8. A hydrothermal system for use at high temperatures and pressures particularly involving temperature gradients. The reaction vessel is mounted vertically by hanging from the rigidly mounted transducer; the furnace slides along vertical guide rods.

The vessel and associated equipment in Figure 8 were designed specifically for use of temperature-gradients to equilibrate solids at conditions up to 600 C, 15,000 psi. Characteristics of the delta-seal and of the transducer have been discussed earlier in this chapter and only the furnace needs comment here. It may be lowered away from the vessel along two guide rods; the counterweight supports the furnace only to the extent that friction against the guide rods prevents movement wherever the furnace is placed. With the furnace lowered, the vessel may be readily connected to, or disconnected from the rigidly mounted transducer without interference. A relatively strong outer shell of the furnace also acts as a safety shield against bursting by directing expansion up or down and not towards the operator.

Insulation in the furnace is purposely thin favoring heat dissipation so that steep gradients can be realized when required. As a consequence, furnace windings must be concentrated both at the top and at the bottom of the vessel (Figure 8) to compensate for the high heat loss in these areas and also to linearize the gradient between the two thermocouple wells. With a linear gradient, the temperature of growth of a solid at a known position in the vessel can be accurately interpolated provided the solid is in good thermal contact with the vessel wall.

B. Sealed Tube Experiments

Control of corrosion is less of a problem when only equilibration of solids is important and sampling of fluids becomes unnecessary. Individual runs can simply be sealed into tubes of glass, gold, or platinum which are supported by external pressure within the vessel, but the type of tube used must be compatible with the run composition. Reactions which inhibit the utility of gold and platinum tubes have already been examined, and glasses also must be used with discretion. Silica glass (at least 99.8% SiO_2) and Vycor glass (96% SiO_2) may be used continuously at 1100 C and 900 C respectively if dry, and both withstand instantaneous quenching into cold water. Pyrex (borosilicate glass) may be used continuously to 500 C if dry, but, in spite of these high working temperatures, all three glasses readily recrystallize at low temperatures in the presence of alkali-rich solutions. Depending on the vagaries of any one run, these glasses may recrystallize and leak within a few days even at temperatures somewhat below 200 C when used with 1 molal NaHS solutions. The rapid rate of deterioration may be reduced by cleaning the exterior of the tube of perspiration and by heating only in an alkali-free environment.

1. Filling of Tubes

Preparation of sealed tubes is very similar to that given in the preceding chapter on dry sulfide systems. One end of the glass or metal tube is welded shut and solids inserted. Precise quantities of deaerated solution are added by syringe directly into metal tubes, or if into glass tubes only after they have been necked down to an inside diameter about twice the diameter of the syringe needle. Prior to sealing and to prevent boiling, the solution can be frozen by immersion in an acetone-dry ice bath or in liquid nitrogen; wrapping the end of the tube containing the solution, with cloth, keeps it frozen and makes handling the tube easier. If necessary, the frozen tube can now be evacuated through rubber tubing connected to the warmer end of the tube. Alternatively, oxidation could be prevented by displacing air in the tube with N_2, H_2, or H_2S. Sealing is completed by welding metal tubes along a crimped portion or by fusing glass tubes at the neck.

2. Support Pressure

The pressure in the tube is fixed predominantly by the percentage fill and the temperature, and can be calculated from the P-V-T properties of the solution if known, or approximated from similar solutions. Metal tubes, being flexible, will deform to equilibrate the internal and external support pressures but only within a narrow range; excessive external pressure will collapse part of the tube often preventing transport of solids, but excessive internal pressure of a few atmospheres will rupture the tube. Glass tubes with dry charges can support, depending on diameter, up to a maximum of 100 atmospheres of internal excess pressure (see Chapter 11), but as pointed out above may be considerably weakened by alkali solutions. Although glass tube runs require less exact balancing of internal and external pressures, one does not know the internal pressure except by calculation. The metal tubes must have virtually the same internal pressure as that read on the gage for external pressure but filling may demand precision within one drop (.04 ml) to prevent collapse or bursting of small tubes, such as a tube 5 mm inner diameter by 5 cm long.

The external supporting pressure is best selected to be slightly greater than the estimated internal tube pressure, particularly if the tube will be nearly filled with subcritical liquid at run temperature. In runs containing simple salt solutions, such as with NaHS, NaOH, or NaCl, the external pressure generated by pure water, at the same percentage fill in the vessel as in the tube, provides sufficient support. If the tube contains a volatile reagent, such as NH_3, H_2S, or SO_2, then a supporting pressure greater than the vapor pressure of water will be necessary below 374 C. One method of controlling these pressures is by connecting two pressure vessels in series with the cooler one filled with liquid water and the hotter with supercritical water to act as a pressure buffer. The exact pressure can be adjusted by adding or withdrawing water from the two bomb system using a manual pressure pump.

3. Productivity

The vessel shown on Figure 8, with nominally 50 ml capacity, internally is 7/8 inches in diameter and about 5 inches deep and can accomodate one large or several smaller tubes simultaneously; seven tubes 5 mm in diameter will fit comfortably. For good contact with vessel walls and resulting control of internal tube temperatures, Pyrex wool should be stuffed in the center of the vessel. It is also beneficial in impeding convection in the vessel, thereby decreasing the gradient across the diameter of each tube from the vessel wall toward the interior.

Limited quenching of the reaction vessel is feasible by simply lowering the furnace away from the vessel. Since this leaves the vessel unshielded, it is recommended that the operator leave the area until the vessel is cool. A limitation to the cooling rate results from the more rapid cooling of the vessel than of the internal tubes. Too rapid quenching is not desirable because the support pressure will fall faster than the internal tube pressure thereby rupturing the tubes. During cooling, overgrowths may form on previously crystallized solids, but it is often possible to remove them by washing with an acid for a few minutes.

In experiments above 300 C with the temperature gradient method using the equipment in Figure 8, growth of sulfide solids (pyrite, pyrrhotite, sphalerite, wurtzite, cinnabar, metacinnabar, argentite, galena, etc.) has been effectively completed sometimes as quickly as in one day and occasionally as long as two weeks were needed. However, runs of longer duration have offered no advantage. The rate of transport is strongly dependent on temperature gradient and has been varied (between thermocouples) from about 5-100 C. The smaller gradient generally leads to growth of fewer nuclei and larger, nearer perfect crystals. Where solubility is low and nearly independent of temperature, — an uncommon phenomenon — the larger gradients are needed for transport.

Because solids grow along the sides of each tube at a series of temperatures and at virtually constant fugacities (fixed normally by equilibria in the coolest portion of the tube), one tube can be used to determine the position of a univariant curve. For example, if a sulfur-rich polymorph grows in the cooler portion of a tube and the corresponding sulfur-poor polymorph in the warmer end, then the univariant curve separating their stability regions has been straddled by conditions in the tube. Interpolating the temperature, at the point in the tube between the two polymorphs, locates the univariant curve for the fugacity in the tube. In this manner, the results from one tube are equivalent to those from many experiments made at individual temperatures and fugacities.

BIBLIOGRAPHY:

Barnes, H.L. (1963). Ore solution chemistry. I. Experimental determination of mineral solubilities. Econ. Geol., 58: 1054-1060.

Barnes, H.L., editor (1967). Geochemistry of hydrothermal ore deposits. New York: Holt, Rinehart and Winston.

Barnes, H.L. and Czamanske, G.K. (1967). Solubilities and transport of ore minerals,in Geochemistry of hydrothermal ore deposits, H.L. Barnes, editor. New York: Holt, Rinehart and Winston.

Barnes, H.L., Romberger, S.B., and Stemprok, M. (1967). Ore solution chemistry. II. Solubility of HgS in sulfide solutions. Econ. Geol., 62:957-982.

Clark, L.A. and Barnes, H.L. (1965). Metastable solid solution relations in the system FeS_2-CoS_2-NiS_2. Econ. Geol., 60: 181-182.

Dickson, F.W., Blount, C.W., and Tunell, G. (1963). Use of hydrothermal solution equipment to determine the solubility of anhydrite in water from $100^{\circ}C$ to $275^{\circ}C$ and from 1 bar to 1,000 bars pressure. Amer. J. Sci., 261:61-78.

Haas, J.L. (1966). Solubility of iron in solutions coexisting with pyrite from $25^{\circ}C$ to $250^{\circ}C$, with geologic implications. Ph.D. Thesis, The Pennsylvania State University, University Park, Pa.

Harned, H.S. and Owen, B.B. (1958). The physical chemistry of electrolyte solutions, third ed. New York: Rheinhold Publishing Corp.

Hoffman, A.N., Montgomery, J.B., and Moore, J.K. (1948). Agitation in experimental rocking-type autoclaves. Ind. and Eng. Chem., 40:1708-1710.

Marshall, W.L. (1969). Correlations in aqueous electrolyte behavior to high temperatures and pressures. Record Chem.

Prog., 30:61-84.

Rogers, H.L. (1968). Hydrogen embrittlement of metals. Science, 159:1057-1064.

Scott, S.D. and Barnes, H.L. (1969). Hydrothermal growth of single crystals of cinnabar (red HgS). Mat. Res. Bull., 4:897-904.

Snyder, J.R., Hagerty, P.F., and Molstad, M.C. (1957). Operation and performance of bench scale reactors. Ind. and Eng. Chem., 49:689-695.

Weissberg, B.C. (1970). Solubility of gold in hydrothermal alkaline sulfide solutions. Econ. Geol., 65:551-556.

ACKNOWLEDGMENTS:

Many of the techniques described herein have been developed and tested during collaborative research at Penn State by R.K. Cormick, D.A. Crerar, J.L. Haas, G.R. Helz, S.B. Romberger, and S.D. Scott. The manuscript has been beneficially criticized by Professors C.W. Burnham and H. Ohmoto.

INDEX